THE SCIENCE OF MIDDLE-EARTH

A New Understanding of Tolkien and His World

ILLUSTRATED BY
ARNAUD RAFAELIAN

EDITED BY
ROLAND LEHOUCQ, LOÏC MANGIN, AND JEAN-SÉBASTIEN STEYER

TRANSLATED BY
TINA KOVER

PEGASUS BOOKS
NEW YORK LONDON

Dedicated to my brother Jean-Claude
(June 7, 1954 to September 7, 1973).

THE SCIENCE OF MIDDLE-EARTH

Pegasus Books, Ltd.
148 West 37th Street, 13th Floor
New York, NY 10018

First Pegasus Books paperback edition April 2022
First Pegasus Books cloth edition April 2021

Interior design by Maria Fernandez

Library of Congress Cataloging-in-Publication Data is available.

ISBN: 978-1-64313-954-8

10 9 8 7 6 5 4 3

Printed in the United States of America
Distributed by Simon & Schuster
www.pegasusbooks.com

CONTENTS

TOLKIEN,
LORD OF THE SCIENCES

 ow are the Elvish language and Old English related? What is the geology of Middle-earth? How does the dragon Smaug fly? You don't need to be a scientist or a Tolkien specialist to answer these questions; all you have to do is open this book.

In 1937, John Ronald Reuel Tolkien left an indelible mark on the history of fantasy with *The Hobbit*, originally written for his children. The multitalented author, Oxford university professor, philologist, and poet would repeat this success with his best-known works, including *The Lord of the Rings* (1954-1955) and *The Silmarillion* (published posthumously in 1977). Passionate about mythology and languages, which he entertained himself by inventing, Tolkien left behind a complex *legendarium*, an imaginary world so rich, fully formed, and elaborate that it serves as an outstanding argument in favor of science as entertainment! That is the goal of this book: to use Tolkien's universe—its history, languages, geography, monsters, and characters—to speak about the human, physical, and natural sciences. We are well aware of how many themes there are to be explored, how many questions raised—but, though we cannot be exhaustive, we hope to be reader-friendly, and to share knowledge with a wide audience. We have no desire to shatter the fantasy of Tolkien's creation, much less to criticize the man himself; rather, we hope to show how the sciences can enrich the world. Welcome to the Middle-earth of science! A cross-disciplinary—and sometimes undisciplined—journey through the land of Tolkien . . . " And if you [come back], you will not be the same!"

—Roland Lehoucq, Loïc Mangin,
and Jean-Sébastien Steyer

TOLKIEN AND THE SCIENCES:
A RELATIONSHIP WITH MANY FACES

ISABELLE PANTIN, historian, École normale supérieure

olkien saw himself primarily as a poet and a creator of myths through words, but he was certainly just as much a scientist, if we keep to a fundamental (and therefore timeless) definition of science: the rational effort made to learn about and understand natural reality, from the organization of the cosmos to human activity and production. If, on the other hand, we limit ourselves to the standard (caricatural?) definition of modern Western science, dominated by models created by and for physics and mathematics, and engaged in a conquering approach based on technology, the analysis demands far greater nuance. We must, then, examine these two approaches separately.

Tolkien had a scientific mind in the broadest sense: not the kind defined by the philosopher Gaston Bachelard, perhaps, or by others for whom the aptitude for "objective knowledge" could not exist before the advent of the modern era, but rather that of Plato, Aristotle, and Roger Bacon (1214–1294), that insatiably curious "admirable doctor" who had preceded him at Oxford, albeit some seven centuries earlier and occupying a different chair. Tolkien loved the real world, and he was curious about it, and wanted to have an unbiased relationship with it; endowed with a sharp critical mind, he maintained a cordial loathing for the sort of convoluted, pretentious discourse on any subject which offered false access to knowledge while simultaneously acting as an opaque and deceptive barrier.

A REFUSAL TO LIE ABOUT REALITY

To express his disgust for pretense, Tolkien often used the word *bogus*, meaning cheap and artificial, and often associated with trickery and fraud. He uses it, for example, in a 1968 letter to *Time-Life International* (*Letters*, no. 302), in which he declines to have a series of photos taken in a work setting, pointing out that he and the agency clearly have different definitions of the word "natural"; as he is never photographed while writing or speaking with someone, a snapshot of him "pretending to be at work would be entirely bogus." The term is used here to refer to writerly practices, but it is applicable to various sciences as well.

In his study of the *Kalevala*, Tolkien points out that this collection of ancient Finnish poems, compiled and arranged from oral sources with great care by the physician and linguist Elias Lönnrot in the 19th century, is unsullied by the "bogus archaism" that marks the work attributed to Ossian (a mythical third-century A.D. Gaelic poet) by the man who claimed to be its translator, James Macpherson.* He uses the same adjective to excoriate the authors of bad historic novels who make their characters speak in falsely "ancient," falsely noble, and frankly ridiculous language (*Letters*, no. 71), as well as the pseudo-scientific arsenal deployed in the clumsy faux-technology of mediocre science fiction ("mere abracadabra in bogus 'scientific' form."†)

Bogus syndrome can also afflict those who affect an ivory-tower purism in matters of botanical terminology, like those pedantic proofreaders of *The Lord of the Rings* who substituted *nasturtium* (the Latin name of the flowering plant also known as "Indian Cress") for its true English name, *nasturtian*. In a letter written in July 1954, Tolkien cites the expertise of Merton College's gardener and rejects nasturtium as "bogusly botanical" and "falsely learned."

The disorder had a far more serious effect on certain champions of the Indo-European hypothesis and its biased application to the study of mythology. Georges Dasent (1817–1896), a leading author of Scandinavian studies who did not hesitate to express in his work his belief in the superiority of the Nordic race (which, to him, included the English), was accused by Tolkien, in his essay *On Fairy-Stories*, of having distorted the true function of legends, armed with and blinded by the theoretical hodgepodge with which he was

---

*   *Tolkien Studies*, 7, 2010, p. 269.

†   Tolkien, "The Notion Club Papers," in The History of Middle-earth, 1983–1996, t. IX, p. 167.

infatuated, that "mishmash of bogus pre-history founded on the early surmises of comparative philology." But the worst offense was committed by the "wholly pernicious and unscientific race-doctrine" propagated by the Nazis, which Tolkien condemns in a July 1938 letter to his editor Stanley Unwin, who had informed him of the demands made by a German publishing house wishing to have *The Hobbit* translated.

In light of the fundamental nature of this refusal to lie about reality, it is clear that the distinctions often made between Tolkien the rigorous scholar and acclaimed philologist, Tolkien the citizen who loved to read newspapers to keep abreast of current affairs, and Tolkien the creative artist allowing his mind to escape to the vast reaches of Middle-earth, have no validity whatsoever. It is absurd to disassociate—and even worse to view as mutually exclusive—a freely creative spirit and the ability to face things as they really are. It was for this reason that, like his friend and Oxford colleague C.S. Lewis (1898–1963), Tolkien fought tirelessly against the prejudices that pitted so-called serious literature, which claimed to bear faithful and accountable witness to current issues (to be "in touch with the world," so to speak), against so-called escapist literature, an easy means of taking one's mind off things with stories as far removed as possible from the dull grind of daily life.

SCIENTIFIC MYTH, SCIENTIFIC TRUTH

The last section of *On Fairy-Stories* (before the epilogue) concerns the purpose of these tales, which deliberately distance themselves from realism and the seemingly ordinary in order to tap, through myth and the creation of imaginary worlds, into a higher form of realism, one that approaches reality at a deeper, more fundamental level. Here, Tolkien refutes criticisms aimed at so-called escapist literature. Fairy tales, he maintains, make a vital contribution thanks to the complete trust they place in the imaginary, which paradoxically makes them a source of richer and deeper knowledge. They make our sight clearer, stripped of the dreariness of habit and liberated from the illusion that things belong to us—so much so, in fact, that we are free to pay attention to them.

> *I do not say "seeing things as they are" and involve myself with the philosophers, though I might venture to say "seeing things as we are (or were) meant to see them"—as things apart from ourselves. [. . .] The*

The Elven army of Rivendell

things that are trite, or (in a bad sense) familiar, are the things that we
have appropriated, legally or mentally. We say we know them. They have
become like the things which once attracted us by their glitter, or their
colour, or their shape, and we laid hands on them, and then locked them
in our hoard, acquired them, and acquiring ceased to look at them.

The most skilled tellers of fairy-stories keep simple, fundamental things
as the focus.

For the story-maker who allows himself to be "free with" Nature can be
her lover, not her slave. It was in fairy-stories that I first divined the
potency of words, and the wonder of things, such as stone, and wood,
and iron; tree and grass; house and fire; bread and wine.

There is no reason to scorn the "escapism" provided by "fantasy." Why
despise the prisoner who tries to escape his prison? In the same section (on
"Escape") of this lecture given in 1939 and revised for publication after the war,
Tolkien even uses a radical analogy: would we call it desertion, or treason, if a
person escaped from the Third Reich or any other dictatorship? The vital function of stories is to respond to humans' deepest desires, the ones most revealing
of their own nature: to push back the limits of one's experience by swimming
with the fish, by flying with the birds, by speaking to other living species, and,
the ultimate in imagined freedom, by crossing the barrier of death.

The love of myths, then, in no way precludes a rational approach. It
does nothing to diminish "the appetite for scientific truth," any more than
it clouds our perception of it. On the contrary, the worth of an imaginary
world depends on the scientific excellence of its creator, on their ability to see
things in the real world for what they are, and to work scrupulously to give their
inventions cohesion, consistency—for it is the task of these creators to construct
a universe governed by a system of natural laws (*On Fairy-Stories*, "Children").

Remembering the sort of reader he was himself as a child, Tolkien emphasizes the fact that he was in no way an exclusive and unconditional lover of
folk tales (a predilection that developed later on, for the most part, with his
study of languages). What he wanted, more than anything, was to "know." He
fulfilled this need by reading stories that transported him to another world,
where he found things that appealed to him (dragons, for example) without
confusing them in any way with reality. These stories also, and even more

frequently, allowed him to discover history, botany, grammar, etymology. A note mentions his early taste for zoology and paleontology, yet his keen interest in these subjects could never have made him focus wholly on science merely to satisfy adult prejudices:

> *In fact, I was eager to study Nature, actually more eager than I was to read most fairy-stories, but I did not want to be quibbled into Science and cheated out of Faërie by people who seemed to assume that by some kind of original sin I should prefer fairy-tales, but according to some kind of new religion I ought to be induced to like science.*[*]

Tolkien would learn, later, how right he had been in his instinctive resistance; there was no need to give up fiction for science. The two are in a constant state of exchange, with the former acting as a source of theories for the latter, and borrowing the knowledge needed to give substance and consistency to its inventions in return. Each, in its own way, imparts a sense of wonder at the physical world, and the means to react to it. Fiction, however, does this through a secondary creation, an imaginary world, seemingly independent and governed by its own laws, but made of elements taken from the first world.

In the latter years of his life, Tolkien spent a great deal of time answering letters from readers so captivated by Middle-earth as to be convinced of its existence, who constantly asked him to expand upon certain details, or to explain some phenomenon or another. Rather than looking down his nose at their naivete, he did his best to answer their questions, willingly admitting that he, too, was under the spell of his own creation. It wasn't simply a world on paper for him, either.

But it would be wrong to see, in this attitude, the dangerous drifting of an imaginative mind caught in the seductive trap of fantasy. Describing Middle-earth was, for Tolkien, a way of fully understanding the planet on which he lived, of feeling more deeply the admiration it aroused in him, and of responding to it with the work of a poet, different yet complementary to that of scientists. He explained this very simply in March 1966, during a telephone interview conducted by Henry Resnick, whose article on him appeared in the July 2, 1966 issue of the *Saturday Evening Post* under the title "The

---

[*]   *On Fairy-Stories*, note D on the section "Children".

Hobbit-Forming World of J.R.R. Tolkien": "If you really want to understand what the foundation of Middle-earth is, it's the wonder and joy given to me by the real world, and the natural world in particular." This wonder and joy, he went on, had first been awoken in him by his discovery, aged three or four, of the countryside around Birmingham. These comments, which confirm the deep-rootedness of Tolkien's creation in his experiences of the real world, contain no thoughts on science *per se*, but they are reminiscent of the way in which Aristotle, in his *Metaphysics*, evokes the beginnings of philosophy and the quest for an understanding of nature, emphasizing the vital role of astonishment or wonder (*thauma*) and the kinship between the love of myths (that is, fictional stories) and the love of science.

> *For it is owing to their wonder that Men both now begin and at first began to philosophize; they wondered originally at the obvious difficulties, then advanced little by little and stated difficulties about the greater matters, e.g. about the phenomena of the moon and those of the sun and of the stars, and about the genesis of the universe. And a man who is puzzled and wonders thinks himself ignorant (whence even the lover of myth is in a sense a lover of Wisdom, for the myth is composed of wonders).*

Aristotle goes on to insist that these early researchers, who often expressed their questions and theories through fictional stories, pursued knowledge for its own sake, and not for utilitarian purposes—an idea equally dear to Tolkien's own heart, which will be touched upon later.

The scientific merit of Tolkien's approach is thus undeniable, and has been spontaneously recognized by many of his own readers, including astronomers, physicists, biologists, and paleontologists, who are in no way put off by the archaic and fantasist nature of Middle-earth. On the contrary, they are appreciative of the way in which its creator has sought to give consistency to the elements of his universe, and has made it live and evolve, rather than simply constructing a mock setting, throughout a patient process during which he seems constantly to have asked himself questions about the delineation of coasts and the courses of rivers, the flora and fauna, the adaptation of beings to their environment, and many other points, before coming up with answers to them. The quality of his observations regarding natural history has not been lost on these scientists, either.

Henry Gee, a paleontologist by training, expresses this recognition eloquently in his book *The Science of Middle-Earth*, in which he begins by developing an analogy between the methods of Tolkien's academic discipline (philology) and those of cladistics (the grouping of related species into tree-shaped diagrams), which, he claims, have done the most to advance our understanding of the evolution of species: whether it is a matter of establishing filiations between manuscripts, languages, or living organisms, tree-shaped diagrams (dendrograms) are always used. Gee then goes on to broaden his survey to include other fields of study, making connections and pointing out reciprocal borrowings.

The nomenclature of Middle-earth—precise, suggestive, respectful of the laws of language, and carefully tailored to fit what it is designating—is not only widely admired; it has sometimes been imitated as well. Entomologists, paleontologists, geologists, and astronomers have tapped into this treasure trove when in need of a name for some recently-discovered insect, fossil, or asteroid which strikes them as being linked, or simply analogous, even very indirectly, to an object or character described by Tolkien. Kristine Larsen, professor of physics and astronomy, has drawn up an inventory of these borrowings.

THE REJECTION OF SCIENTIFIC MATERIALISM

However, this image of perfect harmony between Tolkien's creation and the world of the science fails to take a more complex truth into account, and we must not allow it to make us forget the writer's numerous and strenuous denunciations of what he called "scientific materialism," which he saw as the scourge of the modern world.

This "scientific materialism" in its most diabolical form, which connects actual research to technological escalation and manipulation by financial or political superpowers, is evoked in *The Lord of the Rings* through the empire of Mordor and its satellite, Saruman's Isengard; it is clear that this is a portrait of totalitarianism. Tolkien confirms this in a letter to his son Christopher, dated January 1945, in which he expresses his hope for a sort of millennium—that is, an earthly rule of the "Saints," those who have not bowed to the forces of Evil, specifically, "in modern but not universal terms: mechanism, 'scientific' materialism. The Socialism represented in either of the factions now at war."

Tolkien puts the word "scientific" in quotation marks here, which shows that he was keeping things in perspective; yet this did not keep him from feeling deep and lasting hostility toward the "materialist" tendencies that he felt had long since penetrated the field of science, quietly and insidiously, even in the most seemingly virtuous places, such as the faculties of Oxford University.

His previously cited remark concerning his childhood refusal to enlist in the service of science by abandoning "Faërie," or fantasy, is significant in this respect. It stands alongside the opinions he willingly expressed on developments in the way academic study was organized. Despite the tragedies and horrors of the early 20th century, Tolkien considered himself extremely fortunate to have been young during a period when he was given the time and freedom to find his own way and to broaden his knowledge; in short, to become himself, J.R.R. Tolkien, renowned professor and creator of Middle-earth. In that golden era, a young person had the time to root himself in fields of study having to do with his vocation without risk of damaging his prospects for an academic career; a doctorate was not necessary to obtain a post, as long as one could provide proof of ability and merit.

The farewell speech he gave on June 5, 1959 upon his retirement, published in *The Monsters and the Critics and Other Essays*, further emphasized this theme, painting an ironic portrait of his university as it had become. It was no longer a place of learning, but one of initiation into research. The most talented students were steered inexorably toward doctoral studies narrowly determined according to the policies of their departments, which sought to produce PhD theses as plentiful as they were regularly calibrated, as if they were sausage factories. It was here that materialism insinuated itself into science, playing on both the students' ambition, legitimate as it might be, and on the desire for power of academic institutions. And yet, in 1959 Oxford graduates were not as sought-after by international finance as they are today, and world rankings did not exist to stir up competition among universities.

In a 1966 letter to his grandson Michael George, then beginning graduate studies at Oxford and reflecting on possible research projects, Tolkien reiterated his skepticism and bitterness:

> *There is such a lot to learn first. It is often forced on students after school because of the desire to climb on to the great band-waggon of Science (or at least onto a little trailer in tow) and so capture a little of the prestige and money which 'The Sovereignties and Powers and the rulers of this*

world' shower upon the Sacred Cow (as one writer, a scientist, has named it) and its acolytes.

To better understand this attitude, we must go back to the basics of Tolkien's position, beginning with his religious beliefs. His Christian faith, and his strong sense of the distinction between the desires of the Spirit (or of Christ) and those of the Flesh, to use Saint Paul's terminology, explain the depth of his ethical standards and the forcefulness of his rejection of all "materialist" aspects of science. His concept of nature, and of our way of understanding it, stemmed from this as well. He was in no way a fundamentalist, and each new hypothesis offered by the science of his time was greeted by him with an open and curious mind, but he remained loyal to the idea that "the heavens declare the glory of God" (Psalm 19) and that the scientist exploring nature, like the author of a "secondary world," was a participant in the bringing to light of the beauty of Creation, in the splendid complexity of its organization and its evolution through time.

This feeling can be sensed in the background of a November 1969 letter in which Tolkien thanks Amy Ronald for the gift of a book on the wildflowers of the Cape peninsula, which, he says, opens up a number of perspectives on paleobotany. He draws a comparison with flowers from the first Ages of Middle-earth, *elanor* and *niphredil*, and admits to his fascination for books that introduce him to new plants, especially rare plants whose biological relationships have not yet been established, but which can be linked to known plants: "They rouse in me visions of kinship and descent through great ages, and also thoughts of the mystery of pattern/design as a thing other than its individual embodiment, and recognizable."

FANTASY, MAGIC, SCIENCE FICTION

Tolkien's education—mostly literary, as was standard in his time—counted as well. When he speaks of his school memories, he invariably means Latin, Greek, or other languages and literatures. His scientific literacy, which was considerable, was largely acquired through popular books, his representations of nature were not constructed through the learning of mathematical signs and symbols. To him, one's understanding of nature could be expressed in the same language as the rest of the human experience.

It is no doubt partly for this reason that, in his view, approaching something scientifically did not mean one had to involve paradigms drawn from the so-called "hard" disciplines, or renounce subjectivity. It pained him to observe the irreversible shift in the "human sciences" away from the knowledge of languages, literature, and mythology and toward a subjugation to structuralism, simply in order to claim academic rigor and objectivity. In *On Fairy-Stories*, he dedicates an entire section ("Origins") to dissociating himself from the folklorists and anthropologists striving to found a science of folk tales. These researchers extract "patterns" from the narrative material of mythemes, looking for similarities and differences, mining it like a quarry, the ore extracted from which will then be neatly categorized, labeled, and utilized for various purposes, all authentically scientific. This work, Tolkien grants, would be "legitimate[. . .] in itself" if its practitioners did not claim full ownership of the terrain, subsequently allowing themselves to make utterly flawed arguments due to fundamental ignorance; they do not realize that they are dealing with literary texts, each one of which is unique and must be appreciated as an integral whole, taking all of its elements into account.

Tolkien himself considered a large part of his talent as a linguist to be due to his sensitivity to the aesthetic and musical qualities of languages, and to his ability to listen to his own deepest and most personal instincts. In 1955, he sent the poet W.H. Auden, who was preparing to give a talk on television about him, an outline of his basic character traits. In it, he spoke of his "sensibility to linguistic pattern," which affected him "emotionally like colour or music," and of his "deep response," having to do with his "roots," to certain languages and legends.

He was also extremely wary of any attempts, even well-intentioned ones, to translate the marvels of Middle-earth into scientific terms. He would undoubtedly have agreed wholeheartedly with every page of the aforementioned book by Henry Gee, particularly Chapter 17, which is based on an aphorism coined by science fiction author Arthur C. Clarke, who stated that "any sufficiently advanced technology is indistinguishable from magic."[*] There is nothing, then, to keep the doors of Moria, the lights of Lórien, or the phial containing the star-light of Eärendil, given to Frodo by Galadriel, from being seen as the fruits of some future technology.

---

[*]   Aphorism taken from Arthur C. Clarke, "Technology and the Future," *Report on Planet Three and Other Speculations*, London, Harper Collins, 1972.

Indeed, Tolkien was well-acquainted with the ancient concept of "natural magic," which consists of learning about natural processes in order to reproduce them experimentally, accelerating and intensifying them as needed, so as to create what the naïve and credulous might believe to be miracles.* Certain of his Elves and Dwarves, and even Radagast with his birds and Gandalf with his fireworks, can be seen as "natural mages." But this idea must not be taken too far. In *On Fairy-Stories*, Tolkien differentiates "faërie" from "magic" specifically in order to curtail such interpretations:

> *Faërie itself may perhaps be most nearly translated by Magic—but it is Magic of a peculiar mood and power, at the furthest pole from the vulgar devices of the laborious, scientific, magician.*

In June 1958, Tolkien wrote a detailed (and blistering) critique of a screenplay developed from *The Lord of the Rings* by Morton Grady Zimmerman. Point 22 concerns *lembas* (the light and highly restorative bread provided to the travelers by the Elves), misguidedly replaced with the term "food concentrate." The text of Tolkien's comments is found in his *Letters*. He is clear on this point:

> *As I have shown I dislike strongly any pulling of my tale towards the style and feature of 'contes des fées', or French fairy-stories. I dislike equally any pull towards 'scientification', of which this expression is an example. [. . .] We are not exploring the Moon or any other more improbable region. No analysis in any laboratory would discover chemical properties of lembas that made it superior to other cakes of wheat-meal.*

Fantasy and science fiction were two very different things, as far as Tolkien was concerned. As a reader he loved both of them, even likely preferring the latter, as it inspired more original and imaginative books. He was an avid reader of Isaac Asimov and a sincere, though critical, admirer of David Lindsay's *A Voyage to Arcturus* and the *Space Trilogy* of C.S. Lewis.†

As a writer, on the other hand, he felt at home only in fantasy. In 1936, as revealed in a letter of February 1967, Lewis and Tolkien decided to write

---

* See, for example, Richard Kieckhefer, *Magic in the Middle Ages*, Cambridge University Press, 2000.

† *Out of the Silent Planet* (1938), *Perelandra* (1943), and *That Hideous Strength* (1945).

simultaneous novels on subjects that truly appealed to them, and which they found lacking in contemporary literature: Lewis would tell the tale of a journey through space (which became his *Trilogy*), and Tolkien a story of time travel, with the latter beginning, and quickly abandoning, *The Lost Road*. In 1945–46, a second attempt on a very similar theme (two modern characters travel back to the time of an ancient catastrophe analogous to the sinking of Atlantis) went a bit further; this became *The Notion Club Papers*, which eventually encountered the same problem as *The Lost Road*: as soon as the time-barrier is crossed, the story seems to fall apart.

The completed section of *The Notion Club Papers* is illustrative of the difficulty Tolkien experienced in writing a true science fiction story, as it includes discussions on this very subject between members of the Notion Club, several of whom are writers. Some characters seem to be representing Tolkien's point of view, attacking the artificial nature of the pseudo-scientific explanations their authors feel compelled to give, and the machines they invent. The result, according to these characters, is a literary disaster, the destruction of the story's consistency, its aesthetic unity. Otherwise impressive portrayals of imaginary worlds are ruined by the fact that visitors arrive there aboard ludicrous vehicles. Better, then, to remain faithful to the narrative techniques of fantasy—or to have his heroes travel in time by means of recollections, visions, or dreams. This is the choice Tolkien himself made in *The Lord of the Rings*.

Tolkien's relationship with science was not without its contrasts. Without his avid interest in the cosmology, geography, and geology of our world, and without his knowledge of its plants and the creatures inhabiting it, he would never have had the desire to invent Middle-earth, or the energy and inspiration necessary to make it exist in its full breadth, in the depth of its history, and with the ring of truth in even its smallest details. At the same time, he was well aware of the way science could align itself with power (not only that of a visible dictatorship), imposing its values in an imperialist manner; this unleashed the anarchist tendencies he carried deep within himself alongside his love of order and harmony, inciting him to unyielding resistance.

In any case, science, with the fascination it sparks and the drive that ensures its ceaseless advancement, is part of the real world. It is for this reason that, with its ambivalence, it has a place in Middle-earth. Various ways of practicing it are represented there. Some of the darkest of these have already been touched upon by Tolkien himself: the Elves, a scientific race *par excellence*, ingenious and dangerous researchers ready to sacrifice anything for their passion for their inventions, like Fëanor in *The Silmarillion*, and like the thinkers of Lórien, capable of combining the quest for knowledge with artistic creation:

> *The Elves represent, as it were, the artistic, aesthetic, and purely scientific aspects of the Humane nature raised to a higher level than is actually seen in Men. That is: they have a devoted love of the physical world, and a desire to observe and understand it for its own sake and as 'other'—sc. as a reality derived from God in the same degree as themselves—not as a material for use or as a power-platform.*

This sublime and idealized image was offered by Tolkien in a 1956 letter to Michael Straits, but some of his humbler characters achieve a similar level of understanding, including the hobbit Sam Gamgee, whose botanical know-how and talent for gardening derive from the same love of plants.

1

WORLD-BUILDING

TOLKIEN: SCHOLAR, ILLUSTRATOR . . . AND DREAMER

CÉCILE BRETON, scientific journalist

"Science is what makes me dream."
Mr. Z (personal communication)

How was a professor of philology, so conventional in appearance and domestic in lifestyle, able to create such remarkable fictional worlds? This is the first question that comes to mind when looking at a portrait of John Ronald Reuel Tolkien in his tweed jacket, pipe in his mouth, posing in his cozy library.

We could dismiss this question by listing the many paradoxical literary figures counted among Her Majesty's subjects, but that would be taking the easy way out—and it would also mean forgetting the idea, deeply rooted in our culture, that scientists are necessarily devoid of imagination and feeling, given that their work is devoted to understanding reality. But let us resist the siren call of oversimplification that panders to our intellectual laziness, and, thanks to Tolkien the scholar, illustrator, and dreamer, let us try to show how absurd it is to consider science and imagination mutually exclusive. To do this, we will discuss that talent of his which, according to the prejudice described above, distances him the furthest from rationality: drawing.

Tolkien loved to draw. He would fill nearly any piece of paper with doodles, from his own manuscripts to newspaper clippings. He had little regard for these creations, and was reluctant to let his publishers print them; this was a private pleasure. The image we have of Middle-earth and its inhabitants—rather stereotyped, ultimately—was popularized, well after the publication of Tolkien's work, by illustrators such as Jon Howe and Alan Lee. But what do Tolkien's drawings reveal to us about the way in which he visualized his universe, and about the artists that inspired him? We will focus in this chapter on those sketches of Tolkien's that depict *The Lord of the Rings*, though he also produced drawings from nature (family life and landscapes) and illustrations for children (for his *Letters from Father Christmas* and *Mr. Bliss*), which are deliberately far more simplistic in style and deserve analysis elsewhere.

THE ORIGIN OF THE STORY: MEDIEVAL MAPS AND MANUSCRIPTS

The first edition of *The Hobbit, or There and Back Again*, released in September 1937, included two maps and eight drawings in black ink by Tolkien. These were sometimes colorized by others and republished, as in *The J.R.R. Tolkien Calendars* in 1973 and 1974, and in *The Hobbit Calendar* in 1976, with his son Christopher collecting forty-nine of them in *Pictures by J.R.R. Tolkien*. Later, Wayne Hammond and Christina Scull published around two hundred drawings in their *J.R.R. Tolkien, Artist and Illustrator*. Most of these are landscapes, but the volume also includes pages of calligraphy and decorative patterns.

The Hobbit, a book intended for children, contains far more illustrations than does *The Lord of the Rings* trilogy, the first editions of which include only maps. Concerning *The Hobbit*, Tolkien wrote in a 1955 letter (*Letters*, no. 163) that "... for a long time, and for some years [*The Hobbit*] got no further than the production of Thror's Map." Methodical by nature, Tolkien was obsessive about maintaining temporal and spatial consistency in the story, and his maps were the cornerstone of this. Daydreaming with his nose buried in an atlas was a favorite pastime of this little man born in the 1880s. Tolkien invented languages before he created the people that spoke them; he drew maps in order to imagine those who roamed them.

The cartography of his two main books, produced with the help of his son Christopher, uses the graphic codes common until the 16th century, combining "plan view" and figurative relief in perspective. Waterways are filled in with parallel lines, lakes with concentric lines. Captions, compass roses, and map scales are almost always present, and marine creatures such as dolphins can sometimes be seen at play in the ocean. Here, relieved of the need for realism, there is no desire other than to "create ancientness," to give the impression of reproduction. And, of course, as a linguist as well as an illustrator, Tolkien pays great attention to calligraphy.

As proof of the anxiety Tolkien felt at the idea of representing living beings, his landscapes are usually deserted, their inhabitants absent or reduced to tiny silhouettes. The only beast that seems to find favor in his eyes is the dragon Smaug, whom he shows several times, guarding his heap of gold or attacking Lake Town. This figure is wholly consistent with the common medieval image of the dragon, albeit sometimes touched with an Asiatic influence: the horse-like head, long, reptilian body, and bat wings. We know how much Tolkien loved dragons, from the one that slew Beowulf to Sigurd's Fáfnir.[*]

THE PREHISTORIC IMAGINARY

In Smaug's fiery attack on Lake Town, a quickly-executed pencil sketch, the dragon, hit by a black arrow, comes crashing down on the flaming city, portrayed as a long row of pilings and a palisade. Another image, far more elaborate and titled simply *Lake Town*, shows the lakeside city viewed from the shore. A bridge leads to a vast platform on pilings, atop which an impressive settlement has been built. Several barrels strewn around, a ferryman, and a boat attest to its commercial activity. Tolkien—and the images that moved him. One of the great archaeological discoveries of the mid-nineteenth century was the protohistoric lakeside habitat of the Swiss lakes: the exceptionally cold and dry winter of 1853–54 caused an extreme drop in water levels, revealing veritable "forests of wooden posts." More than two hundred 'villages' were discovered in this way throughout the Alpine region over the next two decades, and the image of these so-called *palafitte* cities,[†] spread through depictions in

---

[*] Beowulf and Sigurd are heroes, the former from Anglo-Saxon epic literature and the latter from Norse mythology. Beowulf is killed by a nameless dragon, while Sigurd vanquishes the dragon Fáfnir, guardian of a treasure hoard. Sigurd is the ancient form of the hero Siegfried from the *Nibelungenlied*, or *Song of the Nibelungs*.

[†] From the Italian *palafitta*, the etymology of which is subject to debate, but which broadly connotes the concepts of 'marshland' and 'wooden stakes driven into the ground'. This was the name given to Neolithic and Bronze Age cities atop pilings when they were discovered.

popular science books, left a profound impression on the popular imagination. These lakeside cities were also naturally represented by the master painter of Swiss pastoral scenes Albert Anker (1831–1910), known for his Rousseau-esque depictions of family and country life, as well as for his many intimate portraits of children. In *Der Pfahlbauer*, a *palafitte*-dweller with a Gaulois mustache lies in wait for prey, bow in hand, stretched out on his belly on a rocky outcrop overlooking his lakeside home. How can we fail to be reminded of the American Indians that populated young Tolkien's books? These archaeological discoveries crystallized in a popular imagination that blended their exoticism with an image of man "in his natural state," freed from the constraints of civilization, with its codified social relationships, giving numerous artists license to depict alluringly wild, half-naked women being assaulted by virile warriors, as in Paul Jamin's (1853–1903)[*] magnificent painting *Rapt à l'âge du Bronze*, or *A Rape in the Bronze Age*.

The boat with its zoomorphic prow (dragon? horse?), which seems to be headed back toward the city in Tolkien's sketch, immediately calls to mind a Viking ship, and *Beorn's Hall*, another of Tolkien's drawings, also points us toward the Nordic world. With its transverse beams and its large, sunken, quadrangular hall, this interior is evocative of mead halls, communal Viking structures used for both political and religious purposes. Hammond and Scull consider this drawing to have been inspired by another, published by his medievalist colleague E.V. Gordon, author of *An Introduction to Old Norse*, in 1927. Thus, we can rightfully see Lake Town as a "Neolithic-Viking"-inspired city, based on romantic interpretations of the archaeological discoveries of the century.

THE ENVIRONMENTAL IMAGINARY

In a letter of October 25, 1958 (*Letters*, no. 213), Tolkien writes: "I am in fact a Hobbit (in all but size). I like gardens, trees and unmechanized farmlands." What is so striking in his drawings, particularly if we compare them to the

---

[*]   French painter of the Academic Classicist school who drew much inspiration from archaeological discoveries, depicting scenes from Roman and Gallic antiquity, touched with patriotism but also serving as a pretext for erotic representations, such as *Le Brenn et sa part de butin* (*Brennus and His Share of the Spoils*), which shows a Gallic soldier entering a *gynaeceum*, or *Le rapt – âge de la pierre* (*A Rape in the Stone Age*), which shows a similar scene to the one in *Le rapt a l'âge du Bronze*, but set in the Stone Age.

battle scenes, monsters, and dark towers emphasized by those who illustrated the books later, is their peaceful, bucolic quality; the deep, dark forests, soaring, craggy mountains, and even the trolls lack any real sense of menace. It is difficult not to make a connection here with his childhood and his love for the green valleys of the Midlands. He was born in South Africa, and it was to protect her children from the heat that his mother decided to return to England when little John Ronald Reuel was a mere three years old. The family settled first in Worcestershire, and then in Sarehole, near Birmingham. Tolkien often spoke fondly of this happy time, which ended five years later when he began attending school in the Birmingham city center.

The landscape of Hobbiton, The Hill: Hobbiton-across-the-Water, may be interpreted as Tolkien's own image of an earthy paradise: hedged farmland, a river, and a hill overlooked by the circular windows of Bag End and its neighboring houses. Tolkien was a hobbit who loved trees above all else, not only for their majestic beauty, but also because he saw in them an illustration of how myths are constructed. He referred to *The Lord of the Rings* in a September 1962 letter (*Letters*, no. 241) as "my own internal Tree." Trees are omnipresent in his work, and a large number of them appear in one of his own favorite drawings, *Bilbo Comes to the Huts of the Raft-Elves*. The style of this piece, even more, perhaps, than *The Elvenking's Gate* or *The Doors of Durin*, is strongly Art Nouveau.

THE SYMBOLIST IMAGINARY

The Parisian metro stations designed by Hector Guimard (1867–1942) are the most famous expression of Art Nouveau, an artistic movement that emerged around 1890. The painters, illustrators, and architects who identified themselves with this movement drew extensively on shapes and forms found in nature, exalting the grace of curves both botanical and feminine. Animal life is also omnipresent in Art Nouveau and, among the forerunners of this anti-positivist movement, we find—paradoxically—a scientist who left his mark on the history of biology, Ernst Hæckel (1834–1919). His famous illustrative plates were used to support descriptions of numerous species, such as the jellyfish and octopi sagely coiling their tentacles in perfectly symmetrical spirals.[*]

---

[*]   Ernst Hæckel, *Kunstformen der Natur*, Olaf Briedbach ed., 1904.

Art Nouveau is considered to be one of the offshoots of Symbolism, an artistic movement with its origins in the mid-nineteenth century that emerged in reaction to Naturalism and Impressionism, postulating that art should attempt to represent a "reality" situated beyond the sensory world. Symbolism itself had its roots in a strictly British movement, the Pre-Raphaelite Brotherhood, whose preferred subjects were medieval mythology, Shakespearean theater, Dante's *Divine Comedy*, and, of course, the Bible. Ophelia, Merlin, Chaucer,* and Saint Agnes mingle on their canvases in an atmosphere blending the exaltation of religious sentiment and a prehistoric/medieval dress code. In a letter dated March 16, 1972, Tolkien wrote: "The great bank in the Fellows' Garden† looks like the foreground of a pre-Raphaelite picture."

It is to the Pre-Raphaelites that we owe the unrestrained use of color and the resurgence in popularity of stained glass (solid areas of rich color framed by line patterns), a style found in many of Tolkien's watercolors, such as *Rivendell*. These artists represent the pictorial expression of the nineteenth century craze for a romanticized medieval era.

Tolkien was also a great admirer of William Morris (1834–1896), that political figure emblematic of British socialism and possessor of many talents: writer, (another) translator of medieval literature, painter, architect, illustrator, and designer of textile patterns. A friend of Edward Burne-Jones (1833–1898), the driving force behind the Pre-Raphaelite movement, William Morris illustrated his own books with striking borders, intertwined botanical designs, and neo-medieval typeface. Many of Tolkien's watercolors feature these motif-embellished borders and captions surrounding the title of the piece. His admiration for Morris is impossible to miss in *Númenórean Tile and Textiles* and *Heraldic Devices*. He also said that while the Dead Marshes owed much to his dreadful experiences in the trenches of the Somme, Morris was the principal inspirer.

To "fill in" and give depth to his drawings, Tolkien uses a number of graphic techniques involving the use of lines and dots, as often seen in his maps as well. Another frequent user of these techniques, drawn from wood-engraving, included Henry J. Ford (1860–1941), illustrator of *The Red Fairy Book* by Andrew Lang, which introduced Tolkien to Norse mythology as a child. The

---

*   14th-century writer who was to Tolkien what the Italian Masters were to the Pre-Raphaelites in the field of literature; that is, the ultimate in literature itself.

†   The Fellows' Garden at Oxford University.

highly popular style, introduced by illustrators such as Aubrey Beardsley, can be seen in the trees and swirling smoke of *The Trolls*, a piece directly inspired by another illustrator of fairy tales, Jennie Harbour (1893–1959).

Symbolism and Pre-Raphaelitism, deemed elitist and backward-looking, have long been condemned to the purgatory of art history. However, they have found inheritors in those artists who deal in what has, since the 1950s, referred to itself as *fantasy*, a field of art long (still?) considered minor—undoubtedly because, like its precursors, it is subordinate to the texts in which it resides.

MAGIC AND MACHINES

The artistic atmosphere into which Tolkien was born was, of course, only one manifestation of a very specific historical and social context. The England of the late 19th century was steeped in the smoke that billowed from the factories that were the fruit of rampant industrialization, smoke that young Tolkien breathed in while living in Birmingham. Engineers were bent on definitively confirming both mankind's subjugation of nature and the supremacy of the British Empire over the rest of the world.

Thus it was in reaction to what they denounced as the "tyranny of reason" that the Symbolists developed a style of art touched with spirituality. Equally paradoxically, it was writers such as Mary Shelley[*] (1797–1851) and Auguste Villiers de L'Isle-Adam[†] who established the foundations for science fiction literature, with the intent of denouncing the way in which technology made man's most ludicrous desires into reality. Readers would do well to remember, incidentally, Tolkien's great admiration for the visionary American science fiction author Isaac Asimov (1920–1992).

The conflict between philosophies is unmistakable, and Tolkien perhaps put it best when he described a parallel between magic and machinery, both human creations intended to alter the natural order of things. In a 1951 letter (no. 131), he wrote:

> *By the last [machines] I intend all use of external plans or devices (appa-*
> *ratus) instead of development of the inherent inner powers or talents—or*

---

[*]   Author of *Frankenstein, or the Modern Prometheus*, 1818.

[†]   Author of *L'Ève future* (*The Future Eve*), among other works.

even the use of these talents with the corrupted motive of dominating:
bulldozing the real world, or coercing other wills. The Machine is our
more obvious modern form though more closely related to Magic than is
usually recognised.

"Progress," for Tolkien, is a Sauron which, through empty promises of power, alienates and distances us from the beneficial effects of the very nature we wish to dominate.

With the first social conflicts and railway disasters of the early twentieth century, the cult of progress took a hit, eventually becoming mired in the trenches of the Great War. The fallen angel of positivism took a collateral victim down with it as well: the scientist, who was suddenly seen as a lackey of power, a man-machine who posed a threat to the arts as well as to all forms of spirituality. And yet it was the knowledge he had acquired that fed Tolkien's inspiration: ancient texts in which knightly values were exalted and dragons still lived; archaeology that revived the image of a vanished Eden in which Neolithic hobbit-men enjoyed the simple pleasures of life in the open air.

Researchers seeking to explain the current rise of religious extremism, or the success of astrology, in a world where one might expect belief to have been weakened by advances in knowledge, point out, in the words of sociologist Romy Sauvayre, that "scientific discoveries and theories that science is attempting to prove are so stimulating to both non-scientists and insiders that they open doors to the imaginary, to the inexhaustible realm of the possible." It is undeniable that science nourishes the imagination. Fantasy, while turned toward a mythical past, and science fiction are two sides of the same coin.

Let us return to our question, a deliberately provocative one: do scientists have imagination? How could a man who never traveled farther from home than Switzerland invent Barad-dûr and Lothlórien? The question is answerable thanks to Tolkien, living proof that his objective understanding of medieval myths and legends, of the way in which they were constructed and subsequently evolved, and of the language that expressed them, nourished and informed his dream. It was his scientific, methodical mind—thought by some to be so devoid of romanticism—that enabled him to construct his complex universe on such solid foundations. Tolkien's work is fantastical without being absurd, and it is in its very consistency and logic that its evocative power lies.

The antagonism between scientific curiosity and an attraction to fantasy, then, exists only in those unhappy minds that remain bound by preconceived notions. Tolkien understood this very quickly. "I was an undergraduate before thought and experience revealed to me that these were not divergent interests—opposite poles of science and romance—but integrally related," he wrote in late 1951 (letter no. 131). Readers of this book will undoubtedly understand it, too.

TOLKIEN AND SOCIOLOGY:
FACING THE LOSS OF A WORLD

THIERRY ROGEL, Associate Professor of
Economic and Social Sciences

TWO PERSPECTIVES ON
A WORLD IN TRANSITION

 peaking at a conference on fairy tales* at the University of
St. Andrews in 1939, Tolkien addressed two common miscon-
ceptions. The first was that fairy tales are intended for children;
he, who had developed an interest in fairy stories through his study of philology, believed that adults could appreciate these tales as much as children,
and that it was only by historic accident that fairy stories had become associated
with childhood. The second common misconception concerned the escapist
nature of these tales; while fairy stories are indeed "stories of escape," Tolkien
said, this is not the escape of the deserter, but that of the prisoner, and it is
not duty that is being fled from, but rather a prison, in order to reach a
desired world. Tolkien considered fairy tales (and fantasy, the first example
of which is generally to have appeared in 1865 with Lewis Carroll's *Alice's
Adventures in Wonderland*) to be a way of understanding the world ("I think

---

*   Tolkien assigned a broader meaning to this term than the common definition: for him, a
"fairy tale" was a story that dealt with fantasy (or magic).

that fairy story has its own mode of reflecting 'truth'," he wrote in Letter no. 181, dated January or February 1956) for in creating a "secondary world," we are not speaking of what is possible, but of what is desirable. Stories do not stand in opposition to knowledge in this respect, and have their own view of the real world.

However, Tolkien did not use stories as a means of criticizing the world, and had no intention of making *The Lord of the Rings* an allegory for the modern world. He regularly dismissed the suggestion that the trilogy was meant to depict the Nazi threat, Communism, or the nuclear bomb. And yet, for all that, he did not reject the notion that the sensitivity of writers to the world around them could be seen in their work. Clear in Tolkien's writings are his rejection of—and even disgust for—the modern era, born during the "long 19th century," which spanned the years from the Industrial Revolution to the First World War, and dying in the 20th.

His rejection of modernity is plainly visible in his correspondence. He loathed the automobiles that were ruining his world: "[. . .] the spirit of 'Isengard', if not of Mordor, is of course always cropping up. The present design of destroying Oxford in order to accommodate motor-cars is a case" (draft of letter no. 181, January or February 1956). He disliked machines in general, with the exception of the typewriter* (he would dearly have loved to possess a model equipped with Feanorian letters, as he mentioned in Letter no. 257 on July 16, 1954): "[. . .] There is the tragedy and despair of all machinery laid bare. Unlike art, which is content to create a new secondary world in the mind, it attempts to actualize desire, and so to create power in this World; and that cannot really be done with any real satisfaction." (Letter no. 75, July 7, 1944). Likewise, he disdained modern cities, and was horrified by "the lunatic destruction of the physical lands which Americans inhabit" (draft of Letter no. 328, Autumn 1971). Modern entertainment found no favor in his eyes any longer either, including what was being produced by Disney, as noted in a letter written in May 1937 (just a few months before the release of *Snow White*, the first feature-length color cartoon in history) as part of an epistolary exchange concerning the possible publication of *The Hobbit* in the United States. Tolkien, speaking of possible illustrations, firmly vetoed "anything from or influenced by the Disney studios (for all whose works I have a heartfelt loathing)" (Letter no. 13 of May 13, 1937), an opinion which had

---

*   Though he was wary of tape recorders . . . Humphrey Carpenter, *J.R.R. Tolkien—Une biographie*, Paris, Christian Bourgois, 1980, p. 193.

not changed years later when he spoke of one story as a "terrible presage of the most vulgar elements in Disney" (Letter no. 234 of November 22, 1961). In short, he did not like the modern world: "Such is modern life. Mordor in our midst," he wrote in Letter no. 135 of October 24, 1952, later describing "the evil spirit (in modern but not universal terms: mechanism, 'scientific' materialism)" (Letter no. 96, January 30, 1945).

This sense of loathing crops up numerous times in books I and II of *The Lord of the Rings*, when, for example, Gandalf explains how Saruman imprisoned him on Orthanc, from whence he surveyed the landscape: "the valley below seems far away. I looked on it and saw that, whereas it had once been green and fair, it was now filled with pits and forges. Wolves and orcs were housed in Isengard [. . .]. Over all his works a dark smoke hung and wrapped itself about the sides of Orthanc." Wouldn't this description apply equally well to England, which was at the forefront of the Industrial Revolution?

In this sense, Tolkien was like Don Quixote, battling an unbearable reality. Another way of reacting would have been to accept the real world and its changes, and to analyze them in order, perhaps, to change the course of things. The social upheavals of the time were also responsible for the surge in popularity of sociology, the vital role of which, according to the German sociologist Norbert Elias, is to "trace the origins of myths."

Indeed, many "classical" sociologists working between 1830 and 1920[*] were struck by the effects of the two great revolutions, the Democratic and the Industrial, wrote Robert Nisbet in 1966. The convulsions that gripped the world were many-faceted: industrialization, and the calling into question of traditional hierarchies; growing rationalization and disenchantment with the state of the world; the displacement of holy sites; the transformation of social bonds, etc. So much change did not come without crises, social conflicts, anomie, and a "leveling of the world." These became focal points of interest for the classical sociologists, with each concentrating on a specific angle of analysis: class and production relationships (Karl Marx), rationalization and disenchantment (Max Weber), the objectification of society and the role of money (Georg Simmel), transformations in social bonds (Émile Durkheim), and so on.

---

[*]   Such as Karl Marx (1818–1883), Alexis de Tocqueville (1805–1859), Ferdinand Tönnies (1855–1936), Max Weber (1864–1920), Émile Durkheim (1858–1917), and Georg Simmel (1858–1918).

Unlike Tolkien, sociologists acknowledged the global changes occurring. As Robert Nisbet wrote in 1966: "Our civilization is urban, democratic, industrial, bureaucratic, rationalized; it is a civilization on a large scale that is formal, secular, and technological. [. . .] The fact that many of us feel a certain malaise, a kind of perplexity, and even a certain nostalgia in viewing the results of these two revolutions changes nothing, and even if a few Don Quixotes attempt, now and then, to tilt against windmills, these results are here, and they are irreversible."

COMMUNITY STEPS

The reading grid that has remained foundational to modern sociology is that of Ferdinand Tönnies, emphasizing the opposition between community and society that can be seen in the background of Tolkien's work. In his book *Community and Society*, published in 1887, Tönnies showed that every human grouping is the result of one of two forms of social bonding: the first, the "communal bond," ensures solidarity between individuals through the depth and warmth of feeling and the recognition of mutual common feelings linked to habit and custom. This is the bond typically found in family relationships and close friendships, or the relations between neighbors in a village. In these cases, we generally belong to the group without having chosen to join it.

But, in order for humans to be able to live together, we must also have "social bonds," deliberately chosen relationships usually based on logic or calculation. Contracts of association, business contracts, and commercial relationships are the most obvious examples of this type of bond; companies are the organizations that come closest to this sort of social interaction, and cosmopolitan cities the prime setting for it.

According to Tönnies, all forms of society result from a combination of these two types of bonds. However, he also specifies that the developments occurring in the 19th century caused a shift from a world in which communal bonds (families, villages, and tradition) dominate, to one where social bonds (commercial relationships, business activity, and social relations in large cities) are paramount. In other words, a shift from *community* to *society*.

Tolkien's descriptions of hobbits and the Shire in the first chapter of book I of *The Lord of the Rings* are thoroughly representative of community as Tönnies meant it, based on families and clans:

The houses and the holes of Shire-hobbits were often large, and inhabited by large families. [. . .] All hobbits were, in any case, clannish and reckoned up their relationships with great care. They drew long and elaborate family-trees with innumerable branches.

Above all, their world is rural: "they love peace and quiet and good tilled earth: a well-ordered and well-farmed countryside was their favorite haunt." They have no love for knowledge for its own sake, or for the unknown that it represents: "A love of learning (other than genealogical lore) was far from general among them, [. . .] they liked to have books filled with things that they already knew, set out fair and square with no contradictions."

And they dislike machines: "They do not and did not understand or like machines more complicated than a forge-bellows, a water-mill, or a hand-loom, though they were skilful with tools."

Most importantly, the community formed by hobbits is quite closed ("[They] meddled not at all with events in the world outside") and, mostly impervious to the passage of time, they "do not hurry unnecessarily." Tolkien was unquestionably describing a lifestyle to which he aspired himself, modeled on an England that was disappearing. Change will occur in the hobbits' world as well, however, caused by two things: the "different" individual, and the outside world.

TOWARD FELLOWSHIP

The sociologist Émile Durkheim, discussing the concept of "mechanical solidarity" in his *Division of Labour in Society*, believed that "community" based its vital cohesion on the conformity of individuals to a collective model, and on their submission to a collective consciousness. Anyone who is different is therefore perceived as a threat by the group, and will be irrevocably rejected. This is precisely what happens to Bilbo; seen as a friend of the Elves, honored by Dwarves and wizards, he is suspected of associating with all sorts of strangers, and his reputation is blackened by it.

However, while deviants are rejected by the group, they are also agents necessary for its evolution, for what is seen as a crime today may be the harbinger of normality to come. Likewise, the change at the center of *The Hobbit* and *The Lord of the Rings* is not initiated by just any individual; Bilbo and

Frodo are very particular hobbits, unmarried at the beginning of the story, each adventurous in his own way, and, in the end, friends with outsiders. Tolkien states this explicitly in book 1 of *The Lord of the Rings* ("Concerning Hobbits"): "The houses and the holes of Shire-hobbits were often large, and inhabited by large families. (Bilbo and Frodo Baggins were as bachelors very exceptional, as they were also in many other ways, such as their friendship with the Elves.)"

Bilbo is also remarkable because he leaves his cocoon and embarks on an adventure, as a thief, with Gandalf and Thorin's Dwarves. It is in this that we see the beginnings of change. In point of fact, the association between Bilbo and the Dwarves is not based on their belonging to the same group, or on mutual feelings, but on converging interests, and has all the hallmarks of a social bond. The symbol left on Bilbo's door (undoubtedly written by Gandalf) attests to this, resembling as it does an application for employment: "Burglar wants a good job, plenty of Excitement and reasonable Reward." Any doubt about the commercial nature of the undertaking is eliminated when Thorin addresses himself to Bilbo the next morning:

> *Thorin and Company to Burglar Bilbo greeting! For your hospitality our sincerest thanks, and for your offer of professional assistance our grateful acceptance. Terms: cash on delivery, up to and not exceeding one fourteenth of total profits (if any); all traveling expenses guaranteed in any event; funeral expenses to be defrayed by us or our representatives, if occasion arises and the matter is not otherwise arranged for.*

That "and Company" is more reminiscent of a trading company than it is of a medieval company of soldiers—indeed, conditions of payment and of compensation in case of loss are clearly laid out. There is no honor or set of values to be defended here; a commercial relationship is indisputably being formed.

The same is true at the beginning of Book One of the Trilogy, when Gandalf returns to ward off the threat posed by Sauron's acquisition of the One Ring. This is certainly not a business association, but neither is it a "communal relationship" as meant by Tönnies. It is no longer a question of remunerating Frodo, but one of saving Middle-earth, which explains the coming together of representatives of the land's various free peoples. Nine companions to face the nine Black Riders: Gandalf for the Wizards (but sometimes closely connected to the Elves), Legolas for the Elves, Gimli for the Dwarves, Aragorn

Gandalf surveying the devastation of Isengard.

and Boromir for the Men, and finally the four hobbits, Frodo, Sam, Merry, and Pippin. The fellowship is not commercial, but voluntary, and destined to destroy the One Ring in order to vanquish Sauron; therefore, it is not a true community. Community or association? It is noteworthy that Gérard Klein, discussing *The Lord of the Rings* in 1969, before its French-language publication, translated the original title *The Fellowship of the Ring* as *The Company of the Ring* and not *The Community of the Ring.*[*]

This association is also a likely harbinger of other connections and alliances, including the friendships between certain hobbits and Elves, and the upcoming marriage of a Man, Aragorn, to the Elf (or half-Elf) Arwen. And so, the diverse peoples that are Elves, men, Dwarves, and hobbits will live their respective lives while following a common path, thus creating a shared sense of belonging. Without going so far as to speak of a "nation," which would be anachronistic, we might borrow the terms employed by the philosopher Ernest Renan in his famous lecture "What is a nation?": "having common glories in the past and a will to continue them in the present; having made great things together and wishing to make them again; these are the essential conditions of being a people. [. . .] A long past of efforts, sacrifices, and devotions [and] a will to continue them in the present." This is what is in the making, throughout the entire journey of *The Lord of the Rings*.

ON THE BORDERS

Above all, a sense of community lies in the awareness of that community's border with the outside, which strengthens its internal cohesion in return, as explained by the sociologist Georg Simmel. The arrival of a foreigner, a stranger, can disturb this border, and the community. Simmel employs the stranger as a sociological concept designating one who is simultaneously inside and outside of the group (traveling salesmen, tourists, travelers, but also those who are deviant, or represent a minority group); for him, a human collective defines itself by the attitude it adopts regarding strangers. And the small world of the hobbits is built, in part, around the fear of those who prowl along the borders:

---

[*]   The first French translation was by Francis Ledoux (1972). In the new translation by Daniel Lauzon (2004), the title becomes *The Brotherhood of the Ring*.

At the time when this story begins, the Bounders, as they were called,
had been greatly increased. There were many reports and complaints of
strange persons and creatures prowling about the borders, or over them:
the first sign that all was not quite as it should be, and always had been
except in tales and legends of long ago.

(From *The Fellowship of the Ring*, Prologue,
"Concerning Hobbits")

We think immediately of Sauron and his accomplices, but the true figure of the stranger is Gandalf. Arriving from elsewhere, he has kept up ancient connections with a few members of the community, and is therefore both a member of the group and an outsider. It is he who will take the home-loving Bilbo out of his familiar world, and lead him to discover other peoples. This will change Bilbo profoundly, as forging alliances with Elves and Dwarves means *understanding* them (in the sociological sense of the word; that is, giving meaning to their customs and their way of life). Understanding may then replace other ways of encountering the other, such as conflict, rejection, and the desire for domination. Wasn't this the challenge Europe faced at the turn of the twentieth century? Whereas, before the nineteenth century, Europeans were interested in the other countries of the world only as objects of curiosity, pagan lands to be converted, or sources of wealth and labor, now they became subjects of scientific interest, with the flourishing of ethnology and anthropology. And while there are no Elves or Dwarves on our Earth, still we have ceased to believe that any of its creatures are monstrous half-humans half-animals, embracing unity of humankind at last.

MYTHOLOGY VS. MYTHOLOGY: TOLKIEN AND ECONOMICS

THIERRY ROGEL, Associate Professor of
Economic and Social Sciences

A PRE-ECONOMIC WORLD

If there is one social science that dominates our world, it is undoubtedly economics. It has always been necessary to produce and exchange in order to be able to consume, but it was only in the 18th century that companies ceased to be founded mainly on the respect due to allies, religious duties, and political issues. Two years stand out in the history of economics: 1776, which saw the publication of *The Wealth of Nations* by Adam Smith (1723–1790), considered to be the seminal work on political economy in the modern sense of the term, and 1834, which marked the abolition of the Speenhamland System,* which, by imposing the creation of a true labor market, ushered in the first example of a market economy in the full sense of the term.

Yet, there are few traces of economics in Tolkien's writing, given that he is describing a time (and an imaginary one at that) when the economy was not at the center of social life.

---

* Also known as the Berkshire Bread Act, an amendment to the Elizabethan Poor Law; this law, in effect in England and Wales from 1975 to 1834, provided the poorest people with a minimum income based on the price of bread and on family size.

Middle-earth is indisputably a "pre-economic" world. Tolkien himself admitted that he had spent little time developing its economic aspect (Letter no. 154, dated September 25, 1954)—but this was not, as he made a point of saying, because he lacked the knowledge. He does list a number of points that could give rise to a description of Middle-earth's economy: Gondor has enough fiefs and parcels of land supplied with potable water to sustain its population, and it possesses numerous workshops; the Shire is close enough to mountains and waterways to be agriculturally fertile, and the Men of Lake Town live mainly by trade. The Dwarves, who work in the mines, deal in metals and weapons, and it is they who seem the most adept at business, even more so than men, and the most driven by what might be called "the economic spirit." Elves, on the other hand, care nothing for metalworking, commerce, or land cultivation, though the Elf-King keeps a close eye on the cargoes traveling along the toll roads, which guarantee him a certain level of affluence. If the history of *The Lord of the Rings* is that of the end of the world of the Elves in favor of the burgeoning world of Men, we might also infer that it is the end of the world of art in favor of the world of commerce, as well.

Tolkien's remarks on economics in his letters imply that he had some knowledge of economic geography, and even economic history, but there is no hint that he ever studied political economy or economic science as they existed in the twentieth century. Modern economic thought is young, dating from the same period as the first Industrial Revolution and the emergence of the other social sciences. The classical economists,[*] however much they differed on other points, situated their economic reflections in the context of a society based on class structure (Adam Smith and David Ricardo spoke of social classes well before Karl Marx), but the real surge came at the end of the nineteenth century. Amid the scientistic atmosphere of the era, a group of economists, of whom Léon Walras is the best known, resolved to make economics into a true "pure and hard" science modeled on physics, and with the objective of establishing universal laws similar to the laws of physics, valid in any time or place. The analysis of concrete cases (applied economics) would rely on the demonstration of these laws. Abstract concepts would be created such as the rational and independent *homo economicus* (so independent, in fact, as to be impervious to all social relationships), and a market structure of "pure and perfect

---

[*]   Adam Smith (1723–1790), Thomas Robert Malthus (1766–1834), David Ricardo (1772–1823), Karl Marx (1818–1883)

competition" which would be equivalent to the perfect gas in physics. The new discipline would, in other words, remove sociology and history from its thought process. This approach became largely dominant in twentieth-century economic research even as it was harshly criticized by Friedrich Hayek and John Maynard Keynes, probably the two greatest economists of the century.

Without knowing Tolkien's thoughts on economics, we might reasonably suppose that he had little interest in this excessively mathematized discipline, which claims to be rational by eliminating all passion from the field of human activity. However, his letters do give us a few hints about his opinions regarding economic doctrine: he disliked socialism "in either of its factions now at war," seeing it as "the evil spirit" along with the machine and "scientific" materialism (Letter no. 96 of January 30, 1945); he was opposed to planning (Letter no. 181, dated 1956), and hated it when anyone talked to him about the government. Yet he was not a reformer, either ("I am not a 'reformer' (by exercise of power) since it seems doomed to Sarumanism," he wrote in Letter no. 154 on September 25, 1954), and the idea that even the slightest political reform would lead irrevocably to Saruman's unjust rule is not without echoes of Hayek's words in *The Road to Serfdom*, when he insists that any government intervention can only end in totalitarianism.

Tolkien wasn't a socialist, and might not have agreed with the economist Karl Polanyi's criticisms of the market system in *The Great Transformation* (1944), but the respective descriptions of the damage done by Saruman by the former, and industrialization by the latter, are curiously similar. When, in *The Two Towers*, Tolkien describes Isengard: "there Saruman had treasuries, store-houses, armories, smithies, and great furnaces. Iron wheels revolved there endlessly, and hammers thudded. At night plumes of vapor steamed from the vents, lit from beneath with red light, or blue, or venomous green," Polanyi seems to reply: "No society could stand the effects of such a system even for the shortest stretch of time[. . .], unless its human and natural substance, as well as its business organization, was protected against the ravages of this satanic mill." When, also in the second volume of the trilogy, Frodo and his friends venture into Mordor, Tolkien writes: "They had come to the desolation that lay before Mordor: the lasting monument to the dark labor of its slaves that should endure when all their purposes were made void; a land defiled, diseased beyond all healing," and Polanyi echoes, "Such an institution [the market] could not exist for any length of time without annihilating the human and natural substance of society; it would have physically destroyed man and transformed his surroundings into a wilderness," concluding: "After a century of blind 'improvement' man is restoring his 'habitation.' If industrialism is not to extinguish the race, it must be subordinated to the requirements of man's nature."

And Keynes? It is hard to imagine Tolkien, ardent Catholic that he was, who struggled to save money all his life and complained about the increased taxes that came along with the success of *The Lord of the Rings*, would have seen much allure in Keynes, a libertine and homosexual whose economic doctrines were based on the benefits of expenditure. However, he might not have disagreed when Keynes wrote in 1930, in his *Economic Possibilities for our Grandchildren*:

> *It is a fearful problem for the ordinary person, with no special talents, to occupy himself, especially if he no longer has roots in the soil or in custom or in the beloved conventions of a traditional society. To judge from the behaviour and the achievements of the wealthy classes to-day in any quarter of the world, the outlook is very depressing! [. . .] I see us free, therefore, to return to some of the most sure and certain principles of religion and traditional virtue—that avarice is a vice, that the exaction of usury is a misdemeanour, and the love of money is detestable [. . .] We shall once more value ends above means and prefer the good to the useful. [. . .] But, chiefly, do not let us overestimate the importance of the economic problem, or sacrifice to its supposed necessities other matters of greater and more permanent significance.*

Believing that the economic issue could be resolved during the century to come, and that three hours of work per day would be enough to satisfy our basic requirements, Keynes worried about the ways in which people might use the additional free time, a challenge humanity had never before confronted: "Yet it will only be for those who have to do with the singing that life will be tolerable and how few of us can sing!" We know how important singing is in *The Lord of the Rings*.

MONEY AND POWER

Gold and money are frequently talked about in *The Lord of the Rings*, but talking about money, or even currency, doesn't necessarily mean talking about

economics. According to the common definition of the term, and to mainstream economists,* currency is merely a neutral object that facilitates exchange
and makes it possible to avoid the difficulties of the barter system. Such is the
pretty tale originally spun by Adam Smith (who could hardly do otherwise,
given the absence of data), but it must be abandoned now, undermined as it has
been by historical and anthropological research. This has shown, in fact, that
exchanges have also been made between groups or within a group as part of
a "gift culture"; it appears that a widespread bartering system did not exist in
traditional societies, and hardly ever appears in modern societies except during
a period of crisis (as explained by André Orléan in 2008). Be that as it may, the
"myth of bartering" has become rooted in our collective imagination, with not
insignificant consequences. It implies that money is not desired for itself, but
solely for the objects it enables us to acquire: this is the idea of "the neutrality
of currency" (or the "veil of money") that lies at the heart of mainstream economic thought, but which runs counter to an even slightly careful observation
of reality. Fortunately, not all social-science thinkers adopted this perspective:
for Keynes, money could be desired for itself, either for comfort in the face
of an uncertain future (in which case it will be kept for as long as possible in
the form of savings), or for the purposes of speculation (making it the basis
of a love of money for money's sake, which Keynes saw as a sickness). In his
weighty tome *The Philosophy of Money*, Georg Simmel insisted that psychological predispositions with regard to money are anything but neutral: one may
desire money for itself (cupidity), and refuse to be separated from this money
without good reason (avarice); one may also desire it wholly for the pleasure
of spending, without even trying to gain from the goods purchased (what
Simmel calls prodigality). In rare cases, one may refuse to possess money (as
in the case of ascetics or individuals stricken with "monetary anorexia"), and
one may desire to possess money out of a simple wish for power or respect.

We see very little of this "veil of money" in Tolkien. The desire for gold in
The Lord of the Rings and *The Hobbit* is rarely driven by the yearning for a better
life (though this is occasionally the case for hobbits), but rather by a hunger
for power. In *The Hobbit*, the Elvenking dreams of nothing but amassing
"treasure, especially silver and white gems," and he is always eager to enlarge

---

*   Mainstream economists are those who participate in the current dominant research trends
      in economics, characterized by significant use of mathematized modeling, to the detriment
      of possible contributions made by other social sciences. Though mainstream economics
      is not necessarily linked to economic liberalism, many observers associate the two.

his fortune, to make himself the equal of the elf-lords of ancient times. The Dwarves, more than gold and silver, wish to possess "truesilver," the Elvish name for which is mithril. Mithril can be hammered like copper and polished like glass; it is both light and strong and does not tarnish over time, and is used to make mail-shirts, helmets, ships, the gates of Minas Tirith, and, of course, jewelry (including Nenya, the second Ring of Power). Its rarity and usefulness help to explain why it is worth infinitely more than gold, but it is also surely because its possession brings great admiration that "all folk [desire] it." More striking still is, of course, Smaug's wish to sleep atop a pile of gold from which he gains nothing; neither enjoyment of objects purchased, nor the esteem of others, nor power. Smaug's desire for gold is desire for its own sake; this is cupidity, pure and simple.

This brings us to the question of the value of things, the determination of which separates economists into two camps. For classical economists and Marxists,* value stems from the work that goes into an object (in this case, the fact that one must dig deeper and deeper to find mithril). For the neo-classicists that succeeded them, an item's value derives from its "utility," the objective usefulness linked, for example, to the various qualities that mithril possesses, but also subjective utility, which has to do with the fact that value is above all a phenomenon stemming from individual psychology. Both of these approaches have the serious flaw of situating value outside of any social bond between humans; thus it predates both exchange and society. However, modern economists such as André Orléan posit that value, even subjective value, does not arise miraculously in the head of each person, but is rather created by interactions between individuals and their desire for imitation, born of social comparison, a theme found in somewhat exaggerated form in Thorstein Veblen's analysis of the "demonstration effect."

Tolkien's views on economics and money are, then, a world away from the tenets of prevailing economic thought but, with regard to money, he is perhaps in closer touch with reality than the mainstream economists who, remember, embraced the famous misconception of the creation of money as an overcoming of the barter system. The idea of this neutral currency is vital to maintaining the image of a rational and desocialized *homo economicus*,

---

*   The term "classical" includes most 19th-century authors (Smith, Ricardo, etc.). They are distinguished from the "neoclassicists" (Walras, Jevons, Pareto, etc.), who emerged at the turn of the 20th century.

the cornerstone of the economic approach prevalent today (even if this *homo economicus* is regularly revised and fine-tuned). Indeed, if money can be desired for itself, then the individual does not regard himself only in relation to the goods he is able to acquire, but also in comparison to other members of society (as we see in the case of the Elvenking), unless his behavior has no rational basis (as in the case of Smaug).

According to the anthropologist David Graeber, the myth of the barter system persists because it is the linchpin of mainstream economics as a whole. So, we may smile at the way in which Paul Samuelson—the author of *Economics*, the bestselling economics textbook in history—explains the genesis of currency: "If we had to reconstruct History according to logical theories, we would naturally suppose that the age of the barter economy had to be succeeded by the age of commodity money." Reconstruct history according to logical theories! No need then, for long hours spent in archives reading about the use of currency in past societies, or for extensive fieldwork aimed at learning what currency represents in a particular traditional society, or understanding the way in which exchanges of money are woven into friendships or family relationships (as is found in the work of historians, ethnologists, and sociologists). Samuelson's thoughts alone are apparently enough to reconstruct history. It's clear that we are not so far from Tolkien here; he wanted to give England a mythology of its own, and this explains the fact that the invention of currency in order to avoid the problems posed by bartering is categorized as a "myth" by many ethnologists and economists. Indeed, it *is* a myth, in the broadest sense of the term: this is a story located outside of time (when did two humans decide to invent money?), featuring powerful heroes (already rational despite being freshly hatched), accepted by most of our contemporaries as a faithful account of reality as it happened, and describing the creation of a world (like all origin myths). Like any foundational discourse, it has the capacity to transform the world, and what a world! Ours, the one that Tolkien despised. The truth is that, if there is hardly any economics in Tolkien, it's because two mythologies cannot exist side by side.

FAMILIES, POWER, AND POLITICS IN *THE LORD OF THE RINGS*

THIERRY ROGEL, Associate Professor of
Economic and Social Sciences

THE LORD OF THE RINGS: A WORLD OF KINSHIP

olkien's writings teem with references to various affiliations and kinships, in a manner reminiscent of the comments on lineage found in the Norse sagas that were among his major sources of inspiration.

Among hobbits, the importance of family is paramount, the most prominent families giving their names to the districts and regions that make up the Shire. The hobbits' society is a patrilineal one[*] (as confirmed by Tolkien in Letter no. 214, dating from 1958 or 1959), and genealogy seems to be one of the few areas of study deemed worthy of interest: "A love of learning (other than genealogical lore) was far from general among them" (*The Fellowship of the Ring*). In *The Two Towers*, Gandalf does not hesitate to warn his listeners: "These hobbits will sit on the edge of ruin and discuss the pleasures of the table, or the small doings of their fathers, grandfathers, and great-grandfathers, and remoter cousins to the ninth degree, if you encourage them with undue patience."

---

[*]   In which descent is traced through the paternal line.

The bond between cousins is particularly important to hobbits. A cousin is both close and distant; distant enough not to be part of the immediate family, but close enough to be trusted, and to make a good ally. It is not surprising, then, to find in *The Lord of the Rings* a number of cousins who have embarked on the same adventure: The hobbits Frodo, Merry, and Pippin, and also the Dwarves Bifur, Bofur, and Bombur. Likewise, Thorin invites the adventurers to contact his cousin Dáin in the Iron Hills. In the French edition of *The Return of the King*, the elf Celeborn goes so far as to say "Cousin, farewell!," which may be confusing to readers; the English word used by Tolkien here is *kinsman*. Are they indeed distant cousins, or does the appellation mark the beginning of a new alliance?

Yet, the role of the cousin is an ambivalent one. He may be a protector; Bilbo, for example, "had no close friends, until some of his younger cousins began to grow up" (*The Fellowship of the Ring*). Likewise, it is thought in Hobbiton that the weakened Frodo has gone to live in Buckland with his cousins, and the Underhills of Staddle, who share Frodo's assumed name, although they have never met him, "[take] [him] to their hearts as if he were a long-lost cousin" (*The Fellowship of the Ring*). But cousins can be hostile as well, as in the case of the Sackville-Bagginses, who are eager to strip the presumed-dead Bilbo of his possessions in *The Hobbit*.

In human societies, cousins are situated at the junction between the family and the rest of society. The cousin's position within the framework of the rules of marriage is significant in this respect, generally poised on the boundary between the permissible and the forbidden; cousin-marriage can, depending on the case, open a group up to the outside world—which is desirable—or close it off, coming under the taboo of incest. For example, as described by the anthropologist Christian Ghasarian, in many populations, marriage to a cross cousin (the daughter of either the mother's brother or the father's sister) is encouraged, while marriage to a parallel cousin (the daughter of either the father's brother or the mother's sister) is forbidden, with the former beginning a cycle of alliances and the latter closing the family group in on itself.

Certain references in Tolkien indicate that some cousin-marriages are prohibited: "the laws of Númenor did not permit the marriage, even in the royal house, of those more nearly akin than cousins in the second degree" (*The Silmarillion*). But this type of marriage does not seem to be forbidden among hobbits, since we are told in *The Fellowship of the Ring* that Mr. Drogo, Frodo's father, married Primula Brandybuck, Bilbo's first cousin on his mother's side, "and Mr. Drogo was his second cousin."

Because cross-cousin marriage is favored in many societies, the position of the mother's brother becomes crucial, as it is this brother who gives up his sister by handing her over to a third party (who is a future ally), becoming a "wife-giver," to use the term coined by Claude Lévi-Strauss. The relationship between a nephew and his maternal uncle (the "avuncular relationship") then becomes a central one, with the son/mother/mother's brother triptych thus forming the "atom of kinship." This avuncular bond, which is key to ethnological analyses of the family, also plays a key role in medieval literature, which contains frequent references to fosterage, a situation in which the maternal uncle is bound to educate his sister's son and to make him an heir of equal standing to his own children.*

Knowing Tolkien's expertise in medieval literature, it is not surprising that the uncle/nephew relationship features prominently in his writings (it would also be contemporaneously analyzed by the ethnologist Alfred Radcliffe-Brown, himself an Oxford professor, in a 1941 article), but not always in enough detail to determine if the relationship in question is an avuncular one or is between a paternal uncle and a nephew. In *The Lord of the Rings*, we learn from Boromir that Isildur instructed his nephew in Elendil in the ways of governance, entrusting him with the rule of the southern kingdom, and Denethor takes an interest in the position of Eoren, the king's nephew. Nephews often succeed their uncles; Tarondor succeeded his uncle King Telemnar, and Eärnil his uncle Tarannon (who reigned under the name Falastur). Succession can also occur via usurpation: Ar-Pharazôn (the Golden) takes the Sceptre of Númenor from his uncle Tar-Palantir and marries his cousin, Tar Palantir's daughter, against her will. But the central relationship in *The Lord of the Rings* involves Bilbo and Frodo. Tolkien does not give us an easy task here; Bilbo is the grandson of Mungo, who is himself the brother of Largo, great-grandfather of Frodo, so Tolkien is accurate both when he writes that Frodo is the oldest of Bilbo's cousins (and his favorite), and when he has Bilbo say, during a banquet, that Frodo is his nephew and heir.

---

* The avuncular relationship is also vital in a third case: the familial relationships of Donald Duck, established by Carl Barks . . . one of the great mysteries of the human sciences.

HOUSE SOCIETIES: THE INTRUSION OF POLITICS INTO KINSHIP

Ethnological theories concerning the family are constructed around two axes, alliance and filiation. However, ethnologists have sometimes found themselves faced with situations stemming from neither one nor the other. Claude Lévi-Strauss was the first to study such cases, which he called "house societies" (*sociétés à maison*). The House is a corporate body ("moral person") that possesses an estate composed of both material assets (land, property, etc.) and immaterial ones (traditions, beliefs, legends, names, rights over certain rituals, etc.). It perpetuates itself "by transmitting its name, its goods, and its titles down a real or imaginary line, considered legitimate as long as this continuity can express itself in the language of kinship or of affinity and, most often, of both," as Lévi-Strauss explained to the anthropologist Pierre Lamaison in 1987. The House, which sometimes lacks any biological foundation, thus makes it possible to reconcile the conflicting states of alliance and kinship; "kinship has the same value as alliance; alliance has the same value as kinship," he wrote in *The Way of the Masks*. While Lévi-Strauss claimed to have found this institution in societies in North America (notably, based on descriptions given by Franz Boas, in the Kwakiutl people of Canada), Polynesia, and even—to a certain point—Africa, he considered it to be part of the broader model of the "noble house" originating in the medieval period. Thus it is no accident that we find a plethora of references to various "houses" in Tolkien's work, for example the houses of Finrod, Brandybuck, Elrond, Elendil, and Anárion (not an exhaustive list), sometimes with a certain irony (as when Merry mockingly introduces Pippin as "Peregrin, son of Paladin, of the house of Took," in *The Two Towers*). In Tolkien, houses gauge their importance by their material possessions and "trademarks," such as "the Star of the House of Fëanor, the elfstone of the House of Elendil," as well as intangible possessions such as stories: "If these old tales speak true that have come down from father to son in the House of Eorl [. . .]" (*The Return of the King*).

Belonging to a House is different than belonging to a bloodline:

> *We of my house are not of the line of Elendil. though the blood of Númenor is in us. For we reckon back our line to Mardil, the good steward, who ruled in the king's stead when he went away to war. And*

that was King Eärnur, last of the line of Anárion, and childless, and
he came never back.

(The Two Towers)

Nor is it synonymous with a people, as becomes clear when Aragorn says to Éowyn in *The Return of the King*: "Farewell, Lady of Rohan! I drink to the fortunes of your House, and of you, and of all your people."

Men belong to their houses, sometimes within complex structures as described in various writings: "The Elfstone, Elessar of the line of Valandil, Isildur's son, Elendil's son of Númenor"; "Tuor was the son of Huor of the house of Nador, the Third House of the Edain." This belonging can be by appointment and is therefore not necessarily based on biology: "Rise now, Meriadoc, esquire of Rohan of the household of Meduseld!" Theoden says when Merry enters his service. The members of a house judge themselves on the strength of their moral qualities: "But I am of the House of Eorl and not a serving-woman. I can ride and wield blade, and I do not fear either pain or death" (Éowyn in *The Return of the King*). However, the resemblance can also be physical, suggesting a biological link, even a fictive one: "But you, Théoden Lord of the Mark of Rohan are declared by your noble devices, and still more by the fair countenance of the House of Eorl"; "They were tall, fair of skin and grey-eyed, though their locks were dark, save in the golden house of Finarfin."

In this blending of alliance and pseudo-biology, Houses reflect the intrusion of politics in the family structure. As Lévi-Strauss puts it in Bonte and Izard's *Dictionnaire de l'ethnologie et de l'anthropologie*, "[. . .] the concept of the house manifests a state in which political and economic interests, which tend to encroach on the social sphere, use the language of kinship while simultaneously subverting it." Likewise, in Tolkien, the existence of a house that counts both Elves and Men among its numbers (that is, two eras in Middle-earth's history) attests to the fact that this affinity has become something to be used for political ends and purposes, contrasting sharply with the sort of kinship around which the hobbits' world is structured.

AFFIRMATIONS OF POWER IN MIDDLE-EARTH

According to the ethnologist Maurice Godelier, it is wrong to presume, as many commonly do, that the family is always the foundational unit of society;

symbolic and political relationships are frequently at least as important. Among hobbits, tradition was crucial as well; hobbits respected common laws and rules but, according to legend, these had been proclaimed by a king who hadn't been seen in more than a thousand years. Social control institutions are virtually absent; "Shirriffs," the closest approximation of our police, were rather in the vein of rural constables, "more concerned with the strayings of beasts than of people" (*The Fellowship of the Ring*), and the mayor, elected for a seven-year term on the occasion of the Free Fair during the summer solstice, was merely a figurehead whose symbolic duties consisted mostly of presiding over banquets. There was also a council, the Shire-moot, led by the Thain, but it met only in times of great or unprecedented emergency, and had not done so for many years by the time of the static Third Age. The administrative structure of the Shire is a clear reflection of Tolkien's own political preferences. He was unquestionably conservative, and not at all democratic, as he stated several times in his letters ("I am not a 'democrat' only because 'humility' and equality are spiritual principles corrupted by the attempt to mechanize and formalize them," he wrote in letter no. 186 in April 1956). His opinions leaned more in favor of anarchy (in the philosophical sense) and non-constitutional monarchy, and he swore to execute anyone who used the word "State" in his presence (letter no. 52, November 29, 1943). Representative democracy and intermediary institutions have no place in his stories; in contrast to the Shire and tradition, Sauron represents a hierarchical power structure and obedience to absolute power. Somewhere in between the two are assemblies such as the Council of Elrond, but these are more meetings between powers than true governmental structures.

The Shire, the Council of Elrond, Gandalf, and Sauron are all sources of power that correspond to the famous typology established by the sociologist Max Weber: traditional power, rational-legal power, and charismatic power. Traditional power, typical of "communities" in the sense of the term put forth by Weber's colleague Ferdinand Tönnies, is dominant in the Shire and among most of the peoples encountered in the story. Charismatic power is the sort possessed by a single individual, a "leader of the masses" or a "providential man," and depends on his own personal qualities or, more precisely, on the relationship he is able to establish with his multitudes of followers. This is obviously the type of power that Sauron enjoys. For Gandalf, the case is less clear-cut; his power, based on magic, can seem charismatic (since, according to Tolkien, the command of magic can only be innate), but it also stems from the mastery of certain technology. In this regard, it comes closer to the third form of power identified by Max Weber, rational-legal power, which is based on recognized expertise, or on an election. But it is clear that rational-legal power and its corollary, representative democracy, carry little weight in the clash between those two holders of charismatic power, Sauron and Gandalf, in which the intent of one is to take possession of Middle-earth and the Shire, while the other is trying to save them—that is, to save a world governed by traditional power.

In their struggle, both Sauron and Gandalf use magic as a key weapon. Classical sociologists and anthropologists such as Max Weber, Marcel Mauss, and James George Frazer regarded magic and witchcraft as being in opposition to science and technology, while also establishing a boundary separating the rational from the irrational (assuming that this distinction is always a simple one to make). Tolkien breaks things down somewhat differently; he is clearly wary of anything that can be applied to and used to transform the world (technology and witchcraft), as employed by Sauron, which he contrasts with means of understanding the world (science and "white magic"). What he fears above all is mankind's subjugation of the environment, as is typical of industrial society, and this fear is not without echoes in the present. For example, the physicist Jean-Marc Lévy-Leblond warned in 2015 that "[. . .] science is being threatened by its own success, and overrun by the very technologies it has generated, giving rise to technoscience. The paradox of this trend is that it is leading to intellectual speculation being overcome by material agency, with transforming the world taking precedence over understanding it." In the same vein, Tolkien associates technology and the machine with witchcraft (*goeteia*), with the taste for power that goes along with magic leading rapidly to the "machine." He obviously finds the idea of a world in which man exists in cooperation with his environment far more to his liking.

FICTIONAL TALES AND REAL SYSTEMS

Tolkien does not hesitate to stage encounters between humans (or near-humans) and non-humans (eagles, wargs) or parahumans (Ents). These encounters, which occur frequently in fantasy stories, help us to identify a developing collective imaginary. For Danilo Martuccelli, professor of sociology, all societies rely on cultural narrative productions to establish a "real system" of their own; that is, to define the limits of the possible and the impossible in their society. The recognition of these limits relies on a central actor (religion, the State) and a specific fear (divine punishment, imprisonment, social shame, etc.), with these elements working in tandem to ensure social control of the population.

The "real religious system" relied on spiritual fear, and was replaced by the "real political system" and the "real economic system," which were based on fear of the other and fear of scarcity, respectively. Some artistic works can be seen as markers of the changeover from one system to another. For example, Cervantés's *Don Quixote* marks the shift from the medieval and magical (or enchanted) world to the modern, objectified world dominated by the real political system. Danilo Martuccelli has theorized that we are currently in the process of transitioning to a new real system, the "real ecological system," which is based on the fear of ecological catastrophe and is reconsidering the oppositional relationship between man and nature that led to the domination of the latter by the former. The discourse of this new system will have to do with mutual support and even symbiosis between the two, a concept beautifully illustrated by the Ents, who play a decisive role in the victory over Sauron, the symbol of dominance over nature, and who are neither Men nor trees but a little of both. As Treebeard puts it in *The Two Towers*: "Some of my kin look just like trees now, and need something great to rouse them; and they speak only in whispers. But some of my trees are limb-lithe, and many can talk to me."

LANGUAGE AND EVOLUTION
IN TOLKIEN

CHRISTINE ARGOT AND LUC VIVÈS,
researchers and educators, Museum of Paris

What would Charles Darwin think of J.R.R. Tolkien's world, a world that is biologically both older than and parallel to our own, and in which a variety of anthropomorphic beings has evolved that is quite distinct from primates in paleontological terms, but which would have existed alongside humans before we were left on our own? Would it be an unnatural history rather than a natural one, in which mutations appear as degenerations, with an increased rate of mutation giving rise to monsters and preventing the species from evolving? For example, the Uruk-hai and the Olog-hai suggest a differentiation in a particular area's orcs and trolls that is not adaptive, but rather forced by means of genetic manipulation and, primarily, the corrupting effect of Evil, which acts as a selection agent.* Tolkien would utilize these biological concepts in an act of creation that was solely linguistic at the outset. Though the German linguist August Schleicher had created a genealogical tree of languages in 1862 based on the biological principle of affinity, it is not certain that Tolkien was familiar with it; however, his own approach was drawn from the same principle. He began by creating two languages spoken by Elves (Tolkien himself described Elvish as a language belonging to "fairy tales"), and then created the history and narrative context

---

* See the chapters "The GMOs (Genetically-Modified Orcs) of Saruman," p. 257, and "The Evolution of the peoples of Middle-earth: a phylogenetic approach to humanoids in Tolkien," p. 263.

in which these languages could exist and develop. His son Christopher writes in *The Etymologies* that the Elvish languages "image language not as 'pure structure', without 'before' and 'after', but as growth, in time."

Reading Tolkien's work—*The Lord of the Rings*, of course, but also *The Silmarillion*, *The Book of Lost Tales*, and *Unfinished Tales*—we get the strong sense that the history of Middle-earth is built on a real temporal structure, one made up of historical chronicles whose complexity encompasses both its languages and the peoples who speak them.

Mythological and philological* discourse influence one another in their contribution to the linguistic register. The language tree prefigures the phylogenetic tree, each branching out with the passage of time and, particularly, in cases of geographic isolation. Tolkien's composition is a mythology of forms, but even more a genealogy of languages, which he accounts for through the dissemination of his peoples; men, like Elves, awaken in the east and disperse mainly toward the west. Tolkien, who sought to evoke for his chronicles one or more languages from eras long since vanished, invented grammar, vocabulary, etymology, declensions, and individual cases for his own linguistic creations. His treatise *Lhammas*, written in 1937 and published in the compendium *The Lost Road* in 1987, includes a description of language trees, and slowly an equivalence emerges between the history of his species and that of the languages they speak; for example, the Dwarves created by the Vala Aulë possess their own language, Khuzdul, directly taught to them by Aulë and originating in Valarin, the language of the gods. Likewise, Morgoth derives Orcish, intended for those who serve him, from a distorted and corrupted form of Valarin.

The case of Elves and Men is more complex. Their languages, created by a single god named Ilúvatar, have been subject to an evolutionary process and to linguistic derivation linked not only to situations of geographic isolation resulting in heightened speciation, but also to the influence of other languages, and to the alteration common to all who live on Earth and suffer the corrupting effects of Evil. For example, after Oromë finds the Elves and some of them undertake the long march to the West and Valinor, a first split-off of the early Quendian languages emerges, something between Eldarin, the language of those who depart, and Lemberin, that of those who refuse to answer the call. With the passage of time, the languages and dialects of the Elves that remain

---

*   Philology: the study of a language based on its writings.

in Middle-earth diverge radically from those who journey to Valinor. A third major schism then occurs with the creation of Telerin, the result of a millennium-long isolation on the solitary island of Tol Eressëa, halfway between the mortal and immortal realms.

Tolkien did not begin by creating early Quendian, however, but rather Quenya, a refined and scholarly language dubbed the "Latin of the Elves," invented by the noblest of the Eldar (the West-Elves who took refuge in Valinor), which emerged only later in the Elvish timeline. This people also developed a language for use in daily life, called Lindarin. Tolkien subsequently created Sindarin, the language of the Grey Elves of Beleriand, which became the "common language" of the Elves and was even occasionally adopted by Dwarves and men. The original "mother-tongue" is thus only revealed by deduction, comparisons, and cross-referencing with the various "daughter-tongues." Tolkien vacillated for a long time before giving Oromë a role in the conception of this original protolanguage, but described Elves as extremely fond of words and gifted at the creation and development of their own languages. Accordingly, their mother-tongue subdivides in the course of their history into various branches that form different language families; for example, Lemberin splits, according to its geographic roots in Middle-earth, into Doriathrin (spoken in Doriath, the Elvish Middle-Earth kingdom protectively encircled during the First Age by the Girdle of Melian) and Ossiriandic, the dialect spoken by the Green-Elves.

We are told that the wars against Morgoth, which led to the obliteration of entire populations of Elves, caused the disappearance of the languages associated with these populations. All that remains by the Third Age is Sindarin, the syncretic language of the Grey Elves of Beleriand, the result of various influences: Doriathrin, the exilic Noldorin spoken in the Elvish city of Gondolin, human and orcish languages (as utilized in Angband, the fortress of Morgoth, with which those Elves who were held prisoner there came into contact).

Tolkien seemingly aspires to the mythical/historical restoration of a *lingua elfica* linked to etymological research aimed at understanding the successive stages of a word in order to uncover, filiation by filiation, the origins of the Elvish languages (and, as we will see in the next section, human languages as well). Tolkien, like a paleographer of the Firstborn (another name for the Elves), makes use of a literary creation that is essentially mytholinguistic and mythographic. Elvish language encapsulates a sort of "primordial purity" and functions as a repository of theological truths, as the only names of the guiding Spirits of the world and emissaries of Ilúvatar that have come down to us are Elvish names, as is the very word that designates them (Valar), and as these deities went so far as to adopt the most sophisticated of the Elvish languages to express themselves. Tolkien, in creating a language that had already evolved and was laden with history, also had the idea of going back to the sources of that language, the primitive languages that served as its foundation. He unfurls a universe of ideal languages, bygone languages to which only he holds the key. As an inventor of symbols he is as much a paleographer (one who studies ancient manuscripts) as he is a prophet. He molds and forges the material of meaning, exploring the poetic potentialities of sounds, tries out new rhythmic combinations, playing on the melodic and incantatory qualities of words. It is clear that he gets deep aesthetic pleasure from working on forms of language, and composes poems in Elvish for the pure joy of hearing these unfamiliar sounds resonate.

THE TREE OF LANGUAGES

In *The Book of Lost Tales* we are introduced to the theory that all the languages spoken by Men in Middle-earth stem, more or less directly, from an ancient Elvish language taught by the elf Nuin to the first pair of humans. The first human language was originally called Taliska(n), and was influenced by the languages spoken by Dwarves, orcs, Lembi Elves, and Grey Elves, but without directly descending from any of these. Next Tolkien created Adûnaic and, in parallel with the history of the Elves, Taliska became the language of the Men who stayed behind, as opposed to those who journeyed to the West and reached Beleriand. The human language subsequently became "sindarinized," as the three main human lines of Bëor, Hador, and Haleth had already become unable to understand one another, and were compelled to use Sindarin as a common language. Adûnaic descended from Hadorian, the language of the third group of humans to enter Beleriand, and was similar to Bëorian, while Halethian, which was relatively distinct, died out before the end of the First Age. Long before departing for the Isle of Númenor in the Second Age, Men intermingled with those Elves that had remained in Middle-earth, as well as with the eastern Dwarves, and borrowed words from these civilizations which then became part of their vocabulary. For this reason, when a human word does not appear to stem from an Elvish language, there is a strong chance that it derives from Khuzdul. Adûnaic was spoken for more than 3,000 years by the inhabitants of Nûmenor, and evolved in three periods (Old Adûnaic, Middle Adûnaic, and classical Adûnaic). The Númenóreans learned to speak Sindarin fluently, to which the most learned of them also added Quenya. The descendants of the island's king, Elros, were long given two names; their true name in Quenya, and a common name in Adûnaic. In Middle-earth exilic Adûnaic was spoken; this was one of the languages that borrowed most from others and became, through this linguistic syncretism, the *lingua franca,* and was also known as Sôval Phârë, Adûni, and Westron; it was the common tongue of all peoples endowed with the power of speech in the Third Age.

There are even multiple versions of Westron itself, as the Westron spoken by hobbits and that adopted by the Elves and the nobles of Gondor are not wholly identical. The language spoken by Elves and noble humans was older, more formal, and more concise. At the end of the Second Age, the exiles of Númenor, who were friends of the Elves, brought back Elvish Sindarin as their mother-tongue. Later, by the end of the Third Age, the balance of power had shifted, and the dominant common language among humans came into wider use while Sindarin was spoken only by some inhabitants of Gondor. Additionally, many isolated groups of humans retained their own languages: Dalian in the region of Dale, also used by the Dwarves of Erebor; Rohirric for the Rohirrim, which is to Westron as Old English is to modern English; in fact, Tolkien "translated" Rohirric into Old English in his books, in par-ticular the names of places and characters including Eomer, Eowyn, Edoras, Théoden, and Shadowfax, or *mearas* (horses). Dunlendish, a possible offshoot of Halethian which declined at the end of the Third Age, coexisted alongside the dialect spoken by the Stoors. Some of these dialects, such as Dalian and Rohirric, are more closely related to Taliska than they are to Adûnaic.

WRITING AND NOTHINGNESS

Should we suppose there to be an inherent relationship in Tolkien's writings between sound, spoken language, and the being that language represents? Can a language match itself to the being that speaks it? It is shaped and molded by that speaker, and undoubtedly shapes and molds the speaker somewhat in return. It is frequently said that orcs, characterized by their ugliness, speak a language that is long-winded, repetitive, degenerate, and coarse. Another example: the language of the Ents, able to be spoken only by them, reproduces the rustling of the forest, a blend of long murmurs and shorter calls strung into a musical, tonal chain that even the Elves themselves refuse to put down in writing. Elves, characterized by their beauty and nobility, speak a language that is considered smooth and harmonious. On the contrary, the Khuzdul of the Dwarves, though taught to them by the Vala Aulë, sounds dissonant and discordant to Elvish ears, and only two Elves, Fëanor and Pengolodh, make the effort to learn it, while the Dwarves learn both Sindarin and later the common tongue with ease. Note, too, that the Dwarves share the fear, with the Valar and the Ents, of revealing their true names, even on their tombs, for they believe these names reveal their fundamental essence and their history, and would give too much power to anyone in possession of them. Naming something gives it life, but it kills it a bit as well. Tolkien is in agreement with Plato and Pythagoras on this point, for whom the incarnation of a soul also diminishes that soul, with the body (*sôma* in Greek) being at once a means of expression (*sêma*) and a prison, a tomb for the soul that has "fallen" into it.

The language of the Valar, which did not evolve and never needed to be transcribed, stands as a pure creation, second only to Ainulindalë, the original creation music of the Ainur. While Elvish and human languages blossom and branch out abundantly, it can be observed that orcs, trolls, and other creatures of Morgoth, and subsequently those peoples under the control of Sauron, use the common tongue only as a means of communication. The decline of languages goes along with the disappearance of the people who speak them, but they are also associated with a degeneration and decay imposed by the way in which they are used by those who corrupt them in adopting them; thus the orcs, dispersed into scattered tribes, quickly become incapable of understanding one another and find themselves forced, in the Third Age, to use the common tongue (Sôval Phârë) to communicate, but it is a distorted, chaotic, and depleted version of the language.

This evolution can also be seen in written symbols, in particular the writing of cirth, which are morphologically similar to runes and were invented by the Elves and subsequently adopted by Men (especially those of Dale and Rohan), orcs, and particularly the Dwarves of Khazad-dûm. All of these races modified these symbols for their own purposes and according to their (in)aptitudes: while the orcs degrade cirth by retranscribing them in an over-simplified manner, the Dwarves enrich them, even going so far as to invent a cursive style of writing them—to the point that their Elvish origins have been forgotten by the Third Age. The second category of Elvish symbols, the Tengwar of Fëanor, brought to Middle-earth by the Ñoldor, were used in the Third Age to retranscribe most of the languages commonly used in the West of Middle-earth, Westron in particular. Even the inscription on the One Ring, in Black Speech but engraved in Tengwar characters, remains elegant, though overly decadent and subtly deformed. When not corrupted, the beauty of Tengwar characters lies in their organic and naturalistic appearance; various embellishments to the bodies and legs of the symbols create a flowing cursive writing; a certain relationship to plants and flowers seems to influence their appearance. Indeed, the Elvish cities of the Third Age, Rivendell and Lothlórien, function as isolated havens in a world in turmoil, preserving species and enabling their endemic evolution. The environment, and the species who live in it and constitute an integral part of it, act mutually upon one another, co-creating each other in a sense. This concept can be seen in Tolkien's work; Elves are linked to flora and stars and beauty, while Mordor and the environs of Angband are desolate places, barren and dead. Those who serve Morgoth fear light (orcs flee from it; trolls are petrified by it, and the Nazgûl lose much of their power in daylight).

Is the designation of things using the Elvish symbols of cirth and Tengwar intended to convey an Elvish representation of the world? The Elves loved words and, through the medium of this creation of the sounds and symbols that translated them, took ownership of the world, leaving their linguistic imprint for thousands of years on Middle-earth (most of the place-names used being Elvish). While a language of human origin would become the common tongue of those peoples endowed with the power of speech during the Third Age, after numerous hesitations on Tolkien's part, it would still be heavily influenced by the Elvish languages. Linguistic animism, grammarian wizardry, genetic syntax—which term best describes this language, which seems to have come back from words as if words were the dead? Thanks to this syncretism, Elvish words survived in the common tongue spoken at the beginning of the Fourth Age, even though most of the Elves had already abandoned Middle-earth.

THE SONG OF THE SIGN

If, in the Gospel according to St. John, "In the beginning was the Word," in Tolkien, in the beginning was the sound as well: an eminently creative sound,

the music of Ilúvatar, in which the Ainur, who created the world, were invited
to take part. The musical themes of Ilúvatar, sung in response to the discords
of Melkor (who had become Morgoth, the Dark Lord), originated with two
peoples—the Elves, or Firstborn, and then the men, also called Aftercomers
or Secondborn. The appearance of the former was linked to the destruction of
the Two Lamps that illuminated the world by Melkor, while the latter emerged
upon the destruction of the Two Trees of Valinor and the theft of the Silmarils.
Each instance of the creation of a new people is thus associated with light, with
its annihilation and subsequent recreation in another form: the destruction
of the Two Lamps and the creation of the stars, and the disappearance of the
Two Trees followed by the creation of the Sun and the Moon. Though
the Ainur forged the stars as compensation for these acts of devastation, they
did not participate in the creation of either Elves or men, strange and free
beings they regarded with fascination, seeing in them a part of the wisdom of
Ilúvatar, who had remained hidden even from them.

The mortality of men, standing in opposition to the immortality of Elves
as it does, is revealed to be a counterbalance to the gift of liberty granted to
them; thanks to this gift, within the limits of the powers and vagaries of the
world, they are given the freedom to devise and shape their lives beyond any
destiny that has been previously determined—that is, sung, from the earliest
beginnings of the world, by the Ainur or Ilúvatar himself.

Tolkien's entire oeuvre teems with exchanges in the context of a proliferation of beings whose forms vary but are mostly anthropoid. Technologies,
knowledge, and languages circulate among the peoples of Middle-earth,
while genes are exchanged only between Men and Elves, at least at the
beginning of their relationships, and on an extremely occasional basis
(three marriages)—so the transmission of symbols between them cannot
have happened genetically, or via descent with modification! There is no
common ancestor linking Elves to men; the two races were born of distinct
acts of creation. Should we consider an evolutionary leap to have occurred
with each act of creation? Whatever the case, each one has necessarily
altered the environment, as well as the equilibrium among species. It is
striking that Tolkien's chronicles reflect a decline of the anthropomorphic species that existed so closely alongside Men during a certain phase
of their history. In this, these peoples illustrate the "necrodiversity" that
would afflict them with the approach of the Fourth Age—or well before,
in some cases.

According to Lamarck's theory, environmental influences affect individual organisms directly. In this sense, Sméagol's degeneration into Gollum and the warping of Men into the Nazgûl are quasi-Lamarckian, if we consider the corrupting power of the One Ring as an environmental factor. For Darwin, on the other hand, the process of selection acts on entire populations, this is what happens in the case of both Men and Elves. Though there is no direct competition between them leading to the gradual disappearance of the Elves, it appears that the acceleration of time plays a role in their vanishing; unlike Darwin, time in Tolkien does not always pass at the same rhythm, since after the destruction of the Two Trees of Valinor and the creation of the Sun and the Moon, time speeds up, which accelerates the growth and the aging of all things. Time is not seen as a development factor for Men or Elves in their internal organization, considering that both awakened in a state of sophistication; rather, and far worse, time is a burden for them, particularly for the Elves, who eventually grow weary of the world and decline. In Darwin, as in Tolkien, man arrives on the scene last, but ends by occupying the sole place of dominance on the planet from the fourth Age onward.

Tolkien's writings, then, involve an examination of the role of the great battles against Evil in the physical aspect of Middle-earth, and the respective influences of chance and destiny in the functioning of the world, the origin of the human species, the place of humans among other living beings, and the relationships among different species. This complex world brings up the issues of origins and determinism and, in a way, offers a moral interpretation of the theory of vital competition, explaining the state of rivalry in which species exist.

The attack on the Trees of Valinor by Ungoliant and Morgoth.

THE DEFENSE AND DEPICTION
OF PHILOSOPHY IN TOLKIEN

MICHAËL DEVAUX, Associate Professor of Philosophy,
Université de Caen Normandie

oes Tolkien philosophize? Might we even call him a philosopher? As is frequently the case, a noncommittal answer to these questions might be the wisest choice. No, he wasn't, because he was a philologist by trade;* yes, he was, like any writer within whom literature drives a philosophy of more or less depth, consciously or not. It remains to be seen whether philology (or anything, for that matter!) can exist without an element of philosophy. Rather than explaining Tolkien's philosophical positions on one theme or another, as has been done by the American philosopher Peter J. Kreeft in *The Philosophy of Tolkien*, let us ask ourselves if Tolkien had a clear personal (philosophical) idea of philosophy, or was it rather just a vague notion? Is there anything more than a cousin-relationship between philology and philosophy for him? Like in a family, the two know each other (but not always well), and resemble one another, to a greater or lesser extent. Since we cannot answer the question of Tolkien's philosophical knowledge fully here, let us instead examine his views on the *regina scientiarum*, the queen of sciences. And since we cannot

---

* From what we know of Tolkien's studies as pieced together by J.S. Ryan, the only evidence of philosophical instruction is an exam on Plato taken on February 27, 1913, and another on Boethius's *The Consolation of Philosophy* ("translated" into Old English). Tolkien also attended at least one meeting of the student philosophical club The Dialectical Society, on January 28, 1913, at which Arnold Toynbee spoke on "The Philosophy of History."

review every single occurrence of the lemma *philosoph** in Tolkien's corpus, given the current publication status of his works (in particular the *Letters*), I will focus on three texts that are not well known, and in any case have never been commented on up to now, even in those rare books that do address Tolkien and philosophy (authored by Peter J. Kreeft, Roberto Arduini, and Claudio Testi).

The initial paradox of this research lies in the fact that Tolkien rarely cited philosophers, and rarely used the term *philosophy* itself, though his children referred to him as "the laughing philosopher," a new Democritus, for such was the nickname of the father of atomism, born in Abdera. Moreover, Tolkien was awarded an honorary doctorate in philosophy from the University of Liège! In a speech given on the occasion of this honorary award, Tolkien's pupil Simonne d'Ardenne called *The Lord of the Rings* "a fairy tale for adults, heavy with philosophical meaning." Despite the filial label and the university honor, the paradox redoubles with Tolkien's own statements, which range from denial ("I am not a metaphysician," he writes in one letter, and "I am not a philosopher," he has his alter ego, Michael Ramer, say in *The Notion Club Papers*[†]) to reluctance, at best: "I do not say "seeing things as they are" and involve myself with the philosophers," he says in *On Fairy-stories*.[‡] In short, Tolkien does not drape himself in the philosopher's toga, but leaves open the possibility for others to do it for him. Finally, in a draft of what became his great theoretical text on fantasy, he writes that "[. . .] fairy-stories are mainly man-made and reflect a philosophy or philosophies concerning the nature of the world. [. . .] Fairy-stories repose on a philosophy or philosophies that are a report or response to the evidence or supposed evidence for fairy phenomenon."[§]

THE MELANCHOLIC PHILOSOPHY OF A YOUNG LECTURER

The first occurrence of "philosophy" in Tolkien's work dates from December 1922, in a Middle English poem he published under the pseudonym "N. N."

---

*   A type of request which searches for all words beginning with "philosoph."

†   See *Letters*, p. 268 and *Sauron Defeated*, p. 178, respectively.

‡   *On Fairy-stories*, Harper Collins, London, 2008, §83, p. 67.

§   *Ibid.*, pp. 260-261.

entitled "The Clerkes Compleinte." Lines 31–32 invoke ethics and the "philosophye malencholyk." The poem concerns a group of students on the first day of the new academic year, some of whom have chosen to study professional subjects (the chemistry of colors and textile engineering, rather than philology or philosophy*).

> *In Leedes atte dores as I sat,*
> *At morne was come into tho halles hye*
> *Wel nigh fyve hondred in my companye*
> *Of newe clerkes in an egre presse,*
> *Langages old that wolden lerne, I gesse,*
> *Of Fraunce or Engelonde or Spayne or Ruce,*
> *Tho tonges harde of Hygh Almaine and Pruce,*
> *Or historye, or termes queinte of lawe;*
> *Yet nas bot litel Latin in her mawe,*
> *And bolde men, alas, ther were yet lece*
> *That thoughten wrestle with the tonge of Grece,*
> *Or doon her hedet aken with etyk hedes*
> *And with philosophye malencolyk.†*

Tolkien may have been thinking of the subjects lectured on by his colleagues at Leeds, Melville Gillespie (who undoubtedly taught classical philosophy) and Harold Hallet (who probably taught ethics, the history of modern philosophy, and English philosophy). Gillespie, a man of letters turned philosopher, had published an article on Hippocrates and was therefore well-acquainted with the theory of humors, one of which was the melancholic humor. As for Hallet, it is likely that he taught Robert Burton's *The Anatomy of Melancholy* (1621), which had marked its tricentenary the previous year. Tolkien would have heard it spoken about and may even have read it, if we go by at least three sections of

---

*   The subjects mentioned by Tolkien in lines 20–32 owe something, no doubt, to the courses of study listed in *The Leeds University Calendar, 1921–22* [online], Leeds, Jowett & Sowry Ltd., 1921. Available at http://digital.library.leeds.ac.uk/11262/; for philosophy, see pages 146 and 165.

†   "The Clerkes Compleinte," *The Gryphon*, Leeds, December 1922, NS 4/3, p. 95. Republished in *Arda 1984*, 4, 1988, pp. 1–2 (with modern English translation) and then in *Arda 1986*, 1990, pp. 2–3 (Leeds manuscript), and (Oxford version) by Jill Fitzgerald, "A 'Clerkes Compleinte': Tolkien and the Division of Lit. and Lang.," *Tolkien Studies*, 6, 2009, p. 50.

text highly reminiscent of his poem that can be found in the chapter entitled "Love of learning, study in excess, with a digression, of the misery of scholars, and why the Muses are melancholy." Here, Burton, like Tolkien three centuries later, speaks of a student's beginning his university studies, and goes on to point out many students' rejection of history, philosophy, and philology, ending by denouncing the misguided emphasis placed by some on enrollment numbers, whatever the students' level of education (Tolkien himself bemoaned his students' lack of proficiency in Latin)!

A PRE-*HOBBIT* DEFENSE OF PHILOSOPHY

In "Philosophical Thoughts," a little-known piece dating from his early years as an Oxford lecturer, Tolkien relates a discussion between his friend and colleague C.S. Lewis and the president of his college, and takes it upon himself to mount a defense of philosophy. There is no reticence to be found here; quite the contrary:

> *Many people are incapable of seeing any good in philosophy, and speak of it as if it were a juvenile excretory process, following which one can settle down in a state of learned adulthood without thinking any further. For example, the president of no less than a college such as Magdalen spoke of Mr. Fausset, the other night at dinner, as if he were an uncouth lout because he philosophized about literature, 'whereas you and I'—he was speaking to Lewis—'have moved beyond all that and left it behind with our Bachelor of Arts degrees.'*
>
> *[. . .] We also spoke of idealist and atheist philosophy—the latter has always been unfathomable to me; I have never understood how anyone can maintain such a position in these times, if it can be called a position.*[*]

Who is being spoken of here? Tolkien is recounting a discussion that took place around 1931 between C.S. Lewis (1898–1963) and George Stuart Gordon (1881–1942). The latter's biographer, Mary C. Biggar Gordon, tells us that: "Philosophy and ancient history held no particular attraction for him.

---

[*]   "Philosophical Thoughts," *Parma Eldalamberon*, no. 20, 2012, pp. 113–115. Arden R. Smith incorrectly standardizes the name 'Fausset' as 'Fawcett', without identifying him.

They held his attention for a time, and his natural vigor and appetite for tackling difficult intellectual problems turned them into a game."[*] Gordon was aiming his barbed words at Hugh l'Anson Fausset (1895–1965), who wrote on philosophy in literature, notably in his 1923 book *Studies in Idealism*, which explains Tolkien's mentioning "idealist philosophy."

What idea(s) regarding philosophy can we take away from this brief snippet?

Firstly, philosophy, Tolkien says, is seen by many as corresponding to age. It is thought to belong properly to adolescence or youth, and is something that warrants being left behind. An adult who has pursued higher education should have no further need of philosophy; at best, it should serve only as a lever for catharsis, to resolve a conflict, for example, with a kind of respectable conformity preferably coming to occupy its place instead. Tolkien himself clearly does not believe this, and takes it upon himself to dispute this reductive concept of philosophy. Tolkien's negative view of this perception, and his clear disdain for the remarks made by Gordon during a banal discussion of university life, are enlightening; between the lines of them we can read a defense of the philosophical training of the mind, and thus of the importance of philosophy.

Secondly, the idealist philosophy Tolkien speaks of is that of Fausset, so it would be useful to understand how the latter defined philosophy: "By 'philosophy', I do not mean any systematic phantom of the schools. Many critics will deny that a poet has a philosophy, because his philosophy is something different from that of Aristotle or Kant or Dr. Bradley.[†] His philosophy is an intuition or a series of intuitions concerning the universe" (*Studies in Idealism*, p. xix). This idealism remains, of course, a quest for reality. It is an "acceptance of things as they are" (*ibid.*, p. xv). Doesn't Tolkien write in his letters that his fairy stories serve as his "comment on the world?"[‡] Isn't this the equivalent of a series of intuitions concerning the universe? And Tolkien undoubtedly shared the idea that a discourse can be philosophical without being systematic.

---

[*]   *The Life of George S. Gordon, 1881–1942*, Oxford, Oxford University Press, 1945, p. 20. [retranslated here from the French translation of the original].

[†]   Francis Herbert Bradley, British idealist philosopher. A professor at Merton College, Oxford, like Tolkien; he died in 1924.

[‡]   "I write things that might be classified as fairy-stories [. . .]. I do this because if I do not apply too grandiloquent a title to it I find that my comment on the world is most easily and naturally expressed in this way. [. . .] I hope 'comment on the world' does not sound too solemn." (Humphrey Carpenter, ed., with Christopher Tolkien, *The Letters of J.R.R. Tolkien*, HarperCollins, 2012).

His work is encyclopedic, with philosophical lineaments, but without being systematized.

Finally, with regard to atheist philosophy, we can look to the last chapter of Fausset's aforementioned book, dedicated to the modern mind in general. The pairing of idealist and atheist philosophy leads more broadly to a discussion of the conflict between idealism and (atheistic) materialism. Materialism and atheism share the affirmation of human pride without the hidden dimension of religion—and so they can only be disqualified by Tolkien. Here again, Fausset's book can serve as the backdrop for this discussion. In it, Fausset attempts to explain what makes Nietzsche a destructive force: he is an atheist, and a conscious advocate of (intellectual) barbarism as an antidote for his spiritual hypochondria, through the separation of mankind from its own humanity, for the benefit of the Übermensch, or Superman. For Tolkien, atheism is untenable, lacking consistency in its ideas. For Fausset, it is only one aspect of a wider vision. The modern mind is presented as arising from an age in which science predominates—which is both a danger and a hope. If science is able to learn the importance of intellectual honesty it will be a good thing; however, if it reduces empathy, it is a danger. It is dangerous to acquire only knowledge, and to cease to understand values. The issue, then, is one of values, and therefore one of intellectual judgment. In the old days sins against the Holy Spirit were sins of the flesh; now, they are sins of the mind. A cult to the principles of science is being created; it is mechanical intelligence that is immoral. A positivist philosophy is an asset if we think of life as a whole, and take its beauty into consideration.

So, in this short piece written by Tolkien around 1931, we find: a defense of philosophy as an important component of reflection, even in adulthood, and as a vital part of one's education; a non-systematic concept of philosophy that is nevertheless made up of a series of intuitions concerning the universe; and, finally, a rejection of atheism as an example of an illogical philosophy in Tolkien's eyes, implying that philosophy is a vital constituent of logical thought.

PHILOSOPHY FOLLOWING THE AMERICAN SUCCESS OF *THE LORD OF THE RINGS*

Toward the end of his life, Tolkien employed the term *philosophy* in a text on Elvish linguistics, *The Shibboleth of Fëanor*, in his answer to the question of the meaning of the names of Finwë and his son Fingolfin (see the text box on the next page).

Fingolfin is also known by the name of Ñolofinwë (Finwë the Wise in Quenya). Ñolo, Tolkien specifies, is the root of the word family attached to the concept of wisdom.[*] In a note, he goes into more detail regarding the meaning of wisdom for the Ñoldor, reflecting on the relationship of wisdom (*sophia*) to philosophy:

> *'Wisdom' —but not in the sense 'sagacity, sound judgement (founded on experience and sufficient knowledge)'; 'Knowledge' would be nearer, or 'Philosophy' in its older applications which included Science. Ñolmë was thus distinct from* Kurwë *'technical skill and invention', though not necessarily practised by distinct persons. The stem appeared in Quenya (in which it was most used) in forms developed from Common Eldarin* ñgol-, ñgōlo-, *with or without syllabic n: as in* *Ñgolodō > *Quenya* Ñoldo *(Telerin* golodo, *Sindarin* goloð)—*the Noldor had been from the earliest times most eminent in and concerned with this kind of ‹wisdom›;* ñolmë *a department of wisdom (science etc.);* Ingolë *(ñgōlē) Science/ Philosophy as a whole;* ñolmo *a wise person;* ingólemo *one with very great knowledge, a 'wizard'. This last word was however archaic and applied only to great sages of the Eldar in Valinor (such as Rúmil).*[†]

On a first, quick reading of the beginning of the note above, we would take "sound judgment" in the sense it is usually meant today; that is, absolutely sure of the certitude of science (positivist). Reading further confirms the conviction that this judgment is based on experience. The basis for this thinking is Kantian: even when one has knowledge considered sufficient to make a judgment, they should not take the risk of *not* basing this judgment on experience. Knowledge can exist only if the conditions of the phenomenality that are space and time contribute something to our understanding which we can then apply. There is no science without experience; anything that falls outside this category cannot be considered science, and no judgment concerning it can be considered sound. Tolkien pits this view of wisdom against what is, for him, true philosophy; he prefers to reconcile the wisdom of science with the wisdom

---

[*]   J.R.R. Tolkien, *The Peoples of Middle-earth*, London, HarperCollins, 1996, p. 344.

[†]   Ibid., note 30, pp. 359–360.

of philosophy that encompasses science. In this he is reminding us that this is an old(er) view of philosophy; the ancients joined philosophy and science together. However, a positivist reading would *reduce* philosophy to science, which keeps us from interpreting this passage in either a positivist or a scientistic manner.

With regard to the conception of sagacity (*phronesis*, also sometimes translated as "practical intelligence"), also referred to in the passage above, Aristotle's is the key philosophical voice, dwelling on this virtue in Book VI of the *Nicomachean Ethics*, which we cannot analyze in detail here (see the text box below).

Let's reflect for a moment, though, on the representation of philosophy and its relationship to science. We'll look at a bit of Aristotle for comparison: it is the *relationship*, more than anything, that is important in Tolkien's note, while for Aristotle, philosophy is above all a theoretical science, one of pure contemplation using the intellect—the supreme branch of knowledge, but a branch nevertheless. From this perspective, most people are Aristotelians, in thinking that philosophy is an occupation, perhaps not exactly like others, but simply an area of knowledge and no more (or better) than that.

The central term for Tolkien, however—the one that lies at the root of the others—is *ñolmë*, which he immediately differentiates from *kurwë* and later from *ingolë*. Philosophy is not its primary subject. He is specifying that the wisdom of the specialist, the wisdom that gives knowledge, the contemplation of a particular domain, the wisdom of the well-informed opinion, stands in opposition to wisdom as such (that is, wisdom in all things).

◦——≺ FINWË AND FINGOLFIN ≻——◦

1. Finwë was married to Miriel, who gave birth to Fëanor. He later married again, to Indis, with whom he had two sons, Fingolfin and Finarfin, and two daughters, Findis and Irimë.
2. Fingolin was the Elven King who faced Morgoth in one-to-one combat and wounded him seven times before being slain by a mace-blow. His brother Finarfin was the only one to remain with his people in Aman, according to the will of the Valar (though his children, including Finrod and Galadriel, followed Fëanor).

We are close to Platonic philosophy here: "Then the lover of wisdom, too, we shall affirm, desires all wisdom, not a part and a part not" (*The Republic*, V, 475b). Does this overarching desire leave any room for the idea of philosophy?

As Monique Dixsaut, a scholar specializing in Plato, has shown, a philosopher, in Plato's view, is a person driven by the desire for wisdom. He applies himself in this quest to matters of the intellect, of course, but that is not the end of it: I am a philosopher; nothing that is of the mind must be unknown to me. It is also for this reason that the philosopher's duty is to be erudite. However, erudition is not an end in itself, but merely a means to get to the truth. The portrait of the philosopher that emerges from all this is thus similar to, and encompasses, that of the scholar, but the former must remain driven by the search for *truth above all*. The object of the philosopher's desire is not knowledge, but essentiality—that is, truth. Likewise, the philosopher's desire has nothing to do with the degree of importance of his subject (one can philosophize about fairy tales, which is undoubtedly insignificant in the eyes of the establishment, but is fundamentally philosophical if the goal is always to understand the truth contained within these tales).

But while Tolkien's views seem similar to Plato's in this respect, he does not attain the peak represented by the Platonist concept of philosophy. For Plato, the scholar seeking only knowledge is a sham philosopher rather than a real one, because truth is the truth of philosophy, not knowledge in itself, or the

⌁ SAMWISE GAMGEE ⌁

The relationship between wisdom and sagacity has made translating the full name of Tolkien's character Sam a challenge. *Samwise* was translated as *Samsagace* by Francis Ledoux, and as *Samsaget* by Daniel Lauzon. Aristotle, in Chapter 12 of Book VI of the *Nicomachean Ethics*, associates *phronesis* with the accumulated knowledge through experience possessed by elderly people. Is Sam's sagacity not that of the Gaffer to whom he frequently refers? Additionally, the only acceptable meaning of "wisdom" for Aristotle in the *Nicomachean Ethics* is that of practical intelligence/sagacity, such as political wisdom (1141*b*23). And isn't Sam, as mayor of the Shire, not effectively endowed with this type of political wisdom, which is included in *phronesis*? Ledoux's Sam is a son of the Gaffer, and Lauzon's is mayor of the Shire but they are both prudent Aristotelians!

Wisdom (Ñolo)		
Knowledge (Ingolë)	Ñolmë	Kurwë
Philosophy including science. Science/philosophy as a whole.	A field of wisdom	Technical skill and inventiveness
Philosopher (ingólemo)	"Specialist" (ñolmo)	Technician/craftsman

amassing of this knowledge. Tolkien says nothing of this relationship to truth. The issue here is that of the part and the whole; the whole of knowledge and part of it, a section, a branch. Likewise, for Plato, the main characteristic of a philosopher is his desire, which is also perseverance. The philosopher seeks to know; he has the taste and the ability for it (we are speaking of the natural philosopher here), but it is still an effort. On this point, Tolkien is silent.

WAS TOLKIEN A PHILOSOPHER AFTER ALL?

Tolkien studied philosophy as part of his undergraduate education in Classics. His poem written from the viewpoint of a young lecturer and his mentions of melancholic philosophy, idealist philosophy, and atheism suggest a certain amount of reading (Plato, Boethius, Burton, Fausset). *The Shibboleth of Fëanor* points toward Aristotle, and Plato once again. Tolkien's work contains Platonic elements; his view of philosophy as an integral whole that encompasses science attests to this. However, the portrait that comes through the texts we have cited here remains one of a scholar, free of a strictly-defined relationship to the truth. The comprehensiveness of the knowledge Tolkien explored flirts with *encyclopedism*, of which Middle-earth is certainly an illustration. Nevertheless, the internal *consistency* present in each argument, while indicative of a sound philosophy, does not commit him to anything beyond a *series* of intuitions concerning the universe, without a *system* being required. In the end, it all bears a strong resemblance to what Tolkien developed as the philosophy of Middle-earth. They may not place him among the ranks of philosophers, but these writings of Tolkien's show that he knew what he was talking about, and had, at the very least, a philosophical view of philosophy gained from his readings.

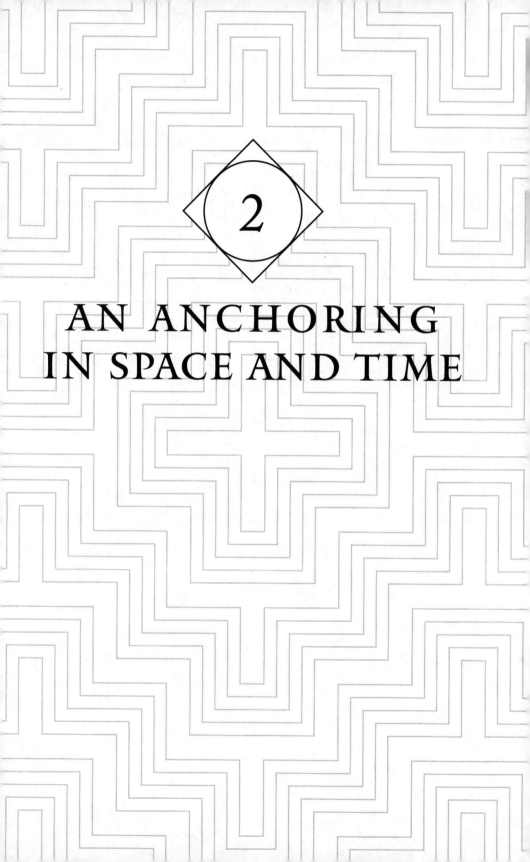

2

AN ANCHORING
IN SPACE AND TIME

ARCHAEOLOGICAL REMAINS AND HIDDEN CITIES

VIVIEN STOCKER, Tolkien specialist

 n a letter to the novelist Naomi Mitchison (1897–1999), Tolkien confessed to being "more conscious of [his] sketchiness in . . . archaeology"* than in other fields, such as economics or music. Nevertheless, as was frequently the case, Tolkien tended to underestimate the real contribution of this discipline to his work, even when he had not consciously incorporated it in his plots, as may have been the case with botany and linguistics.

Tolkien's interest in archaeology is more than supposition. He is known to have found a fossil jawbone (which he speculated had come from a dragon!) during a trip to the seaside at Lyme Regis. The town, to which his guardian, Father Francis Morgan (1857–1935) often took him and his brother on holiday, was famous for its fossil deposits. Later, his various university posts and his interest in historical texts led him inexorably to study the history of the world more closely, and archaeology in particular, which is how he came to contribute in 1932 to a collection of essays on the archaeological site of Lydney Park in Gloucestershire, which included prehistoric and Roman ruins. He also subscribed for a time to the English archaeology journal *Antiquity*, founded in 1927 and still published today, and he possessed at least one volume of

---

* J.R.R. Tolkien, *Letters*, no. 154.

archaeological studies in his library: *Scandinavian Archaeology*, a gift from his colleague Professor E.V. Gordon[*] (1896–1938).

MEGALITHS AND PREHISTORIC ICONOGRAPHY

The year 1932 is a perfect starting point from which to explore the archaeological roots of Tolkien's oeuvre, through one of the most conspicuous reutilizations of prehistoric iconography that occurs in his writing. Since 1920, it had been Tolkien's habit to send an illustrated letter to his children each year that was supposedly from Santa Claus. In the 1932 letter, he relates how the Polar Bear, a friend of Santa, gets lost in goblin caves covered with rock paintings. Tolkien illustrates the story with an overall picture of the cavern, but most notably a full page of cave-wall paintings representing mammoths, bison, and other Paleolithic animals. Though they are intended to be poorly-executed drawings by goblins, the sketches are beautifully done. It is clear that Tolkien has reused representations of cave paintings seen in books he must have had close to hand. Some drawings of animals are copied from the 1928 book *The Art of the Cave Dweller* by the art historian Gerard Baldwin Brown (1849–1932), or inspired by the cave art discovered at Altamira in northern Spain, which he could have seen in the Spanish-language books owned by Father Morgan, who had Spanish roots. But even several years earlier, at the time he wrote the *Lost Tales*, Tolkien must already have been well-acquainted with prehistoric culture when he invented the city of Menegroth, the Thousand Caves, and the underground fortress of Nargothrond, true troglodyte cities.

The Neolithic Age also provided its share of inspirations to the author. In *The Lord of the Rings*, when the Ranger Aragorn, the elf Legolas, the dwarf Gimli, and the wizard Gandalf arrive at the doors of Edoras in Rohan, they discover "many mounds, high and green." This image echoes an incident that occurs earlier in the saga, after the four hobbits, Frodo, Sam, Merry, and Pippin have spent time in the company of Tom Bombadil. When they leave the Old Forest, their path takes them through the Barrow-downs, a region of hills crowned by green mounds and standing stones, the description of which irresistibly recalls the fields of menhirs (standing stones) and tumuli (burial

---

[*]   Gordon and Tolkien are known for having collaborated on a reference edition of the Middle English text *Sir Gawain and the Green Knight*.

mounds) that can still be seen in western France and the south of England. Shortly thereafter, the hobbits are captured and shut up inside one of these burial mounds by a Barrow-wight. The internal structure of the mound is reminiscent of Neolithic dolmens, particularly in terms of their shape and their primary function as collective tombs built of stone and usually covered with earth. One interesting detail is that the tumulus in *The Lord of the Rings* contains a number of objects made of precious metals, principally gold and silver, but also copper and bronze; these materials correspond to two historical periods in which dolmens were constructed and used, the Copper Age and the Bronze Age, which immediately followed the Neolithic Age.

The structure of the dolmen, and in particular its distinctive shape, taking the form of the Greek letter π, seem to have made a strong impression on Tolkien. In his own drawings, both the entrance to the Elves' underground fortress of Nargothrond in *The Silmarillion* and the doors to the palace of the Elvenking Thranduil in *The Hobbit* are depicted in the very noticeable shape of the letter π, with a lintel and two columns. The large stone arch marking the entrance to the Paths of the Dead, traveled by Aragorn and his companions in *The Return of the King*, was also sketched by Tolkien as having this distinctive shape. Later in life, Tolkien referred once again to a π-shaped door in his novel *The Notion Club Papers*, which featured a Great Door "shaped like a Greek letter π with sloping sides" which formed the entrance to an unidentified megalith seemingly situated on the island of Númenor.

NÚMENOR: ANCIENT EGYPT AT THE HEART OF MIDDLE-EARTH

Let us linger on this island for a moment. In 1958, Tolkien wrote in a letter (no. 211) that the Númenoreans, and more specifically those Númenoreans exiled to Gondor and Arnor, were "proud, peculiar, and archaic, and I think are best pictured in (say) Egyptian terms. In many ways they resembled 'Egyptians'—the love of, and power to construct, the gigantic and massive. And in their great interest in ancestry and in tombs."

The Númenoreans were truly accomplished architects, extremely fond of massive, imposing constructions. The city of Minas Tirith and its seven-layer structure is reminiscent of the step pyramids, though a more obvious inspiration may be the Mesopotamian ziggurats, represented by the emblematic

Minas Tirith, the capital of Gondor.

Tower of Babel. The arrowlike tower of Orthanc, seat of the wizard Saruman, may hark back to the shape of Egyptian obelisks. Additionally, the Númenoreans were also renowned artists who did not shy away from sculpting gigantic statues such as those of the Argonath, representing Isildur and Anárion, the first kings of Gondor and Arnor, situated on either side of the River Anduin, between the feet of which the companions of the Fellowship row. The statues of the Argonath are comparable to the two Colossi of Memnon, which guarded the entrance to the mortuary temple of Amenhotep III before its destruction. We might also liken them to the monumental statues erected by Ramses II, such as the colossal seated statue discovered near Memphis in 1820 by the navigator and Egyptologist Giovanni Battista Caviglia (1770–1845). The headless statue of a king Frodo and Sam encounter at the Crossroads on the way to Mordor recalls the famous effigy of Ramses II at the Ramesseum, which was itself decapitated at some point in history.

The history of Númenor also contains numerous parallels to that of Ptolemaic Egypt. In 331 B.C. Alexander the Great founded the famous city of Alexandria on the shoreline of the Nile Delta. On his death in 323 B.C. one of his satraps, the Greek general Ptolemy, governor of the Egyptian province, took the reins of power and made Alexandria the capital of Egypt. Ptolemy gave rise to the Ptolemaic dynasty, which ruled Egypt until 30 B.C., after which Egypt became a Roman province on the death of Queen Cleopatra. Several great Númenorean figures are reminiscent of Alexander. According to legend as recounted by Plutarch in his *Life of Alexander*, as Alexander slept after his Egyptian campaign, Homer appeared to him in a dream and showed him the island of Pharos and the isthmus opposite it as the ideal site on which to build his city of *Alexandria ad Aegyptum* (Alexandria of Egypt). There is no exact parallel to Alexander's dream in the history of Númenor; however, a fairly similar concept appears in the formation of the harbor of Vinyalondë in Middle-earth, at the mouth of the River Gwathló. This port was founded in around the year 800 of the Second Age by the Númenorean prince, the explorer Aldarion. Its existence made it possible for King Tar-Minastir to deal a fierce blow to Sauron's armies in the year 1700 of the Second Age. According to the chapter "Aldarion and Erendis: The Mariner's Wife" in *Unfinished Tales*, Aldarion possessed the gift of foresight, which may have been the source of his decision to build strongholds in Middle-earth, particularly the port of Vinyalondë, in anticipation of the day when the Númenoreans might have need of them.

The cities of those Númenoreans who were exiled to Gondor may also contain echoes of Alexander's Egypt. The first city built in Ptolemaic Egypt was Alexandria, on the Nile delta. In Gondor, long before the great cities of Minas Ithil and Minas Anor were built, the main hub was also a port, Pelargir, constructed on the delta of the River Anduin. Minas Anor and Minas Ithil, built in 3320 S.A. at the time of the exiles' arrival, and the regions these two cities govern, Anórien and Ithilien, owe their names to their respective sovereigns, the princes Anárion and Isildur—exactly like Alexandria, named for the famous conqueror. Tolkien's architecture is not without its share of parallels either; two legendary monuments of Alexandrian Egypt have clear echoes in his work. The most striking example is the existence of a single lighthouse on the whole of the island of Númenor, called Calmindon (Light-tower), also built by Anárion. This lighthouse was erected on the islet of Tol Uinen, in the Bay of Rómenna off the eastern coast of Númenor, a bay that Aldarion built dikes to protect and strengthen. These structures irresistibly recall the famous Lighthouse of Alexandria, connected to the city by the causeway known as the Heptastadion, built by Ptolemy I and his son on the island of Pharos. The other element of Alexandrian architecture that we can connect to a Númenorean construction is the lost tomb of Alexander. On his death, Alexander's body was first taken to Macedonia, his birthplace, before being repatriated to Alexandria by Ptolemy. His great-grandson Ptolemy IV built a magnificent tomb for Alexander, the Soma, which was visited by numerous emperors until it vanished during a massive earthquake in 365 B.C. The tomb has never been located, and there are many theories concerning its possible location. We do know that the Ptolemaic kings had niches built around the Soma for their own remains. In Tolkien, Elendil, king of the exiled Númenoreans, is laid to rest by his son Isildur in a secret tomb on the hill of Amon Anwar, its existence known only to the kings and Stewards of Gondor. When Amon Anwar was ceded to Eorl and the Rohirrim, the Ruling Steward, Cirion, brought the remains of Elendil back to Rath Dínen in the city of Minas Tirith, to the mauseoleums where the kings of Gondor, his successors, were also buried.

The Númenoreans also resemble the Egyptians in their knowledge and their scientific methods of embalming the dead and constructing tombs to their glory. In *The Silmarillion*, Tolkien writes of the Númenoreans:

> [T]hey began to build great houses for their dead, while their wise Men
> laboured unceasingly to discover if they might the secret of recalling life,

or at the least of the prolonging of Men's days. Yet they achieved only the
art of preserving incorrupt the dead flesh of Men, and they filled all
the land with silent tombs [. . .]

These words could just as easily have been written about the Egyptians, whose mummification techniques were unequaled in the ancient world. In Egypt, silent tombs are found in great numbers in the Valley of the Kings, a network of Egyptian royal tombs dug deep in a gap in the mountains bordering the Nile, near Thebes. The valley is overlooked by a rocky summit, then called Ta Dehent (literally "The Peak"), which was considered to be a sacred place, dedicated to the goddess Meretseger, whose name means "She Who Loves Silence." Noirinan, or the Valley of the Tombs, where the Númenoreans buried their dead kings, is particularly reminiscent of the Valley of the Kings. Tolkien placed the valley of the Númenorean kings deep between the ridges of the sacred mountain Meneltarma, the peak of which was dedicated to Eru Ilúvatar, the creator god, where a requirement of absolute silence was imposed.

Moving on from architecture to artifacts, Tolkien added in the same letter previously cited that "the crown of Gondor (the S. Kingdom) was very tall, like that of Egypt, but with wings attached, not set straight back but at an angle. The N. Kingdom had only a diadem. [. . .] Cf. the difference between the N. and S. kingdoms of Egypt." Here, Tolkien explicitly compares the crowns of the two Númenorean kingdoms to the two crowns of Upper and Lower Egypt: the *Hedjet*, the tall, white, closed crown, which merges with the *Deshret*, the open red crown, to form a double crown, the *Pschent*, symbolizing the unification of ancient Egypt's two regions.

There is another type of tall crown that resembles the crown of Gondor even more closely: the *Atef*, a tall white crown (similar to the Hedjet) topped with a sun-globe and decorated with ostrich-feathers on the sides. The resemblance to the crown of Gondor is striking; this crown is also described as being tall and white with "wings at either side [. . .] wrought of pearl and silver in the likeness of the wings of a sea-bird," its top set with "a single jewel the light of which went up like a flame." The common features are obvious: the height, the whiteness, and the presence of a jewel at the peak associated with light, as well as the wings or feathers set on each side.

Going still further, the crown of Gondor displays similarities to the *Khepresh*, the blue crown or headdress worn by the king in battle. Indeed, the Gondorian crown was itself originally a war-helmet worn by the

Númenorean prince Isildur at the Battle of Dagorlad fought by the Last Alliance, and was later replaced by the white crown. Finally, the mithril diadem set with the Elendilmir, a white star-shaped gem, worn by the royalty of Arnor, is very similar to the Egyptian *Seshed*, originally a simple circlet of woven silver ribbon which eventually became a separate diadem, wrought of precious metals and set with gems.

Tolkien's writings also bear traces of the ancient Romans. Again in Númenor, the temple Sauron causes to be built in honor of Melkor has been compared to the Pantheon of ancient Rome. Its structure is "circular," with thick walls surmounted by "a mighty dome . . . roofed all in silver." However, though similar in appearance, the dimensions of this temple bear no resemblance to those of the Pantheon, which was nevertheless the largest dome erected by mankind in its time, and would remain so until the Renaissance. The Pantheon's dome measures 58 meters in diameter, compared to 153 meters for Sauron's temple. Here, again, we find the gigantic proportions typical of Númenorean construction.

A great deal of ink could undoubtedly be spilled discussing the various sources of inspiration Tolkien discovered in his contemplation of our history and the vanished civilizations that left their mark on it. I have deliberately limited myself to antiquity, and to the most obvious influences. Other lines of thought are constantly opening up, some of them more speculative than others; for example, some researchers have laboriously connected the One Ring to the Ring of Silvianus, a Roman gold ring discovered in England in around 1785 and associated with the Celtic god Nodens, based on the fact that Tolkien discusses this name in the article he published in 1932 on the archaeological site of Lydney Park, which I mentioned at the beginning of this chapter.

One thing is certain, whether the influences are clear or not, proven or merely imagined: Tolkien used his archaeological knowledge as a catalyst for his fiction, guiding readers unhesitatingly among the ruined fortifications of Weathertop or the remains of the Seat of Seeing on Amon Hen.

HISTORY AND HISTORIOGRAPHY IN MIDDLE-EARTH

DAMIEN BADOR, Tolkien specialist

OF THE CRIME OF SCHOLARSHIP: ROMANTICIZED HISTORIOGRAPHY

Are history and historiography* intrinsically linked? For J.R.R. Tolkien, the answer would undoubtedly have been yes. The historical narrative and its transmission in his stories constitute an area of study in themselves, unlike what we see with most world-creating writers. Unlike Robert E. Howard (1906–1936), for example, the history of Tolkien's *legendarium* is not related in the form of an objective chronicle that is impersonal and written from an omniscient viewpoint. Nor is it a simple reproduction of a historic record with no other presence than a handful of citations, as is the case with the epigraphs that begin each chapter of *Dune*, by Frank Herbert (1920–1986). Rather, Tolkien acts as if he is an editor working to translate and compile various historical documents that have come into his possession, as he explains in Appendix F of *The Lord of the Rings*. As part of this, he readily describes the texts from which he is supposed to be working, their fictional authors, and even the conditions under which some of these texts were copied, preserved, or adapted.

---

* Historiography is the history of the methods by which the events of the past are studied and written about. It is thus illustrative of the changes affecting subjects, sources, and methods in the discipline of history.

We mustn't forget that Tolkien was a philologist specializing in medieval English literature. The vicissitudes experienced by the sole extant manuscript of *Beowulf** were hardly unknown to him. In several of his university lectures, he openly bewailed the textural uncertainties resulting from the loss of the *Finnesburg Fragment*,† though this loss did not prevent him from reconstructing, in *Finn and Hengest: the Fragment and the Episode*, the legendary history of the Jute lord Hengest and the Frisian king Finn, by cross-checking the rare allusions to these two names that appear in medieval English manuscripts. It would have been impossible for Tolkien to undertake the reconstruction of ancient history without addressing the issue of the transmission of stories, and without taking an interest in the often-anonymous copyists who saved them from oblivion. What is true for the high Middle Ages is obviously even more true for events supposed to have occurred several thousand years ago, perhaps even before the last ice Age, as seen in *The Lost Road*. Indeed, it is amid the mists of our planet's distant past that Tolkien places his history of Middle-earth.

FROM GODS TO ELVES: THE GENESIS OF THE SAGA AND ITS TRANSMISSION

Tolkien's history begins with a fantastical origin story inspired by traditions attached to the Old Testament:

> *There was Eru, the One, who in Arda is called Ilúvatar; and he made first the Ainur, the Holy Ones, that were the offspring of his thought, and they were with him before aught else was made.*
>
> (The Silmarillion)

This account of the creation of the universe is called the Ainulindalë, the "Music of the Ainur." It relates how The One first created a vast angelic cohort and assigned them a demiurgical role by asking them to sing a Great Music.

---

*   Anglo-Saxon epic poem written between the 7th and 10th centuries A.D., narrating the exploits of the Scandinavian hero Beowulf, and his battles with the monster Grendel and a dragon in particular.

†   Fragment of an Anglo-Saxon epic poem from the high Middle Ages, recounting the tale of a battle between the Danes, the Jutes and the Frisians which is alluded to in *Beowulf.*

However, Melkor, the most powerful of the Ainur, revolted out of a desire for power. Breaking the harmony of the Great Music, he sought to mold creation according to his own designs, taking other Ainur with him. In reaction to this attempt, Eru Ilúvatar introduced a new and unexpected musical theme, in which was conceived the existence of his children, Elves and Men. This redemptive act proved insufficient, however, which forced Eru to cease the music prematurely. However, he then decided to transform this great work and give it life, thus creating Ëa, "The World that is." Some of the Ainur decided to enter Ëa to complete the work of shaping the universe; they became the Valar, or the Powers of Arda.

The Ainulindalë is said to be the work of the elf Rúmil, written in Valinor on the instructions of the Valar.* This tale opens *The Silmarillion*, the story of the Elder Days, which preceded the events of *The Hobbit* and *The Lord of the Rings*. Next comes the *Valaquenta*, an anonymous Elvish text that describes the mightiest Valar, their subjects, and the particular attributes of life for which they were responsible. For example, Manwë, the king of the Valar, commands the air and winds, while his spouse Varda is creator of the stars, and his brother Ulmo is Lord of the Waters. Tolkien specifies that these relationships of kinship or marriage are purely spiritual, and have to do with the mutual affinity that some Valar have for one another; nonetheless, the *Valaquenta* is clearly deliberately reminiscent of Hesiod's *Theogony*,† and of some of the *Homeric Hymns*. Eru Ilúvatar's actions no longer manifest directly; now, the Valar, as his representatives, act within the material world, each in a specific and limited manner, with the exception of Melkor and his cohort, who supply the chaotic factor that gives rise to the world's troubles.

---

* *The Silmarillion* was not finished during J.R.R. Tolkien's lifetime. He entrusted his son Christopher with the task of publishing his writings, which were numerous and sometimes disorganized. Christopher Tolkien chose to publish a polished piece for the purpose of reader accessibility; this meant removing most of the passages in which the texts of *The Silmarillion* are attributed to an Elvish author. The Ainulindalë is thus included without a fictional author being identified. The subsequent publication of Tolkien's manuscripts in the twelve-volume *The History of Middle-earth* (only the first five volumes of which have been translated into French to date) enabled Christopher Tolkien to show his father's writing process, and restored elements to their proper context that had been removed from the published *Silmarillion*, including the names of the Elvish authors.

† 8th-century B.C. Greek mythological poem recounting the origins and genealogies of the Greek gods.

The Siege of Angband in Beleriand.

Chronologically, the *Valaquenta* is followed by the *Annals of Aman*, most of which is also the work of Rúmil, though they also include a digression on the reckoning of time, taken from the work of another Elvish loremaster, Quennar Onótimo.[*] These annals recount the beginnings of Arda, the home of the Valar within Eä. They describe their first battles against Melkor, who also entered the realm with the intention of taking it for himself. After a series of devastating clashes over Creation, the Valar took refuge in the realm of Valinor in western Arda, where they created the Two Trees, the principal source of light during this Age of the world. In contrast, Melkor and his followers relocated to an underground fortress in the north of Middle-earth.

The appearance of the Elves in the east of Middle-earth led the Valar to confront Melkor directly, to take him prisoner, and to destroy his fortress. However, they did not manage to capture his main lieutenants. The Valar then invited the Elves to join them in the realm of Valinor where they could live in peace, far from the lands devastated by Melkor and the monsters he created, but some Elves preferred to remain near Cuiviénen, where they first awakened. Others became discouraged by the arduousness of the Great March to the eastern coast of Middle-earth and stopped along the way. The Sindar, or Grey Elves, remained in Beleriand when their king fell in love with a servant of the Valar. After crossing the Great Sea to reach an island detached from the shore by the Valar, three elf-tribes—the Vanyar, the Ñoldor, and the Teleri—finally reached the end of their journey.

The long sojourn of the Elves in the Undying Lands led them to broaden their talents and to develop their arts and sciences, as in the case of Rúmil, who invented the very first alphabet, perhaps for the purpose of writing his chronicles. However, the liberation of Melkor after his term of imprisonment soon wreaked new havoc; Melkor convinced some of the Ñoldor that the Valar were holding them captive on Valinor so they would not come to dominate Middle-earth. The dissension then spread due to the machinations of Fëanor, the oldest son of the Ñoldor king, a particularly gifted elf, but who was also haughty and arrogant. Taking advantage of a festival, Melkor destroyed the Two Trees that brought light to the realm of Valinor and made off

---

[*]   In his (justified) desire for homogenization, Christopher Tolkien merged the *Annals of Aman* with the text of the *Quenta Silmarillion*, which will be discussed later, and so it does not feature in *The Silmarillion*. However, the earliest versions of these annals were published in *The Shaping of Middle-earth* and *The Lost Road*, while subsequent reworkings are included in *Morgoth's Ring*.

with the Silmarils, three immense gems crafted by Fëanor, killing Fëanor's father in the process, and then fled to Middle-earth. Fëanor swore vengeance and the majority of the Ñoldor followed him, despite the Valar's opposition to such impulsive action. Unable to produce a fleet to cross back over the Great Sea, the Ñoldor, led by Fëanor, opted to take by force the ships of the Teleri, their kin, which led to a massacre that roused the anger of the Valar. Those Ñoldor who remained bent on revenge were banished from Valinor, while the small number that repented was pardoned. The *Annals of Aman* end by telling how the revolting Ñoldor eventually split into two factions that journeyed separately to Middle-earth, while the Valar created the Sun and the Moon to compensate for the loss of the Two Trees.

Though their substance is strongly impacted by myth, the *Annals of Aman* bear more resemblance to a history, and are reminiscent of the chronicles of Antiquity or the Middle Ages. Taking his view from these types of chronicles, Tolkien makes frequent reference to other writings that allegedly go into more detail about the events he is describing. For example, the Noldolantë, said to be a tragic poem by the minstrel Maglor relating the fall of the Ñoldor, is cited during the discussion of the Ñoldor's attempt to take the ships of the Teleri by force. The Valar soon cease to be the main protagonists, and the Elves take their place.

The chronicles of the Elder Days are followed by the *Grey Annals*, jointly produced by the Sindar and some of the Ñoldor.[*] They recount the ancient history of the Elves and the settling of the Sindar in Beleriand. These Elves were soon joined by the Dwarves, a people created by the Valar Aulë the Smith. Other Elves who had stayed behind came to Beleriand as well, while the monsters created by Melkor began to proliferate once again. Melkor's return to Middle-earth plunged Beleriand definitively into war, with the Ñoldor, in pursuit of him, establishing numerous kingdoms of their own in the region. A number of battles were fought, and Melkor was besieged in his new fortress, which remained impregnable. Several tribes of Men also reached Beleriand at this time, fleeing the violence instigated by Melkor; some of them allied themselves with the Ñoldor and supported them in their fight. Despite the Elves' valor, Melkor ended by gaining the upper hand. Producing a relentless army

---

[*] As with the *Annals of Aman*, this text was merged with the *Quenta Silmarillion* in view of the publication of *The Silmarillion*. The first version of these annals is also included in *The Shaping of Middle-earth* and *The Lost Road*, while a later reworking is featured in *The War of the Jewels*.

of new monsters, he also joined forces with renegade humans and destroyed the Elvish kingdoms one by one. Finally, the last remaining Elves and Men were forced to take refuge in the forests or on an island off the coast of Beleriand. Fortunately, Eärendil, born of a human father and an Elvish mother, succeeded in his daring attempt to cross the Great Sea, evading the traps set by the Valar, and pleaded the cause of Elves and Men directly to the Valar, who, moved to pity, sent a formidable army that defeated Melkor once more, though the fiery battle caused the sinking of Beleriand.

The subject matter of both these series of annals is revived in the *Quenta Silmarillion*, a continuing narrative compiled by Pengolodh, an elf born in Middle-earth. In this piece, the chronicle-style recitation is frequently enhanced and clarified, but not systematically. Some passages in the *Grey Annals* are cited as sources for the *Quenta Silmarillion*, attesting to a certain complexity of narrative tradition. Additionally, Pengolodh is named as the author of numerous other writings on various subjects, ranging from a treatise on linguistic history to a description of marital customs among the Elves.[*] Other stories are brought in to flesh out certain events in the wars of Beleriand. For example, the Lay of Leithian tells the tale of an improbable love between a man and an Elvish princess, which leads them to undergo all sorts of peril in order to be together. Likewise, the *Narn I Chin Húrin* (or *Tale of the Children of Húrin*) recounts the sorrowful lives of a courageous and proud brother and sister whose many exploits end in tragedy. This poem was written by the minstrel Dírhaval, who died at the very end of the Wars of Beleriand.

All of the accounts cited above are supposed to have been translated into Old English by the navigator Ælfwine, a distant descendant of Eärendil, who miraculously managed to reach the Undying Lands well after they had been hidden from the eyes of the mortal world. Ælfwine is also said to have added many clarifications to these texts taken from other writings, such as the geographical treatise *Dorgannas Iaur* by Torhir Ifant, and met with elf Pengolodh, who answered certain questions for him. Tolkien never specifies what became of Ælfwine after he returned to his English homeland, but it is clear that he had time to write down all of the stories with which he had been entrusted. It is through him that many of the tales of the Elder Days

---

[*] These manuscripts posed a complex problem in view of the publication of *The Silmarillion*, as the annals and the *Quenta Silmarillion* became highly repetitive. Consequently, Christopher Tolkien prioritized the latter while excluding non-narrative essays that were too far outside the scope of the chronicle.

are supposed to have come into the hands of Tolkien, who represents himself simply as a translator. Aware of the dangers of intertwining reality and fiction too tightly, however, Tolkien refrains from telling us exactly how he came to possess the texts. In this, he is drawing freely on the vagaries of transmission of ancient and medieval writings. It is easy to draw parallels with the many Greek or Germanic legends that remain extant only through citations or summaries published by later compilers.

FROM ELVES TO HALFLINGS: HISTORY PRESERVED BY THE HUMBLE

Tolkien did not limit himself to the mythological and epic legends that make up most of *The Silmarillion*. The history of Middle-earth stretches over several Ages, witnessing the resurgence of new forces of evil and the completion by Men of their dominance of the world, to the detriment of the Elves. After the Wars of Beleriand, the Men who had allied themselves with the Elves were rewarded, and received the isle of Númenor on which to settle. They developed a splendid civilization there, thanks to their proximity to the Undying Lands. They also participated in the civilization of a part of Middle-earth that had relapsed into barbarism and successfully repelled an invasion led by Sauron, a former lieutenant of Melkor. However, their increasing pride eventually caused their downfall, and most of their creations were lost when Númenor sank beneath the sea. Only a few loyal followers led by Elendil managed to escape the catastrophe. Unsurprisingly, the parallel with Plato's Atlantis was mentioned by Tolkien himself. A few fragmentary records of this era survive in the archives saved by Elendil, including *Indis i-Kiryamo* (*The Mariner's Wife*), an account of the tragic love between a prince of Númenor and the daughter of a provincial squire. Moreover, Elendil himself wrote the *Akallabêth*, a recounting of the fall of Númenor.

After having escaped the submersion of Númenor, Elendil and his sons founded the kingdoms of Arnor and Gondor in the west of Middle-earth. They defeated Sauron again, causing him to lose the One Ring, which held much of his demiurgical power. Arnor and Gondor then experienced an era of splendor followed by a slow decline, accelerated by a number of conflicts covertly unleashed by Sauron. When the events of *The Hobbit* begin, Arnor has long since vanished, and the last heir to its throne has been raised in Rivendell, in the home of Elrond, the half-elven son of Eärendil. The whole plot of *The Hobbit* is said to be derived from the travel journal kept by Bilbo Baggins, narrating his adventures in the company of the wizard Gandalf and a group of Dwarves led by Thorin Oakenshield, on a quest to recover the treasure stolen from the Dwarves by the dragon Smaug. From this point on, the destiny of all of Middle-earth will be changed by this race of little people that are the hobbits: in the course of the journey, Bilbo comes into possession of the One Ring of Sauron, and it is Frodo, his adopted son, who completes the quest unwittingly begun by his relative. After many adventures, Frodo manages to reach the volcano of Orodruin, where the ring is destroyed in the fire that forged it, which finally enables the defeat of Sauron. One of Frodo's friends, the Ranger Aragorn, is revealed to be the last heir of the kings of Arnor, and restores the kingdom of his ancestor Elendil.

As a sign of their newfound importance, the hobbits now assume responsibility for the preservation of knowledge. Bilbo Baggins takes advantage of his long stay in Rivendell to write *Translations from the Elvish*, a historical record of the Elder Days. Likewise, Meriadoc Brandybuck, one of Frodo's companions, is the primary compiler of *The Tale of Years*, a general history of the time from the destruction of Beleriand to the restoration of the kingdoms of Arnor and Gondor. He also writes a comparative essay on the different calendars used by Elves and men, as well as *Herblore of the Shire*. Frodo himself writes the history of the War of the Ring; this manuscript is subsequently expanded with accounts taken from Gondorian writings including Elendil's *Akallabêth*. Known as the *Red Book of Westmarch*, this version goes through several editions, some of which include further additions such as *The Book of Thain*, which contains *The Tale of Aragorn and Arwen* and is said to be among the sources used by Tolkien for *The Lord of the Rings*. Like the good medievalist that he was, Tolkien had a deep familiarity with manuscripts containing various texts with very few logical connections to one another. Here, the title of Frodo's book is a clear reference to the *Red Book of Hergest*,* an important Welsh manuscript that includes the mythological stories known as the *Mabinogion*† and a large number of bardic poems.

---

*   An important Welsh manuscript written around 1382–1410, containing mythological, epic, and historical texts as well as court poetry and a collection of medicinal remedies.

†   *The Four Branches of the Mabinogi* are four epic medieval stories that make numerous allusions to ancient Celtic mythology.

After the publication of *The Lord of the Rings*, Tolkien planned to write a new novel set a century after the death of King Aragorn. Intended to relate the story of the resurgence of demonic plots, it would have presaged the subsequent fall of Gondor. However, Tolkien eventually wrote only a few pages of this story, finding the idea of a thriller-like narrative "sinister and depressing," and there is nothing in his few surviving notes on the subject to indicate how he would have claimed to come into possession of this text.

WHAT ABOUT TOLKIEN? THE PARADOX OF ETERNALLY STARTING OVER

More than a history of a world, Tolkien invented the way in which a world describes itself and transmits its knowledge to subsequent generations. The dual nature of this task, nearly insurmountable in its complexity, explains in part why Tolkien did not manage to publish a large part of his writing during his lifetime. But there was another hindrance as well. Tolkien was always a meticulous tinkerer, never satisfied with his own stories. Each new version built on the preceding one, adding new details and events. These additions then meant that adjustments had to be made to other parts of the history, sometimes including changes to certain fundamental aspects of the work. Eventually these corrections permeated the whole oeuvre and Tolkien was forced to start over again. By his own admission, Tolkien would never have published *The Hobbit* if the manuscript hadn't been found by chance on the desk of the publisher George Allen & Unwin, and he was even more reluctant to publish *The Lord of the Rings*, bowing only to the encouragement of his friend C.S. Lewis and the perseverance of his publisher.

The saga of the Elves narrated in *The Silmarillion* remained unfinished. It predated *The Hobbit* by many years, having been started by Tolkien in a hospital bed where he was recovering from trench fever contracted late in World War I. Tolkien drew inspiration from the novels of Lord Dunsany (1878–1957), Andrew Lang (1844–1912), and William Morris (1834–1896) to create a fantasy tale linked to the history of the colonization of Great Britain by the Anglo-Saxons. In this version, the navigator who came into contact with the Elves is named Eriol; he finds himself embroiled in the Elves' final battles, and bears witness to their final disappearance, a narrative that can be read in *The Book of Lost Tales*. Dissatisfied with this first version, feeling that the Elves bore too strong a resemblance to Victorian fairies, Tolkien reworked his stories, including narrative poems in some of them. Other, prose versions followed, in which it became clear that the stories had been transposed to a distant, fictional protohistory. The intrusion of *The Hobbit* into this world in gestation made numerous modifications necessary, which led Tolkien to develop the history of the period that followed the Wars of Beleriand.

Twice, Tolkien attempted to find a solution other than the device of the solitary navigator to explain the transmission of these tales to modern readers. In the 1930s he began working on a story entitled *The Lost Road*, in which a father and his son immerse themselves in visions of the earlier eras in which their ancestors lived, until they travel back to the time of the sinking of Númenor. During the writing of *The Lord of the Rings*, Tolkien outlined a novel based on the minutes of a fictional Oxford University club's meetings. One of the club's members, Arundel Lowdham, experiences lucid dreams of a people that speaks Elvish languages. With one of his friends who has been having similar dreams, he finds himself embarking on an adventure in which dreams blend with current reality. The complexity of these plots, and Tolkien's lack of interest in writing a novel set in the modern day, explain his abandonment of both projects.

Ultimately, the overarching plot of Tolkien's narrative, a key aspect for him, remained incomplete, marginalized by stories of the Elves and their unceasing evolution. Undoubtedly, Tolkien had no real desire to wrap up a story that had been part of his life for so long. Assigning a definitive role to Ælfwine or his possible replacements would have meant bringing to a close the great historical chronicle of the Elder Days. Isn't it likely that the prospect of continuing to find new chapters to rework seemed infinitely more appealing? And as for Tolkien's readers, it seems that the blank spaces left around his stories have excited as much thrilled curiosity in them as any Anglo-Saxon specialist might feel when researching the unexplained legends alluded to in *Beowulf*. What more beautiful literary legacy can Tolkien have hoped to leave than that?

LINGUISTICS AND FANTASY

DAMIEN BADOR, Tolkien specialist

PHILOLOGIST AND WRITER

Tolkien disliked novels that tended toward autobiography, though he did not dispute the fact that an author has no choice but to use his or her own experiences in writing fiction. *The Lord of the Rings* is most assuredly *not* an allegory for the 20th century, nor are any of his protagonists a reflection of Tolkien himself. Yet, if there is a domain inextricably intertwined with the life of our author, it is linguistics: comparative philology, to be precise. For Tolkien, language and literature necessarily go hand in hand; this is the only way to ensure proper understanding of a text, particularly in the case of ancient texts. Tolkien conveyed this point of view in his analysis of *Beowulf*, published in *The Monsters and the Critics, and Other Essays*, which combined philological rigor with literary appreciation at a time when critics generally saw the epic poem merely as a source of historical information distorted by myth.

The importance of language is easily discernible in Tolkien's obsession with finding the perfect turn of phrase, even if it meant reworking certain sentences countless times. His preoccupation with linguistic detail also found its way into his stories themselves, focusing on the languages spoken by the various characters. No one who has read *The Lord of the Rings* can fail to have been struck by the passages in Quenya or Sindarin, the two main Elvish languages, and in perusing the novels' appendices, it becomes clear to the reader that these are true languages, each with its own specific grammar and vocabulary, and that Tolkien also paid close attention to the evolution of these languages, and to their relationships to one another. Indeed, the words spoken by the hobbit Meriadoc Brandybuck and the herb-master of Minas Tirith show that many of Tolkien's characters possess great linguistic sensitivity. The Elves are no exception; it is among them that we find the first guild of linguists, the Lambengolmor. From here it is no stretch to qualify Tolkien's inspiration as "fundamentally linguistic," particularly as his letters show a deep interest in the subject.

A LIFELONG UNDERTAKING

By his own admission, Tolkien's fascination with languages began very early in life. His mother Mabel introduced him to Latin, French, and German. Later, he studied ancient Greek, and then Old English, Norse, Gothic, Welsh, and Finnish. Tolkien's first job after World War I consisted of writing definitions for the Oxford English Dictionary, then being prepared for re-edition, and where he became known for his thorough etymological research. In 1920, he obtained an associate professorship in English at Leeds University; he would teach English and the ancient Germanic languages throughout his academic career. However, he had also nursed a passion for the invention of imaginary languages since childhood. After his young Incledon cousins introduced him to the private language they had created, which they called Animalic, Tolkien collaborated with them to invent the more sophisticated language of Nevbosh. Next he created Naffarin for his own amusement, developing its vocabulary and grammar as well as its phonology. The discovery of Gothic impelled him to invent a fictive Germanic language, but his attention was soon fully taken up by Finnish. Strongly attracted by this language and the legends of the *Kalevala*, he threw himself into his first attempt at a novel, an adaptation of the tragic story of Kullervo. This endeavor also represented an initial effort to bring together his two favorite subjects; in it, he began to develop a specific nomenclature based on Finnish.

At the same time, Tolkien was also writing various fantasy poems, which gradually gave rise to a fictional universe in which the Elves met a tragic fate. His forced 1917 convalescence in a military hospital bed provided him with the time to draft several of these stories; they were the earliest of the *Lost Tales*, the first seeds of *The Silmarillion*, and the first tales centered on Middle-earth. Most of Tolkien's creative energy was soon being poured into this work. He invented two languages in succession which were purported to be related, Qenya and Gnomish. Each with their own grammar and phonology, these languages encompass several dialects, and Tolkien took care to outline certain aspects of their semantic and phonetic evolution, enabling him to explain, for example, how the Qenya word *apaire* ("victory, conquest, subjugation") corresponds to the Gnomish *abair*, and how the Qenya *héru*, "lord," is linked to the Gnomish *hîr* (care, anxiety).

As with his books, Tolkien revised his linguistic inventions endlessly. His languages were soon joined by several alphabets, which could be used to transcribe both English and the nascent Elvish tongues, not to mention Old High German! During the 1930s, the initial conceptual framework became insufficient, as Tolkien began to take an interest in the languages of the Valar, orcs, Dwarves, and even humans. It was at this point that he began writing *The Etymologies*, a dictionary tracing the roots of primitive Elvish and listing a large number of derivatives in various languages. He also wrote a history of languages entitled *Lhammas*, said to be the work of the elf Pengolodh. What could be more logical? Of course a people as advanced as the Elves would count linguists among their numbers. The books themselves further influenced the languages, and Tolkien reviewed his own creation more than once, which occasionally led him to remove a language family completely, to modify relationships among languages, or simply to rename them. He also began exploring the narrative avenues offered by the wide variety of languages in more and more depth; thus a king's anger forces the Elves of Ñoldor to abandon the common use of their mother tongue, while the Dwarves' mistrust of strangers manifests itself in a certain reluctance to teach their language to others. Conversely, knowledge of ancient languages helps to resolve the mystery of the secret door to Erebor.

The publication of the appendices to *The Lord of the Rings* forced Tolkien to stop altering the major lines of his linguistic history. These appendices contain a significant amount of information about the languages spoken in the west of Middle-earth at the time of the War of the Ring. The two main alphabets used are described in detail. Tolkien would have liked to add a linguistic glossary that would define all of the invented words used in the novel; however, this plan never came to fruition, as it would have taken too long, and threatened to delay the publication of the final volume of the trilogy, and was never finished. For all that, Tolkien retained his habit of revising his grammatical concepts and continued to develop certain aspects of his two principal languages, now called Quenya and Sindarin, and very different from what they had been forty years earlier. This included playing with those parameters that did not hinge on the forms published in the Elvish poems that pepper *The Lord of the Rings*. The 1965 re-edition allowed him to alter some narrative passages and to make certain linguistic changes; for example, the Quenya *omentielmo*, "of our meeting," became *omentielvo*, a seemingly minor modification, but one behind which lay a complete revision of possessive suffixes and verb endings.

THE LINGUISTIC LANDSCAPE
OF THE WAR OF THE RING

Paradoxically, the Elvish languages to which Tolkien devoted so much thought occupy a fairly limited place in *The Lord of the Rings*. The Elvish population of his Middle-earth is in steep decline, with many leaving for the Undying Lands forever. Quenya, the language of the High Elves, has become nothing more than a sort of "Elvish Latin"; it is rarely used by the Elves in everyday life; and few Men are learned enough to understand it. Sindarin, the language of the Grey Elves, is still spoken in the Elvish enclaves of Eriador and Rhovanion. Its pronunciation differs enough between these two regions to confuse the unprepared speaker, as in the case of Frodo Baggins, but poses no difficulties of comprehension for the Elves. Because of their friendship with the latter, some of the Dúndedain, such as Aragorn, learn Sindarin in their youth, and it is still commonly spoken by some in Minas Tirith and the surrounding lands. Its use among hobbits is exceptionally rare; Frodo is an unusual case, motivated by the remarkable friendship between his adoptive father Bilbo and the Elves of Rivendell.

The former power wielded by the kingdoms of Arnor and Gondor in the west of Middle-earth resulted in the widespread knowledge of the Númenorean language by numerous peoples, who use it as a *lingua franca*. Enriched with Elvish and local words, it is generally known by the name Westron, or the

Common Speech.* The hobbits have adopted Westron as their native language, though their dialect includes terminology related to the language of Rohan, which they formerly spoke. The Rohirrim themselves have continued to use their ancestral language, which is also distantly related to Westron. However, the use of the Common Speech has become customary among their elites, partly as a result of the influence of King Thengel, father of Théoden. The Men of Rhovanion speak languages related to Rohirric, but use the Common Tongue when dealing with outsiders. The Druédain and the Men of Dunland are another story entirely; very few of these mistrustful people speak any foreign language at all. And the Men under Sauron's control speak still other languages, unrelated to those spoken in the west of Middle-earth.

Other peoples have linguistic customs of their own; the Dwarves speak Khuzdul among themselves, a language that has been passed down, virtually unchanged, since their creation. The Ents have created a language so complex that no other being has ever been able to learn it or duplicate it correctly in writing. Each orc tribe has its own dialect, which is generally unintelligible even to a neighboring tribe; this forces the orcs to use the Common Tongue in order to make themselves mutually understood, which greatly aids the survival of Frodo and Sam in Mordor. Sauron has tried, in the past, to propagate his own language among his servants, but by the time of the War of the Ring, only the Nazgûl and some of Sauron's own creations, such as the Olog-hai, the trolls of Mordor, still use Black Speech.

FROM FËANOR TO MERIADOC BRANDYBUCK

Just as Tolkien required fictional historians to retrace the history of his peoples,† so he needed linguists to document the languages of Middle-earth and even put forth linguistic theories. Hence, the Elves, Firstborn of Ilúvatar, gifted in the arts and sciences, were also the first philologists; indeed, Tolkien makes a point of specifying that their fondness for the invention of new words is innate. Some Elves, particularly among the Ñoldor, are especially creative,

---

* In *The Lord of the Rings*, Tolkien uses English to represent Westron, and related Germanic languages to represent those languages related to Westron. For example, the language of Rohan is replaced by Old English, and the language of the Dale of Erebor by Old Norse.

† See the chapter entitled "History and Historiography in Middle-earth," p. 69.

and go so far as to propose deliberate phonetic changes in order to make their language more beautiful and harmonious. Yet, the rules of human linguistics are also partly applicable to Elves. Though they do not die of old age, the Elves witness the evolution of their language through involuntary and unforeseeable phonetic changes, which leads some of the first Elvish linguists living in the Undying Lands to attempt to reconstruct their original language. However, most of these efforts prove to be in vain, due to a lack of opportunities for comparison with other Elvish peoples.

Upon contact with the Valar, the Elves discover a language wholly different from their own—though few of them study it in detail, with the exception of the historian Rúmil, who consequently becomes the first linguist whose name we know. His philological observations have mainly come down to us via Pengolodh, who makes use of them in his *Lhammas*. Rúmil's most significant linguistic contribution is his invention of the first alphabet, called the Sarati. However, this alphabet has fallen into disuse even before the return of the Ñoldor to Middle-earth, supplanted by Tengwar, an alphabet invented by Fëanor. Both the Sarati and the Tengwar are created as abugidas; that is, alphabetical systems in which only consonants are represented as separate letters, while vowels are denoted via diacritical marks.[*] However, the Tengwar forms a more regular system, as the phonological relationships between the different letters are indicated by regular variations in shape, with the dental [t] considered the reference sound. For example, the voicing corresponds to a doubling of the loop of the corresponding letter, while a change in place of articulation may be represented by the addition of a horizontal line (for labial consonants) or the inversion of the letter (for velar consonants):

	[t]	[nd]	[p]	[k]	[n]	[m]	[ŋ]
Sarati	ꙇ	ꙮ	ꙭ	ꙩ	ꙡ	ꙣ	ꙕ
Tengwar	p	pɔ	p	q	ɩɔ	ɯ	ᴄᴄ

---

[*]   This principle is employed in various historical alphabets, including the ancient Indian Brahmi script and its derivatives.

Fëanor is unquestionably the best known of the Elvish linguists, though his main achievements all date from his youth. In addition to his alphabet, he also founded the linguists' school of Lambeñgolmor. His insightful analyses enabled a better delineation of the evolution of High Elvish in relation to Primitive Elvish; however, he did not share with his disciples all of the knowledge entrusted to him by the Valar. Later, he abandoned linguistics to dedicate himself entirely to the crafting of the *Silmarilli* and the *palantiri*, his most famous creations. One of Fëanor's sons, Curufin, inherited some of his father's talents. After his return to Middle-earth, he was one of the rare Elves to forge a friendship with the Dwarves, which made it possible for him to study their language, Khuzdul, and It is to him that we owe much of our knowledge of this tongue. But the greatest Elvish linguist in Middle-earth remains Pengolodh, compiler of the *Lhammas*, a history of the Elvish languages, and who was interested in the languages of Dwarves and Men as well. Gifted with a prodigious memory, he saved much of his predecessors' work from oblivion, particularly that of the minstrel Daeron de Doriath, who had invented a runic alphabet still in use among the Grey Elves of Beleriand before the return of the Ñoldor.

At a certain point in Middle-earth's history, Men gradually begin to take over from the Elves as linguistic guardians, though their efforts focus more on preserving knowledge than on new research. King Elendil is credited with having saved Pengolodh's work *Eldarinwe Leperi are Notessi* during the flooding of Númenor. Indeed, Elendil is renowned for his love of the Elves' languages and history. Surviving linguistic texts are kept in Gondor for the most part, as is the case with the fragmentary, anonymous essay *Onondóre Nómesseron Minapurie*, a study of Gondorian place-names. With the decline of the Númenorean kingdoms, linguistics declines as well; however, the discipline enjoys a certain renaissance after the War of the Ring ends, with the hobbit Meriadoc Brandybuck following the example of his illustrious predecessors and composing a short treatise on the dialectical words of his homeland, entitled *Old Words and Names in the Shire*—yet another example of knowledge passing into humble and discreet hands.

AN ESSAY ON "AESTHETIC LINGUISTICS"

In one of his letters, Tolkien calls himself a "*pure* philologist," who found one of his greatest sources of joy in creating phonetically correct languages whose meaning and form were in harmony with one another. A perfectionist, he wanted his Elvish languages to be refined and logical in their design, while also introducing a number of irregularities meant to render them more plausible. He once filled an entire page with his attempts at a satisfactory translation of the English expression *try harder*, determined both to be idiomatic and not to imitate the English. The very scope of his creation proved too challenging even for him at times, and some phonetic changes no longer fit with the form and meaning of certain terms. At these moments he was forced to retrace the entire etymology of a word, crafting ancient compound words, imagining dialectical variants, or reinventing the origin of the word. In some cases, such as that of the name of the mountain Halifirien, these linguistic contemplations even gave rise to the account of an important historical event in Middle-earth.

It is no surprise, then, that Tolkien's philological creations encompassed his whole life and work, so much so that the latter might fittingly be called "logo-fiction." More than forty years after his death, passionate devotees continue to publish his linguistic writings in the journals *Parma Eldalamberon* and *Vinyar Tengwar*, and the trove of as-yet-unpublished material appears to be sizable. Tolkienian linguistics is a rich discipline, but one that remains unpretentious, with new publications sometimes calling certain long-established views into question—yet this continuous state of re-evaluation has not affected the popularity of Tolkien's invented languages. Numerous dictionaries and grammar texts for the Elvish languages have been published in multiple languages, and conferences dedicated to the languages invented by Tolkien are regularly held; not to mention the countless amateur poets who write quatrains in Quenya or Sindarin. One enthusiast has even taken the trouble of translating the entire New Testament into Quenya, not hesitating to invent neologisms when the vocabulary is lacking. Tolkien wrote in one letter that he considered Esperanto to be a dead language, for the reason that its creator never invented a legend in Esperanto. Without prejudging the future of international auxiliary languages, the current popularity of Tolkien's invented languages tends to confirm the importance of the links between novelistic and linguistic fiction.

THE LORD OF THE RINGS:
A MYTHOLOGY OF CORRUPTION
AND DEPENDENCE

THIERRY JANDROK,
doctor of psychology and psychoanalyst

"Power tends to corrupt, and absolute power corrupts absolutely. Great Men are almost always bad men."

—Sir John E.E. Dalberg, Lord Acton

he characters that people the history of Middle-earth are as much symbols as they are metaphors. They condense the characteristics of themes carved into the body of a legend.

Since the birth of this mythological universe, the subjectivity of its characters and the dynamics of their personalities have been organized according to musical arrangements. Songs follow one another, giving a sense of order to the divisions of this world and its inhabitants. Evil impresses its haunting rhythms on the substance of a totalitarian imaginary formerly lulled by well-being. This portrayal is present in all of Tolkien's stories, from *The Silmarillion* to *The Lord of the Rings*, *The Hobbit* to *The Book of Lost Tales*. Good or evil, every character goes back to a melody stemming from the songs of Ilúvatar, the creator god.

SONGS OF CORRUPTION

All of the evil characters in Middle-earth are typified by their psychic dependence on a protective controlling figure. Morgoth and his lieutenant, Sauron, are the main examples of this. They are devious minds, able, like the original gods, to implant their own thoughts in the hearts of inferior species. Their objective is always the same: to seduce. Through various offerings, they attempt to pass themselves off as benevolent entities in the eyes of those they wish to bend to their will. In the mythology of Middle-earth, Morgoth and Sauron are representations of the "devil child" or "demon seed"—one who, feeling ontologically wronged, lashes out endlessly at his parents and siblings.

The devil child feels humiliated, thwarted in his desire for power, like the victim of an injustice on the part of his guardians and siblings. He consequently seeks out a surplus of love, along with boundless admiration and recognition, unlike that received by those who represent the otherness whose very existence is a threat to his quest for satisfaction. The object of his desire belongs to his prehistory, to a time before the songs of his father, the impossible memory of a merging into a unit both matricial and universal.

The devil child shapes his own psyche around his hatred for his fellow beings. He is held hostage to a totalitarian imaginary. He makes use of symbolic connections in his conflictual relations with others. The devil child does not hold himself to his word; he prefers the pleasure and fulfillment of his fantasies of control and domination.

HATRED AS AN INHERITANCE

In Middle-earth, orcs and other inferior species, such as rats, spiders, ravens, and the powerful dragons ("foul worms") have aligned themselves with the silent promise of the forces of darkness. All the suffering and discrimination to which misshapen beings are subjected further aggravates their sense of humiliation. By this logic, darkness is experienced as illumination, and light as acid-like burning. For them, vengeance, mass murder, and degradation are acts of courage. The quest for peace, on the other hand, is seen as unspeakable violence. The more the servants of the demon meet with failure, the stronger their determination grows. In their differences, the devil's soldiers resemble one another. Their skins are as dark as their souls. In Tolkien's mythology, the agents of Evil, Morgoth and Sauron, are their nurturing fathers. In Middle-earth, Evil involves a radical challenging of the rules of affiliation, and the installation of a totalitarian order of merciless ferocity. For those enslaved by Evil, the only way out is death.

THE MELODY OF THE ONE RING

All that is gold does not glitter, Not all those who wander are lost.
—J.R.R. Tolkien

The Ring of Power is the quintessential corrupting object. Its form reminds us that its destiny was forged in the circularity of an infinite exchange. This object possesses a will of its own. It passes from hand to hand at the whim of the impulses of the desire from which it has been created. Conceived in the bowels of Mount Doom, it can only be destroyed in the fires of this ancient volcano. In Middle-earth, this symbol of absolute power fits anyone . . . like a glove!

The function of the Ring is to entrap corruptible minds. It is a snare as onerous as it is fascinating. Yet, even when licked by the flames of a brazier, it remains as cold as death. The interior space of the Ring is occupied by the desire of its creator. It is the surface of a mirror of desire whose invisible surface is an open channel to the lidless eye of the Beast.

The Ring of Power is the shadow of the desire that lies in each of us, the memory of infantile dreams aborted in the mists of education and Oedipal tragedy. Such is the power of this corrupting object that it can awaken fantasies of omnipotence in mortals, and the mad hope of escaping their inevitable death. All those who come into contact with the Ring are immediately corrupted. It separates and divides, before irrevocably severing the subject's bonds with his fellows. It acts in the shadows of consciousness, prolonging the wearer's years of life.

With the passage of time, the wearer eventually has no choice but to flee his community. The Ring also confers the gift of invisibility—actually more of an alternative reality. Wearing the Ring projects the subject to the other side of the mirror, into a dimension in which the totality of the Master is apparent.[*]

---

[*] See the chapter "Invisible to the eyes of Sauron?," p. 182.

Morgoth, formerly called Melkor.

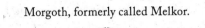

There, the subject is penetrated by a lidless eye, finding himself transported into a hellscape, from the skies of which a terrifying god, rather than asking "Who are you?," as in Genesis, states: "I see you!." The mythological inversion here is one of exceptional violence. Sauron is the dialectical opposite of the Jewish people's Eternal God; he is his dark face, obscene and immoral in his lack of eyelids. Anyone who wears the ring has sinned against the Beast by stealing away his own Goodness. Nevertheless, they are not truly thieves, because the Ring will not allow itself to be stolen. It merely passes from hand to hand, and no one can presume to stop it in its path in order to recover that missing part of them.

PART FOR THE WHOLE

The Ring is thus always a part, already lost, of its creator. It is its destiny, born in the volcano of the same name. The Ring is transient. Faced with the promise of its power, the only sensible thing to do is turn away from it. Gandalf and Galadriel, queen of the woodland Elves, have understood, of course, that given the elusive nature of this object, no one can take possession of it. The Ring is not an object that one can have mastery over. It is an object of desire. As evasive as it is insidious, it is so powerful that even Sauron himself has no control over its presence in the world. Like any object that causes desire, the Ring of Power is revelatory. It offers the subject a channel toward total satisfaction.

Sauron lost his pseudopod to the blade of Isildur, king of Arnor, who later fell victim to an ambush when he refused to destroy the Ring. It was a hobbit, Déagol, who then discovered it, buried in river-mud, near the Gladden Fields where Isildur had fallen. Sméagol, a friend of Déagol, was immediately entranced by the Ring's diabolical aura. He attacked his companion and murdered him in cold blood, becoming a slaughterer, a thief, a pilferer, a manipulator, and a master singer, blithely taking advantage of the power of invisibility conferred upon him by the Ring. His fellow villagers had soon had enough of him, and Sméagol was driven into exile. During the course of his wanderings, his body changed. He soon became unable to tolerate light and took his place deep inside a mountain. There, in the dim depths of this stone tomb, his mind unraveled (*dementia*, in Latin). As the centuries passed, Sméagol transformed into Gollum,* his mirror alter-ego, his shadow, and the

---

*   See the chapter "Gollum: The Metamorphosis of a Hobbit," p. 226.

incarnation of his basest desires. * Gollum incarnates the madness distilled by the One Ring. Sméagol changes from a spoiled child to a demon seed, before sinking into the darkness of the underworld. There, he sheds his socialized veneer. The mountain becomes a new womb for him. Infantilized by his dependence on the object of his desire, Gollum becomes enclosed within himself to the point of amnesia. A shadow among shadows, he becomes one with the archaic darkness of his desire. Becoming a puppet of his own destiny, Gollum will lose the ring in his turn when, continuing on its way, it falls into the hands of Bilbo Baggins.

To possess the Ring is to be held hostage by desire, of which the wearer cannot rid himself except through death or the true loss of the object. The Lord of the Rings is not Sauron, but the One Ring. As the text states, "One for the Dark Lord on his dark throne." The Ring is for the Beast, and not the other way around. "One Ring to rule them all. One Ring to find them, one Ring to bring them all and in the darkness bind them in the Land of Mordor where the Shadows lie." The text is clear.

DISPOSSESSION THROUGH POSSESSION

The War of the Ring is a tragedy, that of desire. At the core of the legend, the worst fate that anyone can meet is to be reduced to an invisibility of being. This is the curse inflicted by Sauron's rings. While the hobbits emerge relatively unscathed, Men fall under the spell of the nine rings that transform them into wraiths.

The Nazgûl, also called Black Riders, are phantoms. The orcs call them "the Shriekers" because they communicate via harsh cries in the Black Speech. Their intimate relationship to Sauron's rings has caused them to pass through to the other side, where they have been reduced to the living dead, or the undead. No weapon wielded by a man can wound them; only Elvish blades are able to penetrate them and to tear through the veil of their diaphanous appearance. They are so dangerous that the simple act of being near them can cause mortals to be poisoned by their breath. Moreover, the weapons they carry

---

* The patronym "Gollum" is formed from the two main syllables of Sméagol. Reversing these gives "Golaems," pronounced "Gollûm," with the 's' of Sauron silent, and emphasis on the second syllable.

are interdimensional, with their poisoned blades infecting anyone they strike. The wound remains open until the whole body passes into shadow. The stain of the Nazgûl, like that of the corruption of the Ring, is an indelible one.

The Black Riders have merged with their master's desire. They are not priests, but officers. The servants of the Ring are under a curse that no spell can undo, with the possible exception of one word, *Elbereth*, one of the names given to Varda, the queen of Arda. This word of liberation and disintegration suggests the mythological equivalent of *Emeth* ("truth"), which was written on the forehead of the Golem of Prague.* We can imagine the Nazgûl mythologically as Golems, slaves tethered to the omnipotence of a master. Like the Ring-wraiths, the Golem is a domestic creature. Like it, the servants of the Ring were originally Men of clay, as Genesis describes the first humans, formed from the dust of the ground. Thus the Nazgûl are identified with the hell from which they come.

The only tools of resistance against the temptation of omnipotence and the curses that come with the Ring are love and the honoring of one's ethical commitments. While hate satisfies itself in the presence of others, love is the eternal response to its absence. Images of power are not fought against with weapons, but with metaphors involving the symbolic desire for accomplishment. This is how Aragorn, a remote descendant of Isildur, lives his life.

THE LIGHT OF THE FATHER

In the story, Aragorn represents the Good man who, rather than resigning himself to slavery to the Ring or to cowardly exile to the land of the Elves, dedicates himself to overcoming his own inheritance. Starting from nothing, the child becomes a man only by conquering the metaphors that define him without his knowledge. The poetry of existence lies in constructing the poem from which the subject comes. It is not fate that decides! Precisely because they are weak and fragile, humans owe it to themselves to push toward the light, to fight against the seductive forces that lure them away from their brothers. Fulfillment does not lie in possession, but in transmission.

---

* Each night, the rabbi who had created the Golem erased the first letter of the name written on his servant's forehead. At that moment the body of the clay servant became immobilized and returned to its shapeless state. The word *meth*, created by erasing the first letter, means *death* in Hebrew.

In the story of the king of Gondor, if his ancestor's sword is re-cast, it is by the grace of its possessor's fraternal commitment. That is how Narsil ("red and white flame"), the sword of King Elendil, becomes Andúril ("flame of the west"). Re-forged in order to fulfil the desire that gives it meaning, Andúril is the sword of a monarch, but that monarch must prove himself worthy of it. Nobility does not lie in military exploits, but in the way in which the subject keeps his word and guides his own destiny. The royal role is to liberate, while that of demons is to shackle. The prince's weapon is his promise, but it is also the light that pierces the darkness woven by the evil forces at work in Middle-earth. In this Aragorn embodies the principle laid down by Goethe in *Faust*: "What from your father you've inherited, you must earn again, to own it straight. What's never used, leaves us overburdened, but we *can* use what the Moment may create!"

This is how *The Lord of the Rings* brings together the *Faust*s of Marlowe and Goethe. Beyond the warlike setting, we are essentially being offered a lesson in wisdom by this British saga. Tolkien wanted to create an English myth, a story that would become etched on the souls of his contemporaries like the *Iliad* and the *Odyssey*. A product of its reference culture, the myth aspires to travel through space and time to guide humans to rise above, to go beyond themselves, their doubts, and—even more—their certainties.

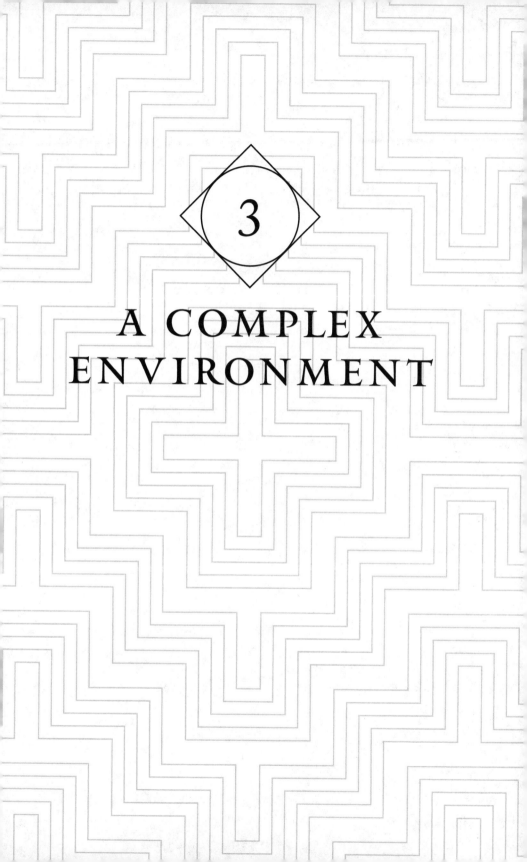

3

A COMPLEX
ENVIRONMENT

LANDSCAPES IN TOLKIEN:
A GEOMORPHOLOGICAL APPROACH

STEPHEN GINER, geomorphologist

Rohan, Mordor, the Shire . . . evocative places that resonate in our ears as if they really existed! With his descriptions, maps, and topography, Tolkien paid particular attention to the geography of his fantasy-world, and it provided true support for his stories. Thus, these places belong to both the fantastical and the real. What can we say about them, in light of our current knowledge? Since it would be impossible to discuss them all, we have chosen to describe certain emblematic places. Have a nice trip through the land of Tolkien!

THE UNDYING LANDS

Let's begin with the Undying Lands, a sacred continent off the coast of Belegaer, the Great Sea, west of Middle-earth. It was in this place described in *The Silmarillion* that the Valar settled during the First Age: a vast "island-continent" bounded by a crown of mountainous terrain, including the famous Pelóri, "the highest mountains on earth," which acted as a sort of boundary-fence on the eastern coast of the island. In the real world, Australia, bounded by prominent landforms (the Great Dividing Range to the east, the Kimberley plateau to the northwest, and the Hamersley Range of mountains to the west), is evocative of the Undying Lands even though its mountains do not form a continuous band.

The Undying Lands also display a mosaic of biotopes; the Woods of Oromë, for example (named after the Huntsman of the Valar and friend of the Elves), form an immense forested massif situated on the eastern part of the continent, just behind the southern Pelóri. The dominant winds here, we are told in *The Silmarillion*, come from the east. Behind a mountain range, winds normally undergo what is called orographic lift: passing above the mountain peaks, they come into contact with layers of colder air, which causes water vapor to condense and to produce heavy precipitation on the exposed slope. Once they have passed the peaks, these winds descend and warm up again; having been emptied of their humidity, they dry up the land behind the mountains. This phenomenon, called the "föhn effect" after a dry, hot wind that occurs in the Alps, is what gives rise to so-called coastal deserts such as the Atacama, situated between the oceanic trench and the Andean mountain range. With this type of eastern winds pummeling the Undying Lands, a vast desert behind the southern Pelóri would have been more reasonable than an immense wooded region . . .

BELERIAND AND THE NORTHERN LANDS

Another of the interesting places in the universe created by Tolkien is Beleriand, a veritable lost world once located in the north of Middle-earth. During the First Age it was the battlefield on which Men and Elves fought against Morgoth, before sinking partly beneath the waves. A map of Beleriand is shown in *The Silmarillion*; this region is bordered by mountain ranges (including the Ered Luin, or Blue Mountains, to the east), and bounded in the west and south by the Great Sea of Belegaer.

Northwest of Beleriand, the Hithlum Plain was occupied by the Grey Elves, who were Sindar. Open to the northern winds, it is bounded in the south, east, and west by mountainous terrain; this landscape is relatively common in reality, as in the case of the Zapadno-sibirskaya Ravnina, or West Siberian Plain, bounded by the Ural Mountains, the Kazakh Mountains, and the Siberian plateau. The southeastern part of the Hithlum Plain, called Mithrim, is traversed by watercourses that empty into the eponymous central lake, as often happens in real regions bounded by mountains. This place served as a refuge for the Ñoldor, the Elves of the Tatyar clan, the mountains providing shelter and the lake water. This type of geography is reminiscent of many real lakes such as Lake Victoria in Africa, which is surrounded by mountains.

In reality, this type of lake forms in an unstable region, in which a fault zone creates a depression which is then filled by the watershed, forming a lake. In Tolkien, this tectonic context is confirmed by the angular position of the mountain ranges of Hithlum, suggesting the presence of shearing faults, which geologists refer to as strike-slip faults. These localized tectonics, coupled with erosion, create landscapes. Erosion, in Tolkien, is perceptible to the west of Hithlum, where the river Annon-in-Gelyth flows beneath the Ered Lómin ("Echoing") Mountains, which must therefore be karstic; that is, composed of partly soluble carbonate-containing rocks.

The region of Nevrast in eastern Beleriand displays a similar landscape, with a central lake, Linaewen ("lake of birds"), ringed by mountains. However, the triangular configuration of these mountains is geologically impossible to explain. The absence of a watercourse feeding into this marshy lake suggests that the water source is a karstic resurgence and is actually filling a foiba, a sinkhole resulting from the dissolving of the calcareous terrain typical of a karstic landscape.

The north corresponds to the frozen lands of Dor Daidelos, bordered on the south by the Iron Hills, which are hills only in name; in fact they are an imposing range of mountains mined for their iron ore by the Dwarves in the First Age. On the south slope towers Thangorodrim, a group of three immense volcanoes comprising the heart of the realm of Morgoth. Volcanoes are extremely significant in Tolkien's work.[*] Those of Thangorodrim, "made of the ash and slag of [Morgoth's] subterranean furnaces," are "black and desolate and exceedingly lofty; and smoke issued from their tops, dark and foul upon the northern sky" (The Silmarillion). This is a typical description of explosive (or gray) volcanoes, which are themselves accumulations of ash and slag. The black smoke suggests volatile ash, and therefore a Plinian eruption; did Pliny the Younger's description of the 79 A.D. eruption of Mount Vesuvius that destroyed Pompeii inspire Tolkien? It is worth noting that this type of volcano is usually found in the midst of a range of mountains, in a geological plate collision zone (as with the Andean mountains), and not at the foot of a mountain range. However, the description of the great plain of Ard-Galen, at the foot of Thangorodrim, fits that of a plain devastated by one (or more) gray volcano(es): Tolkien mentions heavy falls of ash as well as "flame that ran

---

[*] See the chapter "Volcanoes, sources of magic and legend," p. 118.

down swifter than Balrogs"; that is, pyroclastic flows (rapid flows of gas and volcanic material). Might he also have drawn on the eruption of Mount Pelée, which ravaged the city of Saint-Pierre, Martinique, in 1902? Tolkien describes "great rivers of flame," which is also reminiscent of the liquid lava that flows from Hawaiian volcanoes; he commingles various types of volcanoes to increase the drama of his tales.

The Blue Mountains are a range that borders Beleriand to the east. They feed several streams that run parallel to one another and perpendicular to the mountains. This geological layout is common in plains at the foot of accentuated slopes, where movable sediment is cut through by rectilinear watercourses whose high outflows prevent the formation of meanders, as is the case in Provence, where the Vaucluse Mountains are drained by perpendicular watercourses emptying into the Calavon river.

In the center of Beleriand, a northwest/southeast-oriented line of hills called Andram ("long wall") evokes a buckled, or Appalachian, topography, of which only a ridge remains. This would make Beleriand a peneplain, of which Amon Ereb, a small mountain east of Andram, is an outlier. This range of hills is cut through by the "mighty river Sirion, renowned in song" (*The Silmarillion*). Downstream, the river anastomoses, its flow slowing to take on a so-called Durancian appearance. This is unrealistic, since "mighty falls" are mentioned beyond that, yet water cannot stagnate or undergo anastomosis upstream of falls like this. On the other hand, the marshy delta at the estuary of the Sirion is a topography found in real geography, as in the case of Camargue with the Rhone delta, or Louisiana with that of the Mississippi.

NÚMENOR:
TOLKIEN'S MYSTERIOUS ISLAND

From Jules Verne to H.G. Wells by way of Hergé, islands have long been a source of writerly inspiration. At the end of the First Age, Tolkien causes to arise from the waves a fantastic island in the shape of a five-pointed star: Númenor (Figure 1).

Númenor's "emergence from the waves" recalls the very real volcanic islands which, through accumulation of material, form quite rapidly in some cases: the island of Surtsey, for example, off the coast of Iceland, was created by an eruption that began around November 10, 1963 at 130 meters below sea level, and emerged on November 14, 1963, continuing to rise until 1967. It is also notable that the five-pronged shape of Númenor is reminiscent of the volcanic island of Sulawesi in Indonesia, and its northern arm, called Forostar, features numerous mountains, as does the northern arm of Sulawesi, which has numerous active volcanoes. Númenor also bears a resemblance to ancient depictions of the peninsula of Cadiz off the Andalusian coast, of which Tolkien was aware.

The highest peak on Númenor stands at its center: Meneltarma, or "pillar of heaven," described as "a mountain tall and steep"; this is certainly a volcano, since, when Sauron attempts to take possession of the island, "a groaning as of thunder underground" is heard. Furthermore, Meneltarma's summit is sacred and forbidden, as in numerous legends where volcanoes are dwellings of the gods. Logically, Meneltarma is composed of accumulated lava flows, a landform referred to by geologists as a stratovolcano. This composite volcano is surrounded by large volcanic plateaus (presumably basaltic), or planezes. In reality, this type of landscape is found in the Cantal Mountains, for example.

But Númenor, like all mythical islands, meets a tragic end; when its old King, Ar-Pharazôn, tries to invade the land of the Valar in a quest for immortality, the island is laid waste and engulfed by the waves, in an episode Tolkien himself admitted was inspired by the legend of Atlantis.

MIDDLE-EARTH:
COMPOSITE LANDSCAPES

Middle-earth, the setting of *The Hobbit* and *The Lord of the Rings*, remains the best-known part of Tolkien's world, though not necessarily the best understood.

This region is composed of a mosaic of landscapes. Here are a few key points for understanding them:

Lake Evendim, which lies north of the Shire, is an uncommon body of water in that, instead of being fed by rivers or streams, it is the source of a river called the Brandywine by hobbits due to its brownish waters. This color suggests that the lake is fed by karstic resurgence (defined above), as is the case with Ranco Lake in Chile, which is a source of coastal rivers.

The Gap of Rohan is an erosion zone between the Misty Mountains and the White Mountains, through which two rivers flow, the Isen and the Anduin. The strong outflow of the latter will have hollowed out the gap following the subsidence of the eastern part of Rohan. In our world, such gaps can be seen in the Atlas Mountains in Algeria, where an erosional notch links the Chott el Hodna in the north to the Chott Melrhir in the south.

The River Anduin empties into the Bay of Belfalas where it forms a strange, concave delta. Deltas, accumulations of sediment in oceans and seas, are usually convex, except in the event of rapid marine transgressions (rises in sea-level). Did Middle-earth experience a rise in water levels during its history?

Mordor, Sauron's dwelling-place, is surrounded on three sides by the mountain ranges Ered Lithui and Ered Dúath, an enclave reminiscent of the Taklamakan Desert in central Asia, nicknamed the "Sea of Death" due to its extremely harsh conditions. In addition to arid regions, Mordor contains ashy wasteland in the northwest, on the volcanic plateau of Gorgoroth, and fertile terrain in the southern region of Nurn. As specified in *The Lord of the Rings*, the dominant winds in this area blow from the east, and the river-veined terrain here is realistic; these rivers flow into the inland sea of Núrnen, which is reminiscent of the Aral Sea in the 1960s (before its virtual disappearance).

To the west of Mordor lies the kingdom of Gondor, founded by Isildur, who cut off Sauron's finger. This region begins in the north at the confluence of two rivers, the Entwash and the Anduin, whose junction bizarrely appears in the form of a delta—an anastomosis, or large interweaving, would be more logical, as deltas form more frequently off coasts. Further upstream, the Anduin cuts through the mountain of Emyn Muil, carving deep labyrinthine escarpments in which Frodo and Sam get lost before reaching Mordor; this topography of gorges confirms the subsidence of Rohan mentioned earlier, a phenomenon geologists call erosion by superimposition. The Anduin then hurls itself into the vertiginous Falls of Rauros, which empty into the Dead Marshes. This vast, marshy plain full of corpses—it is the scene of many ancient battles—seems poorly positioned from a hydrodynamic point of view, since watercourses located near the bottom of waterfalls like Rauros normally flow so rapidly that, in reality, the water cannot stagnate.

On the other hand, the Forest of Fangorn, refuge of the Ents since the First Age, is situated on the good side (the windward slope) of the Misty

Mountains; here the orographic lift (described earlier in this chapter) caused by the eastern wind generates enough rain to enable the extent and density of this great forest massif.

Let us finish our walk in the north of Middle-earth in the Grey Mountains, occupied by the Dwarves since the First Age, which split into two ranges surrounding a desolate plain, the Withered Heath, inhabited by dragons—here Tolkien remains faithful to the tradition of medieval cartographers who depicted dragons in the distant and unexplored regions of their *mappae mundi*. This branching is somewhat similar to that of the Colombian Andes, which split into three ranges: the Eastern range, the Central range, and the Western range—with the slight difference that no dragons soar above these mountains.

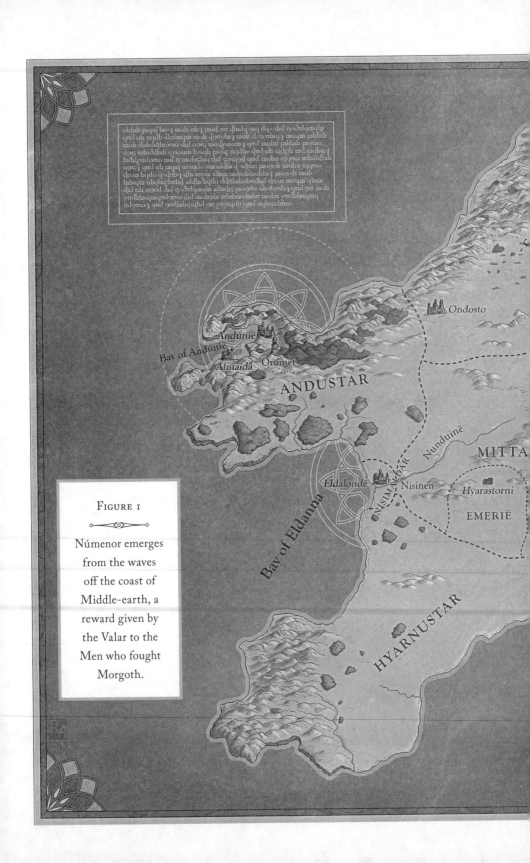

Ondosto

Andúnië

Bay of Andúnie

Almaida   Oromet

ANDUSTAR

Nunduinë

MITTA

Eldalondë   Nísinen

Hyarastorni

Bay of Eldanna

EMERIË

HYARNUSTAR

FIGURE 1

Númenor emerges
from the waves
off the coast of
Middle-earth, a
reward given by
the Valar to the
Men who fought
Morgoth.

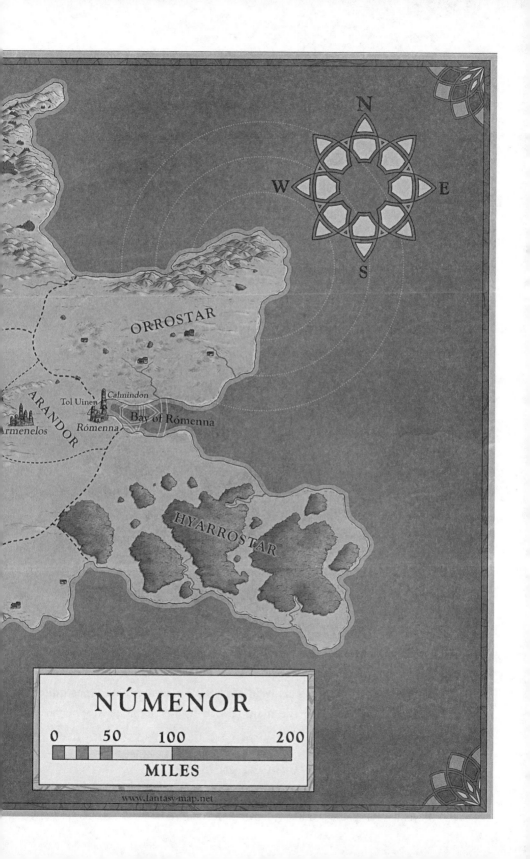

ORROSTAR

Calmindon

Tol Uinen

ARANDOR

Armenelos

Rómenna

Bay of Rómenna

HYARROSTAR

N

W

E

S

NÚMENOR

0   50   100   200

MILES

A GEOLOGICAL STROLL THROUGH MIDDLE-EARTH [*]

LOÏC MANGIN, scientific journalist

In *The Hobbit*, published in the United Kingdom in 1937, and *The Lord of the Rings*, published between 1954 and 1955, John Ronald Reuel Tolkien introduces us to a crowd of characters who roam throughout Middle-earth in every direction. The geology of this land, we can see, is torturous; for example, in *The Hobbit*, Bilbo, a hobbit,[†] in the company of a wizard and thirteen Dwarves, sets out for the Lonely Mountain of Erebor, an ancient volcano where a dragon[‡] guards a hoard of treasure. In *The Lord of the Rings*, another volcano is the ultimate destination of two other hobbits, Frodo Baggins and Samwise Gamgee, accompanied by Gollum, who must destroy the One Ring in the place it was forged, on Mount Doom in the heart of Mordor, the territory of Sauron. That's two volcanoes already!

The detail and verisimilitude with which the landscapes navigated by the heroes of both epics are described are great enough for us to make an attempt at reconstructing the geological history of Middle-earth. Tolkien's depictions—mostly sketches done in the 1930s that are sometimes quite difficult to decipher—and the maps drawn by his son Christopher are

---

[*] This text is drawn from an article by the same author, "Promenade géologique en Terre du Milieu," published in *Pour la Science*, no. 423, December 2012.

[†] See the chapter "Why Do Hobbits Have Big Hairy Feet?," p. 211.

[‡] See the chapter "Smaug, Glaurung, and the Rest: Monsters for Biologists, Too," p. 327.

also extremely helpful; these are available in *The Shaping of Middle-earth*, published in 1986). So, let's set off now on a geological stroll through Middle-earth!

AND MIDDLE-EARTH
BECAME ROUND

First, let's determine just what the term *Middle-earth* encompasses. It includes all the lands of Arda, stretched east from the Great Sea of Belegaer. We know little about the geography of Arda, except that the Great Lake lies at its center, from the middle of which rises an island. Called Almaren, this was the residence of the Valar (divine spirits, Ainur of the highest order). In the first significant geological upheaval, this island disappears when one of the fifteen Valar, Melkor, who will become the first Dark Lord, destroys the Two Lamps that lit the world. The astonishing consequence of this change in the landscape is the appearance of two seas, Helkar to the north and Ringil to the south.

During this period, Middle-earth is sandwiched between two oceans to the east and west, and traversed by great mountain ranges on a north-south axis, with each designated by a color: the Blue Mountains, the Red Mountains, the Grey Mountains, and the Yellow Mountains. After the awakening of the Elves, the Valar wage war against Melkor to free Middle-earth. The field of battle, located in the northwest, will become Beleriand. At the end of the First Age, a second conflict between the Valar and Melkor, who has now become Morgoth, results in the submersion of Beleriand except for its easternmost part, which becomes Lindon, and several islands including Tol Fuin and Himling.

At the end of the Second Age, the Númenoreans, led by Ar-Pharazôn, a usurper of the throne, attempt to invade Valinor. Eru, or Ilúvatar, the creator god who gave rise to the Ainur, does not stand for this, and retaliates by opening a chasm at the bottom of the seas of the West; the depression thus created swallows up the enemy fleet. The consequences of this are considerable. The world, which was originally flat, suddenly becomes round! (There are no Flat-Earthers in Middle-earth!) The coasts of Middle-earth, however, are altered far less significantly. Now that we have set the stage, we can focus more closely on Middle-earth.

A FAILED INITIAL ATTEMPT?

The first attempt at a geological reconstruction of Middle-earth was made by Margaret Howes, who gleaned all the necessary information from Tolkien's work. In an article entitled "The Elder Ages and the Later Glaciation of the Pleistocene Epoch," published in 1967, she retraces the evolution of the geography of Middle-earth following the fall of Morgoth, at the end of the First Age, and up to the last ice ages of the Pleistocene, which corresponds to a period from 95,000 to 65,000 years ago. Her intentions were good, but the critics did not hold back, reproaching Howes for straying too far from Tolkien's texts in order to link the geological history of Middle-earth with that of our own Earth.

The research of Robert C. Reynolds, published in 1974 in an article entitled "The Geomorphology of Middle-Earth," met with more success, and remained the authoritative text for a long time. In his analytical study, he manages to reconstruct the plate tectonics of Middle-earth, identifying four plates: Eriador in the west, Rhovanion in the north, Harad in the south, and Mordor in the east, sandwiched beneath Rhovanion and Harad (see Figure 1).

Also visible on the map drawn by Reynolds is Anduin, the river that flows through a rift valley formed between two faults that the author viewed as transform faults. This type of fault corresponds to a sliding boundary between two lithospheric plates; the San Andreas Fault in California is a famous example of this in our world. According to Reynolds's interpretation, the region south of Emyn Muil, a rocky maze through which Frodo Baggins and Samwise Gamgee are guided by Gollum, constitutes, along with the northern part of the White Mountains, a tectonic basin. To the northeast of this formation, Reynolds says, Rohan, the kingdom of the Rohirrim, or Horse-lords, is located on a craton, an area of long-term stability unaffected by tectonic movements.

A GOOD MAP

Reynolds's description seemed consistent and solidly established, but that did not keep it from being questioned in 1992 by William Sarjeant (1935–2002). Sarjeant, a professor of geology at the University of Saskatchewan in Canada, proposed a new geological history of Middle-earth. To do this, he built on the work of Robert Reynolds while taking advantage of almost twenty years of progress in our understanding of the processes involved in plate tectonics—and a new interpretation meant a new map (see Figure 2).

Sarjeant identifies six tectonic plates in Middle-earth, two more than Reynolds. The oldest of these are Forlindon and Eriador, whose collision resulted in the formation (the geological term is orogenesis) of the Blue Mountains. The Forlindon Plate then underwent subduction; that is, a sliding beneath another plate, which drove it deeper down into the mantle; only two pieces remained, the regions of Forlindon and Harlindon, in the northwest. As they moved northward, these two plates created a network of normal faults, atop which the river faults formed. A normal fault consists of an inclined plane between two blocks of continental crust; the sliding of these blocks along this plane pushes them away from one another.

To the south, the collision of the Eriador Plate with the Rhovanion and Harad Plates caused the upthrust of the Misty Mountains and the White Mountains, respectively. The three plates meet at the craton of Rohan, a common point in the two geological scenarios proposed by Reynolds and Sarjeant. In the north, another coming-together of three plates (the Eriador, the Rhovanion, and the Forodwaith) formed another range, the Grey Mountains.

The most "recent" geological event in Middle-earth is the collision of the Mordor Plate with the Harad Plate, to the south, and the Rhovanion Plate, to the north. The plate is bounded to the north and south with transform faults. These movements had the major consequence of creating the rift in which the Anduin River flows. This geological formation is, according to Sarjeant's theory, bordered by normal faults, rather than transform faults, as Reynolds believes. In this rift, the complex interplay of plates has upthrust three rocky blocks, two of which, Emyn Muil and Emyn Arnen, mark the northern and southernmost points of the Nindalf fen.

There was undoubtedly a great deal of volcanic activity in the past, but by the time of the adventures recounted in *The Hobbit* and *The Lord of the Rings* it is confined to the Mordor Plate, whose movement has created numerous fissures conducive to effusions of lava. The valley of Udûn, in northeastern Mordor, was once probably a giant volcano, of which nothing remains but a caldera, a vast crater resulting from an extremely intense eruption, similar to that of Krakatoa in 1883 and Tambora in 1815.

ERIADOR PLATE

Misty Mountains

RHOVANION PLATE

Craton
of Rohan

Rhûn
Basin

White Mountains

Emyn Muil

Nindalf
Basin

Ash Mountains

Mount Doom

Gorgoroth Basin

MORDOR
PLATE

Nûrnen Basin

Anduin

HARAD
PLATE

Belfalas
Basin

E p h e l   D ú a t h

▲ ▲ ▲   Edges of rift

●━●━●   Raised margins of
tectonic basins

FIGURE I

The tectonic plates of
Middle-earth according to
Robert C. Reynolds.

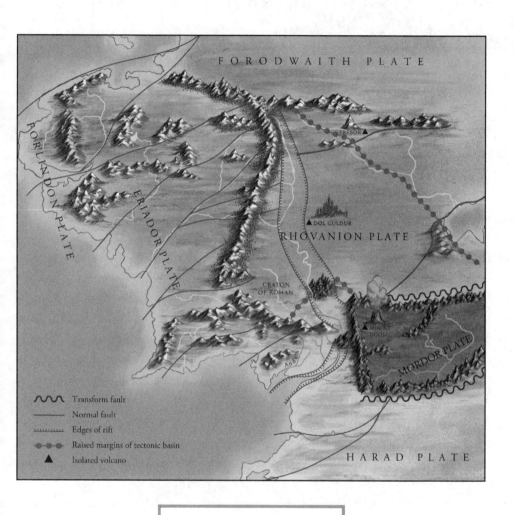

FORODWAITH PLATE

FORLINDON PLATE

ERIADOR PLATE

EREBOR ▲

▲ DOL GULDUR

RHOVANION PLATE

CRATON
OF ROHAN

▲ MOUNT
DOOM

MORDOR PLATE

Anduin

HARAD PLATE

∿∿∿ Transform fault
—— Normal fault
⊥⊥⊥⊥ Edges of rift
●━●━● Raised margins of tectonic basin
▲ Isolated volcano

FIGURE 2

The tectonic plates of
Middle-earth according to
William Sarjeant.

A DENTIST IN THE MINES OF MORIA

The only active volcano in this region is Mount Doom, where the climactic events of *The Lord of the Rings* take place. The three others are Dol Guldur, Orthanc, and Erebor, the destination of Bilbo the hobbit and his companions. The dragon's cave in Erebor is undoubtedly a lava tunnel. These cavities are initially formed by the surface cooling of a lava flow, and then when the molten rock stops flowing within the space thus created. Such tunnels are found in Hawaii, Reunion Island, the Azores, the Canaries, and elsewhere.

Marges surélevées de bassins tectoniques = Raised margins of tectonic basins. Moria is an important region, particularly for its ore deposits. These lodes are situated at the intersection of the Tharbad and Hollin faults with the Misty Mountains, a convergence that probably pushed toward the surface veins of elements formed deep in the earth at high temperatures and pressures.

The most sought-after of these elements is mithril, a metal buried deep in the heart of the mines of Moria. This fictional material has the peculiarity of being both very light and very solid; *The Lord of the Rings* features a shirt of mail, or mithril-coat, woven by the Elves, while in *The Hobbit* it is used in a striking form, ithilden, visible only by moonlight and starlight. According to William Sarjeant, mithril may be an alloy of platinum and palladium. These two materials, which belong to the same group on the period table, are, in fact, often found in combination in actual ore veins, mostly in Russia and South Africa, which themselves account for 84% of the world's production of these metals. Moreover, alloys such as this are in reality extremely hard; some photographers use them for pictures with higher contrast and greater stability. Your dentist might also help himself to the contents of the mines of Moria, for use in making dental crowns.

Tolkien took great care with the world he created, inventing a dozen languages that add to the verisimilitude of the whole. Is Middle-earth as consistent from a geological perspective? It would be of interest to examine possible similarities to our Earth's geography.

FEET ON THE GROUND?

In his prologue to *The Lord of the Rings*, Tolkien places the Shire of the hobbits "in the "North-West of the Old World; east of the Sea." Should we take this to mean northwest of Europe, with Europe being the Old World? Not necessary, since the author notes in a letter to the English historian Hugh Brogan, on September 11, 1955 (Brogan, then aged nineteen, was only twelve when he began corresponding with Tolkien!): "As for the shape of the world of the Third Age, I am afraid that was devised 'dramatically' rather than geologically, or paleontologically." And three years later, in an October 14, 1958 letter to Rhona Beare, a young Englishwoman who went on to become a Tolkien specialist: "All I can say is that, if it were 'history', it would be difficult to fit the lands and events (or 'cultures') into such evidence as we possess, archaeological or geological, concerning the nearer or remoter part of what is now called Europe; though the Shire, for instance, is expressly stated to have been in this region. I could have fitted things in with greater versimilitude, if the story had not become too far developed, before the question ever occurred to me. I doubt if there would have been much gain."

In a 1967 letter to the writers Charlotte and Denis Plimmer, he specifies that "the action of the story takes place in the North-west of 'Middle-earth', equivalent in latitude to the coastlands of Europe and the north shores of the Mediterranean. But this is not a purely 'Nordic' area in any sense. If Hobbiton and Rivendell are taken (as intended) to be at about the latitude of Oxford, then Minas Tirith, 600 miles south, is at about the latitude of Florence. The Mouths of Anduin and the ancient city of Pelargir are at about the latitude of ancient Troy."

It is notable that one of the maps published by Christopher Tolkien in *The Shaping of Middle-earth* displays a likeness to the terrestrial regions mentioned above—for example, a vast continent south of Beleriand is suggestive of Africa, while further northeast, two peninsulas show outlines similar to the Arabian and Indian peninsulas. The cartographer Karen Wynn Fonstad, who specializes in atlases of fictional worlds, used this drawing in her attempt to produce general maps of Middle-earth, maps which did not always agree, however, with some of Tolkien's writings.

While Tolkien did not take his inspiration from real-world geography, the opposite may be true. In fact, if you take an (underwater!) stroll in the Rockall Trough (a sedimentary basin off the west coast of Ireland), you will encounter a number of geological features that have been officially named after places in Tolkien: Eriador, Rohan, and Gondor Seamounts, Fangorn and Edoras Banks, Lórien Knoll, and Isengard Ridge.

The cinematic adaptations of *The Lord of the Rings* and *The Hobbit* were filmed in New Zealand. Would the landscapes of this country have fitted with those in Tolkien's imagination, and the geological history compiled by William Sarjeant?

VOLCANOES, SOURCES OF MAGIC AND LEGENDS

LAURENT STIELTJES, volcanologist
and geological risk consultant

THE ROLE OF VOLCANOES IN TOLKIEN

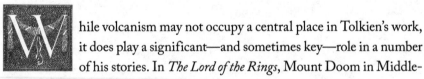

hile volcanism may not occupy a central place in Tolkien's work, it does play a significant—and sometimes key—role in a number of his stories. In *The Lord of the Rings*, Mount Doom in Middle-earth is a pivotally important volcano into which the One Ring of Sauron must be thrown. Then there is the valley of Tumladen, at the heart of which nestles the magnificent city of Gondolin; the description given of this valley in *The Silmarillion* corresponds to a caldera, an open, cauldron-shaped crater several kilometers in diameter that results from the collapse of a volcano's center. Still on the theme of caldera, but resulting from a collapse in the sea this time, the dormant volcano Meneltarma, at the center of the island of Númenor, becomes active during the submersion of the island (*Unfinished Tales*). Finally, there is the Lonely Mountain, inhabited by the fire-breathing dragon Smaug, where he jealously guards the treasure of gold and precious stones stolen from the Dwarves (*The Hobbit*).

Born in South Africa, a specialist in ancient languages, philology, and Anglo-Saxon literature as well as an ardent enthusiast of medieval linguistic traditions, Tolkien never went near, or saw, a volcano in his life. However, he had a solid literary knowledge of them, first through the Norse legends and mythology from which he drew so much inspiration in describing Middle-earth—a term borrowed, incidentally, from Midgard, literally "Middle-earth," the land inhabited by Men in Norse mythology. His knowledge of classical literature and mythology, and of volcanic countries across the globe, also impacted his work. His depictions of the world of volcanoes refer mainly to Viking legends handed down by the Icelandic people, for whom volcanoes and ice traditionally held great power, and were linked to the inevitable downfall of the gods: according to the legend of Ragnarök, which describes the end of the world, Thor will perish in battle with the serpent-dragon, while the flames of the fire-giant Surtr will set the world ablaze, with Earth being swallowed by the sea.

FISSURES, EARTHQUAKES, AND VOLCANOES: SOURCES OF LEGEND AND MYSTERY

To geologists, a volcano is the above-ground manifestation of an eruption of magma from the depths of the earth's crust. For a volcano to form, a cluster of fissures must first form deep underground. This idea of fissures and faults is a prominent one in *The Lord of the Rings*, which was inspired by the myth of the Ring of Gyges (an allegory employed by Plato in *The Republic*), which makes its wearer invisible; in this tale, a Lydian shepherd finds the ring on the finger of a corpse discovered in a cave revealed in a mountainside during an earthquake and violent storm. This link between a ring of invisibility and telluric forces can also be found in the Eastern folk tale *Aladdin* (later associated with *The Book of One Thousand and One Nights*), in which the hero discovers a magic ring stuck in a rock when the earth trembles and smokes and then splits open; it is this ring which will later enable him to obtain the magic lamp. Finally, and above all, Tolkien borrowed these themes from ancient Icelandic texts such as the 13th-century *Snorri's Edda*, from which he took the names of the Dwarves and Gandalf the wizard, and the *Völsunga Cycle*, also dating from the 13th century, which includes a dragon and its ring of power. There are also tempting parallels with Roman mythology; Sauron, the incarnation of Evil in Tolkien, forges the rings of power in the heart of a volcano, in the same way as Vulcan, the Roman god of fire, produces the weapons and trappings of the gods of Olympus in his three forges, the volcanoes of Etna, Vulcano,

and Stromboli. The cycle of the Ring of Sauron ends when it is thrown into
the flames of Mount Doom, the site of its creation.

THE VOLCANO:
A MOUNTAIN LIKE NO OTHER

Tolkien avoided using the term "volcano," preferring to employ more evocative and poetic names such as Mount Doom and the Lonely Mountain. In so
doing, the author was adhering to a tradition found in most of the myths and
legends of volcanic countries around the world. In Antiquity, volcanoes generally appeared to be ordinary mountains, designated by the substantives *oros*
in Greek and *mons* in Latin. To convey their possible volcanic nature, most
authors associated them with fire via periphrasis. For example, Vesuvius and
Etna were associated with "burning" lands by Pliny and Solon; that is, "where
the earth looks burned," and on which nothing can be cultivated. There are
also names meaning "mountain of fire" in Japanese (*ka san*) and Indonesian
(*gunning api*), "smoking mountain" in Nahuatl (*popoca-tepelt*) and Afar (*erta
ale*), and "mountain of God" among the Masai (*Ol Doino Lengai*), in Cameroon
(*Mongo-mo-Ndemi*, the name for Mount Cameroon) and Vanuatu (*Yasur*), and
among the Maya (*Huelhuel*). Tolkien, that great connoisseur of mythologies,
simply adopted this universal practice.

THE VOLCANO, A SOURCE OF BELIEFS
AND MYTHS

Whether they are inhabited by gods or goddesses, genies or dragons; whether
they emit smoke or flame, these "mountains" have a specific origin and a specific destiny. The spitting of fire by their gaping mouths with their mephitic
odors inspires fear and fascination due to its association with the realm of the
dead. Likewise, the devastation and destruction volcanic eruptions cause are
sources of myth in their own right, harbingers of doom (as for the Incas and
Aztecs) or heroic actions (as for the Maya and Incas), and even monotheistic
religious symbols. The underworlds of Greek (and later Roman) mythology,
through which flowed the river Phlegethon ("flaming"), fueled and inspired
the concept of Hell in Judaism, Christianity, and Islam. In the 6th century,

the visionary pope Gregory I said about Vulcano and Stromboli that "the mouths of these volcanoes widen further each day, for, the end of the world approaching, these places of torment must be able to accommodate the growing number of those who will be damned." In the Middle Ages, in *The Book of Miracles* (1180), the monk Herbert compares the great Icelandic volcano Hekla to Etna, which is "merely a little stove in comparison with that gargantuan furnace." He was probably retranscribing the observation made by the abbot Saint Brendan in *The Voyage of Saint Brendan the Abbot*, who, sailing in the North Atlantic, came near to a black island spitting flames and pitch, and saw in it "the wide-open gates of Hell and a hundred devils of fire."

VULCANOLOGICAL KNOWLEDGE IN TOLKIEN'S DAY

Knowledge of vulcanism in the 1930s, '40s, and '50s was far from what it is today. Based mainly on descriptions of eruptive phenomena, the discipline was still in its infancy, and could not yet really be called vulcanology. Until the last thirty years of the twentieth century, volcanoes were seen as a living geological object; fascinating, certainly, but considered as a sort of terrestrial acne, without major geological significance. The origins of magma were nothing but hypothesis. Volcanoes did not come into vogue in the geosciences until the 1970s, and anyone who studied them before then was frequently ignored, and sometimes even criticized. It was only in southern Italy, Martinique, Japan, the Dutch East Indies, and Hawaii that anyone took the trouble locally to study and monitor the pulsations of these monsters, sleeping with one eye open for fear of the terrible threat they represented. The theory of plate tectonics didn't yet exist, and would not be proposed until several decades after Tolkien's writings were published. It was not until the 1970s that the vital role of volcanoes in global tectonics was recognized, not least because there was no television or film footage of an erupting volcano, largely because most eruptions occurred deep in tropical jungles, sweltering deserts, or frozen expanses, which were generally inaccessible except via long journeys by boat and then on foot. Writers and the public at large were forced to content themselves with the stories, drawings, and sketches of naturalists, explorers, and a few rare geologists, and to let their own fertile imaginations do the rest.

TOLKIEN'S SOURCES ON VULCANISM

A reading of Tolkien's works reveals that he relied mostly on three sources: the legendary volcanoes of Iceland, the colossal 1883 eruption of Krakatoa in Indonesia, and that of Mount Pelée in Martinique in 1902.

The volcanoes of Iceland inspired Norse mythology and, by extension, Tolkien's fiction. The heart of the plot of *The Lord of the Rings* hinges on the most important active volcano in Middle-earth, the famous Mount Doom. The descriptions of Mount Doom given in *The Return of the King* are typical of active basaltic volcanoes, of which the volcanoes of Iceland are archetypal examples:

> [T]he furnaces far below its ashen cone would grow hot and with a great surging and throbbing pour forth rivers of molten rock from chasms in its sides. Some would flow blazing towards Barad-dûr down great channels; some would wind their way into the stony plain, until they cooled and lay like twisted dragon-shapes vomited from the tormented earth.

This volcano's lava, then, emerges directly from the Cracks of Doom. And the dramatic climax of *The Lord of the Rings* emphasizes the deep abyss of its crater, into which the Ring is thrown.

The devastating explosion of Krakatoa in 1883, followed by the collapse of the caldera at its center, triggered a tsunami. This eruption, the most famous of the second half of the nineteenth century, was for many generations the model for a volcanic island that exploded and sank beneath the sea. Krakatoa made it possible to understand how calderas form, and to interpret the subsidence of the island of Santorini in around 1600 B.C., the biblical source for the parting of the waters of the Red Sea to allow the safe passage of the Hebrews, and a source of inspiration for Plato's legend of Atlantis. Widely reported and commented upon in newspapers, magazines, conferences, and academic texts across the globe, Krakatoa stood for a century as a typical example of an explosive volcanic eruption. Taking up residence in Western popular imagination, it cropped up frequently in fiction and continues to do so today; Tolkien was no exception, echoing the theme of the caldera collapsing into the sea with the submersion of his island of Númenor when the volcano Meneltarma awakens. Another depiction of a caldera floor is that of the valley of Tumladen, the perimeter of which is distinctively ringed by the Encircling Mountains, surrounding the central hill of Amon Gwareth, the very image of the steeply sloping cone of Krakatoa. Shortly before the publication of *The Lord of the Rings* in 1954, Hergé also used the example of Krakatoa for his comic book *The Eruption of Karamako*, in 1953, even playing on the similarity between the names of the two volcanoes.

The eruption of Mount Pelée in 1902 stunned the entire world, with the annihilation of the city of Saint-Pierre and the resulting loss of nearly 30,000 lives. The Barnum & Bailey Circus toured the globe with a survivor of the catastrophe, exhibiting the man's burned body. Tolkien drew on the image of Pelée's fiery clouds in *The Lord of the Rings*: "great rolling clouds floated down its sides and spread over the land." He describes the desolate landscape of the Gorgoroth plateau, which was "still a waste of soft mud [. . .] pocked with great holes, as if it had been smitten with a shower of bolts and huge slingstones." To this, Tolkien adds descriptions of phenomena quite typical of explosive eruptions of this type: "There was a brief red flame that flickered under the clouds and died away"; "there was a shimmer of lightnings under the black skies"; "the air was full of fumes; breathing was painful and difficult".

Profoundly inspired by Norse mythology, Tolkien's fiction makes great use of volcanoes and their fire, in an echo of Ragnarök: worlds burn, and are then swallowed up by the sea. The open fissures, tubes, and lava flows of Iceland's volcanoes were the principal source of these images, with the collapse of the Krakatoa caldera and the fiery clouds of Mount Pelée completing the picture. And, aligning himself with the tradition of volcanic legends of peoples the world over, Tolkien never uses the word *volcano*, using evocative periphrases to designate these flaming mountains instead.

The devastation of Númenor.

SUMMER IS COMING:
THE CLIMATE OF MIDDLE-EARTH

DAN LUNT, climatologist,
University of Bristol

ABSTRACT

In this paper, I present and discuss results from a climate model simulation of the "Middle-earth" of elves, dwarves, and hobbits (and not forgetting wizards such as myself). These are put into context by also presenting simulations of the climate of the "Modern Earth" of humans, and of the "Dinosaur Earth," when dinosaurs ruled the Earth 65 million years ago.

Several aspects of the Middle-earth simulation are discussed, including the importance of prevailing wind drection for elvish sailing boats, the effect of heat and drought on the vegetation of Mordor, and the rain-shadow effects of the Misty Mountains. I also identify those places in the Modern Earth which have the most similar climate to the regions of The Shire and Mordor.

The importance of assessing "climate sensitivity" (the response of the Earth to a doubling of atmospheric carbon dioxide concentrations) is discussed, including the utility of modelling and reconstructing past climate change over timescales of millions of years. I also discuss the role of the Intergovernmental/Interkingdom Panel on Climate Change (IPCC) in assessing climate change, and the responsibilities placed on policymakers.

I. INTRODUCTION

Computer models of the atmosphere, land surface, and ocean are routinely used to provide forecasts of the weather and climate of the Earth. They are based on our best theoretical understanding of fluid motion, physics, chemistry, and biology, written in the form of equations, and then converted into a form which can be solved by a computer.

Climate models and models used to make weather forecasts are very similar to each other, except that climate models typically simulate longer periods of time than weather models (years to centuries as opposed to days to weeks), and therefore, due to limits on computer time and power, make predictions at a lower spatial resolution (typical scale of hundreds of kilometers as opposed to kilometers). Climate model predictions are an integral part of political and societal planning for the coming decades to centuries, and the recent report from the Intergovernmental Panel on Climate Change (IPCC) summarises many of these future predictions (IPCC, 2013a).

Because climate models are based on fundamental scientific understanding, they can be applied to many situations. They are not designed solely for simulating the climate of the modern Earth, and, in theory, the same underlying science should apply to any time period in the past. The only caveat is that in order to simulate climates different from modern, the user must provide some "boundary conditions"—maps or variables which are not predicted by the model. Examples include spatial maps of the height of the global terrain (topography) and ocean depth (bathymetry), characteristics of rocks and soils, and concentrations of key atmospheric constituents, such as ozone and carbon dioxide (CO_2). In addition, key parameters such as the strength of the sun, and the radius and the rotation rate of the planet, also need to be provided to the model.

Adapting the model to simulate past time periods is potentially very powerful because, in theory, we can know the "right" answer from observations, and test the performance of the models by comparing their results with these observations. For time periods prior to humans making careful observations of the weather, we rely on indirect observations of many aspects of past climates, such as information from tree rings and ice cores, and fossils of plants and animals. However, provided that we understand the uncertainties and errors in these "proxy" records of past climate change, and provided we also understand and account for uncertainties in the boundary conditions we apply to the model, we can make use of past periods going back millions of years, to time periods when the Earth looked very different from the modern.

In addition, by varying the topography/bathymetry, the rotation rate and radius of the planet, and density of the atmosphere, we can, in theory, use climate models to simulate any planet, real or imagined.

In this paper I present three climate model simulations, of the Modern (pre-industrialised) Earth, of the Dinosaur Earth (a time period called the Late Cretaceous, about 65 million years ago, just prior to the extinction event which killed off the dinosaurs), and Middle-earth—the land of hobbits, elves, dwarves, wizards, and orcs.

The aims of this paper are threefold:

1. To demonstrate the flexibility of climate models, arising from their basis in fundamental science.
2. To present the modeled climate of Middle-earth, and provide some lighthearted discussion and interpretations.
3. To discuss the strengths and limitations of climate models in general, by discussing ways in which the Middle-earth simulations could be improved.

2. MODEL DESCRIPTION

I use a climate model developed at the UK Met Office, "HadCM3L," which is capable of simulating the atmosphere, ocean, and land surface. In common with most climate models, HadCM3L represents the world in "gridbox" form, with a 3-dimensional network of boxes covering the surface and layered to extend up to the top of the atmosphere and down to the bottom of the ocean depths. The size of each box is 3.75 degrees of longitude by 2.5 degrees of latitude, with a height dependent on the distance from the Earth's surface-boxes situated near the surface of the Earth which have a smaller height than those at the top of the atmosphere or bottom of the ocean. This results in a "matrix" of boxes covering the world, with 96 boxes in the West-East direction, 73 boxes in North-South direction, 20 boxes deep in the ocean, and 19 high in the atmosphere (a total of more than a quarter of a million boxes, although not all are used as some are effectively below the sea floor).

In this matrix, the fundamental equations of fluid motion in the atmosphere and ocean are formulated and solved, with the additional complication that the Earth is spinning on its axis. Energy is added to the system due to absorption of light and heat radiation emitted by the Sun, and energy leaves the system through emission of heat or reflection of light radiation into space. All variables in the model can be considered as average values over the volume of each gridbox, and so the climate model can not provide any information at a spatial scale smaller than one gridbox (so, for example, although it makes sense to talk about the modelled climate of the UK, or Mordor, the model can not give information about Bristol, or Bree). However, in reality there are many processes which occur at a finer spatial scale than that of a single gridbox. As such, models include "parameterisations" of sub-gridscale processes, such as cloud formation, and small-scale atmospheric turbulence, or eddies in the ocean. It is the representation of these sub-gridscale processes which brings uncertainty into climate modelling (the equations of fluid motion and thermodynamics themselves have been known and understood for several centuries). As well as the atmosphere and ocean, the model includes a representation of the land and ocean surface, including processes associated with sea-ice, soil moisture, and, in our particular version of the model, the growth and distribution of vegetation.

The model is given an initial state of all the variables which it predicts (for example temperature, pressure, wind speed, snow cover, ocean density), and then the model is "run" forward in time, in steps of typically 10 to 30 minutes. Weather systems develop and evolve, rain falls, the seasons come and go, and years of "model-time" pass (for my model, one year of model-time typically takes about 2 hours of "real-time"). Finally, the weather predicted by the model in the final years or decades of the simulation are averaged, resulting in a model-predicted "climatology"—the climate, or average weather, predicted by the model.

HadCM3L is a relatively complex model, known as a "General Circulation Model," or GCM. However, it is not a state-of-the-art model, and includes less processes and has fewer gridboxes (*i.e.* runs at lower resolution) than more recent models, such as those used in the most recent IPCC report (IPCC, 2013a). However, it is useful for my purposes, as its relative efficiency of computation means that it can be run for a sufficiently long time to reach an equilibrium, given that the initial state I put the model into may be quite different from the final predicted climate.

3. EXPERIMENTAL DESIGN

In this paper I present results from 3 climate model simulations using HadCM3L. The first is a simulation of the Earth during the period prior to large-scale industrialization (for sake of argument, the period 1800–1850). I call this simulation "Modern Earth." The second simulation, "Dinosaur Earth," is of the period just prior to the extinction of the dinosaurs (the Late Cretaceous, ˜65 million years ago), and the third, "Middle-earth," is of the climate of Middle-earth.

The model setups for these three simulations are very similar. The only important differences are the boundary conditions.

For Modern Earth, I use the standard pre-industrial global boundary conditions, provided by the UK Met Office, which are derived from observations of the modern continental configuration of the Earth, topography, bathymetry, and land-surface characteristics. This simulation uses a CO_2 concentration of 0.28% (280 parts per million, or "280 ppm") of the total atmosphere, a value which is obtained by extracting bubbles of pre-industrial atmosphere from 200-year-old ice below the surface in Antarctica. It is worth noting that "ice-cores" (long tubes of ice extracted from kilometers into the ice) allow us to reconstruct the CO_2 concentration of the Earth back to 800,000 years ago. This shows that during "natural" cycles of CO_2 and climate variation, between ice ages and warmer "interglacials" such as the Earth has been in for the last ˜6000 years, the CO_2 concentration varies between 180 and 280 ppm. The current CO_2 concentration on Earth is 397 ppm, which is well outside this natural range, a result of the burning of fossil fuels and deforestation by humans.

For Dinosaur Earth, the starting point is a relatively high-resolution (0.5 degrees of latitude by 0.5 degrees of longitude) map of the topography (mountain heights and positions) and bathymetry (ocean depths) of the Late Cretaceous (˜65 million years ago), provided by the geological consultancy company, Getech (www.getech.com). They specialize in making detailed studies of the scientific literature, taking into account information from fossils, rocks, and ancient deep-sea sediments, in order to reconstruct many aspects of the past state of the Earth, including topography and bathymetry. From this basic topographic map, all necessary boundary conditions were created, at the spatial resolution of the model. As well as the topography and bathymetry, these include several land-surface characteristics such as the soil reflectivity ("albedo"), the capacity of the soil to hold moisture, and the capacity of the soil and underlying rock to store heat. The land-surface characteristics for Dinosaur Earth were set to constant values over land points, typical of global-average modern values.

For Middle-earth, all the spatially varying boundary conditions are derived from maps and manuscripts from the exensive archives in Rivendell (Tolkien, 1954, 1986; Fonstad, 1991). The starting point is a map of topography and bathymetry, derived from a map of the Northern Hemisphere during the Second Age of "Arda" (of which Middle-earth is a relatively small region; henceforth, we use "Middle-earth" interchangeably to mean the whole world of Arda, or the region of Arda known as Middle-earth). Considerably more effort could be made to faithfully reproduce the drawings in these manuscripts. Time constraints meant that this was not carried out as fastidiously as was

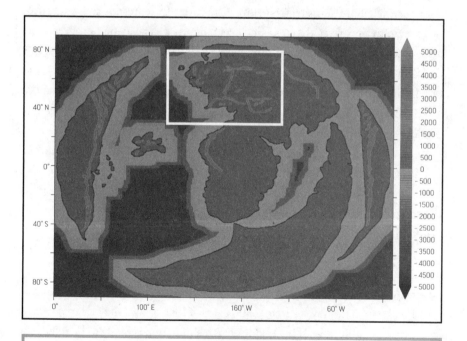

FIGURE I

High resolution map of topography and bathymetry for the "Middle-earth" simulation, shown in units of meters. This world is known as "The Second Age of Arda." The model runs over the whole world, but results are discussed only for the region known as 'Middle-earth' (shown with a white rectangle).

perhaps required (and I had misplaced my spectacles that day . . .). In addition, there is a degree of uncertainty in the mapping of the original maps onto a spherical world, and it is likely that other equally-valid, or better solutions could be found, which could influence the results. Finally, many of the legends associated with Middle-earth (such as the discovery and final destruction of the One Ring) actually took place in the Third Age—here I assume that the climates of these two Ages were similar.

Each surface gridbox in the model is assigned a height above sea-level, calculated from the average topography within the region of that gridbox. The foothills of mountainous regions were assigned a height above sea-level of 2000 meters, the mountains themselves a height of 4000 meters, rivers a height of 100 meters, and all other continental regions a height of 300 meters. In addition, a small random value was added to the assigned height above sea-level of each surface gridbox, in order to represent small features not drawn in the original maps. The ocean depth was deepened progressively away from the coast, to a maximum depth below sea-level of 4000 meters. A decision had to be made of how to wrap the apparently flat, circular world of Middle-earth onto the sphere required by the climate model. I chose the relatively simple solution of a straightforward direct mapping of the circular Middle-earth onto an equal-latitude/longitude grid. The missing corner regions were set to be oceans. The resulting global map, at a resolution of 0.5 degrees longitude by 0.5 degrees latitude, is shown in Figure 1. I set the initial vegetation distribution to be shrubs everywhere on Middle-earth; this distribution evolved during the simulation (see Figure 4 for the final vegetation distribution in the simulation).

The global topography, at the resolution of the climate model, for all three simulations, is shown in Figure 2.

The Modern Earth simulation was run for more than 1000 years. The Dinosaur Earth simulation was run for ˜500 years, which allows the surface climate to approach equilibrium. The Middle-earth simulation is likely to be relatively far from equilibrium, having been run for only 70 years.

In the Dinosar Earth and Middle-earth simulations, the first 50 years of the simulation have a CO_2 concentration of 280 ppm, and the remaining years are at 1120 ppm (4× pre-industrial levels). The high CO_2 in Middle-earth could be interpreted as accounting for greenhouse gas emissions from Mount Doom.

Finally, it should also be noted that I assume that the radius of the Earth and Middle-earth are the same, that they spin at the same rate and thus have the same day-length, and that they spin in the same direction (from my window here in Middle-earth I can see the sun setting over the mountains to the west, so this is a valid assumption). Also, I assume that the strength of the sun, combined with the distance from the Sun to the Earth, and the tilt of the Earth on its spin axis, results in an identical amount and seasonal variation of sunlight reaching the top of the atmosphere in Middle-earth as it does in Earth.

4. RESULTS

The model-predicted annual-average temperature, precipitation (rain and snow), and winter wind speed/direction and pressure, in the region of Europe and the North Atlantic (for the Modern Earth and Dinosaur Earth simulations) and Middle-earth, are shown in Figure 3.

4.1 PRE-INDUSTRIAL SIMULATION, "MODERN EARTH"

The climate of the model of Modern Earth (see Figure 3(a,d,g) for the temperature, precipitation, and winds and pressure) has been compared to recent meteorological observations and to other models in the assessment reports of the Intergovernmental Panel on Climate Change (IPCC). The model does a good job of simulating many aspects of climate, although compared to more recent models taking part in the latest IPCC report (IPCC, 2013a), its simulation of many aspects of climate is relatively poor. This is not surprising given its relatively low resolution compared to many models, and given recent improvements in the representation of subgridscale processes, such as those associated with clouds.

4.2 LATE CRETACEOUS SIMULATION, "DINOSAUR EARTH"

The temperature, precipitation, and winds/pressure in the Cretaceous simulation is shown in Figures 3(b,e,h). There has been significant continental drift and reshaping of the ocean sea floor and other tectonic change since 65 million years ago, and so the continents in the region of the North Atlantic are only just recognizable. This simulation is just one in a series of simulations that scientists at the University of Bristol have carried out, covering the last 150 million years. There is a project, being led at the University of Bristol, and funded by the Natural Environment Research Council (NERC) to investigate these simulations in more detail, and carry out additional work to understand the confidence in their predictions, and how well they compare with the proxy climate data from the geological record that exists from this time period.

In particular, scientists at Bristol are interested in the relationship between past and future *climate sensitivity*. Climate sensitivity is a measure of how much the Earth warms given an increase in atmospheric CO_2 concentration. Climate sensitivity is a useful parameter to know because it summarizes in simple terms the susceptibility of the Earth to global warming. Climate sensitivity is usually defined as the global average surface temperature increase (in degrees Centigrade) that would occur if the atmospheric CO_2 concentration were doubled, and the Earth "equilibrated" to a new temperature (the Earth may take many centuries or even millennia to fully equilibrate, but most of the warming is likely to occur in the first few decades following the CO_2 doubling). For more information on climate sensitivity, see Section 5.

4.3 MIDDLE-EARTH SIMULATION

The temperature over the land of Middle-earth is shown in Figure 3c.

The relationship between temperature and latitude is clear, with the more northerly regions being relatively cold (*e.g.* the annual average temperature is below freezing in the far Northern Kingdon of Forodwaith), and the more southerly regions warm (*e.g.* the annual average temperature is higher than 30°C in Haradwaith). In this respect, to a first approximation, the climate of Middle-earth is similar to that of Western Europe and North Africa. This is unsurprising because I have assumed that Middle-earth is part of a spherical planet (Saruman tells me this is a ridiculous concept); regions near the equator on a spherical planet face directly toward the sun, whereas more poleward regions are tilted at an angle, and therefore receive less sunlight averaged over the year.

Middle-earth, just like Earth, spins on an axis which is tilted relative to the orbit of the Earth around the sun, so the land of Middle-earth also has seasons like the Earth.

High altitude mountainous regions (for example the Misty Mountains) are colder than the surrounding low-lying regions. This is because air temperatures reduce with increasing altitude, as rising air has to give up heat energy in order to expand as it rises.

East of the Misty Mountains, the temperature decreases the further eastward one travels. This is because, just as in the European regions of the Earth, the farther from the ocean the greater the "seasonality"—*i.e.* winters become colder and summers become warmer. But winters cool more than summers warm, and so annual average temperatures in general decrease away from the ocean.

The model-predicted precipitation (rain and snow) over the land of Middle-earth is shown in Figure 3f.

The most striking effect is that the highest precipitation occurs over and to the west of the mountainous regions; *i.e.* there is a "rain-shadow" to the east of the mountainous regions (for example, to the east of the Misty Mountains). This is because the prevailing wind brings moist air from the western oceans onto the continent of Middle-earth. As the air rises over the mountains it cools, causing the moisture it was carrying to condense from vapor into a liquid, to start forming clouds, and eventually fall as rain or snow. On the eastern side of the mountains the air has lost much of its moisture and there is less precipitation.

The regions in the far South of Middle-earth, in southern Mordor and Haradwaith, are very dry. This is because these are "subtropical" regions, similar to the desert region of the Sahara in the Earth. The subtropics are dry because they are in a region where air tends to descend from high altitudes toward the surface. This is part of a large scale atmospheric circulation called the "Hadley Cell," in which air rises in the equatorial regions, moves away from the equator, and descends again in the subtropics. The descending air supresses rainfall for the same reason that rising air enhances rainfall—the descending air warms, and can hold more water in vapor form as opposed to liquid form.

The surface windspeed and direction, and mid-atmospheric pressure over the land of Middle-earth is shown in Figure 3i. It can be seen that there are strong westerly (*i.e.* coming from the west, towards the east) winds in the coastal southern regions of Middle-earth, in particular in the Bay of Belfalas. Conversely, there are easterly winds in the north of Middle-earth. This may explain why ships sailing to the Undying lands to the West tended to set sail from the Grey Havens, situated in the region of these easterly winds.

The vegetation component of the model allows us to examine the model-predicted vegetation of Middle-earth. It is important to realize that the vegetation model does not take account of disturbance of vegetation from its "natural" state (such as forest fires caused [inadvertently or otherwise] by dragons, deforestation by dwarves, the growing of pipe-weed by hobbits, or the wanton destruction by orcs). This is the same for the simulation of the Earth, where the model does not take into account the activities of humans in modifying vegetation, such as the deforestation of the Amazon.

The model-predicted vegetation of Middle-earth, shown in Figure 4, depends strongly on the model-predicted rainfall and temperature. As a result, deserts are found in the warm dry regions of the far South, and low-lying shrub is found over much of Mordor. Shrub and cold desert is found on the peaks of the Misty Mountains and the Iron Hills in the North, and Blue Montains in the West. Much of the rest of Middle-earth is covered with forests. This is consistent with reports I have heard from Elrond that squirrels could once travel from the region of the Shire all the way to Isengard.

An interesting question in relation to these results is to ask where in the Earth is most like a certain place in Middle-earth. For example, one may be particularly interested in knowing where in the Earth is most like The

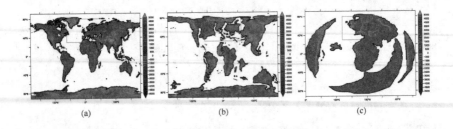

(a)   (b)   (c)

FIGURE 2

Model resolution topography (in meters) for the simulations of (a) Modern Earth, (b) Dinosaur Earth, and (c) Middle-earth. The black lines show the outlines of the continents. White regions are ocean. The rectangles show the position of the regions in Figure 3.

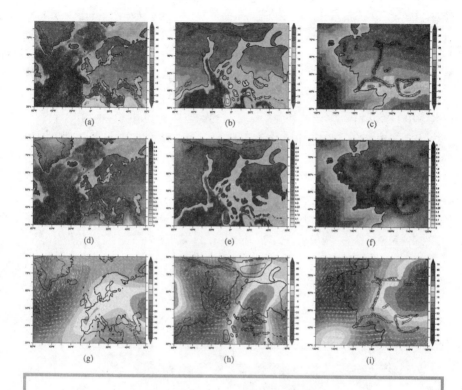

FIGURE 3

Contours of annual-average modelled temperature (top, units of °C), precipitation (middle, units of mm/day), and surface winds and mid-atmosphere pressure ("geopotential height" at 500 mbar, geopotential height in units of meters) in Western Europe (left), the Cretaceous North Atlantic (middle), and Middle-earth (right). The location of these regions in their respective worlds is shown in Figure 2. The annual-averages are calculated over the final 30 years of the Modern Earth simulation, the final 50 years of the Dinosaur Earth simulation and the final 10 years of the Middle-earth simulation. The different lengths of averaging periods should not greatly affect the results. The thick black lines show the outlines of the continents; thin black lines are contours of topography. In the top and middle plots, the ocean is shown in shades of gray which represent depth.

Shire. The climate model predicts that the annual average temperature of
The Shire is 7.0°C, and that the annual rainfall is 61 cm per year. Figure
5(a) shows the regions of the Earth which share this annual-average climate
(regions in light gray have a temperature within 2°C of that of the Shire,
regions in dark gray have a rainfall within 6 cm/year of the Shire, and regions
in mid-gray are most Shire-like—matching both temperature and rainfall). By this metric, eastern Europe has the greatest concentration of Shire-
like climate, in particular Belarus. It is interesting to note that in the UK
(Figure 5[b]), the most Shire-like region is centered around Lincolnshire
and Leicestershire. In New Zealand (Figure 5[c]), north of Dunedin in
the South Island might be considered the ideal location to film a motion
picture based in the Shire . . .

The same question can be asked about Mordor (Figure 5[d,e,f]). Los
Angeles and western Texas are notable for being amongst the most Mordor-
like regions in the USA, and in Australia, much of New South Wales, as
well as Alice Springs, have an annual average climate very similar to that
of Mordor.

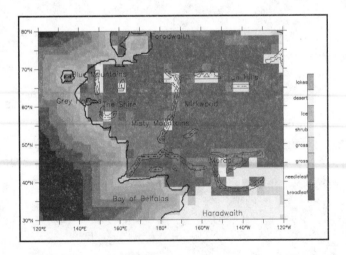

FIGURE 4

Model-predicted vegetation for the Middle-earth simulation. The approximate
position of places discussed in the text are also shown for reference.

5. MORE ON THE IMPORTANCE OF CLIMATE SENSITIVITY

The Intergovernmental Panel on Climate Change (IPCC) currently estimate, with "high confidence," that climate sensitivity (the temperature response of the Earth to a doubling of atmospheric CO_2 concentrations, see Section 4.2) is in the range 1.5 to 4.5°C, with values less than 1°C being extremely unlikely. In reality, CO_2 concentrations are increasing over time, not instantaneously doubling, but to put the numbers in context, at current rates of human CO_2 emissions, the Earth could reach a doubling (560 ppm from a baseline of 280 ppm) by the year 2050.

The actual value of climate sensitivity is uncertain, because our climate models are not perfect representations of reality. In particular, the importance of many "feedbacks," which can amplify or decrease the magnitude of climate change, are uncertain, because the processes which govern them are so complex, and involve the interactions of many parts of the whole "Earth system." These interactions involve, for example, the biology of living plants and animals, the chemistry of atmospheric particles, the ice sheets of Greenland and Antarctica, and the complex circulation patterns of the atmosphere and ocean.

However, it is posible to obtain information on climate sensitivity from studying past climates. In particular, if past CO_2 concentrations and past temperature changes can be estimated, in theory climate sensitivity can be estimated. Climate scientists at the University of Bristol are aiming to assess the utility of the past climates of the last 150 million years for estimating climate sensitivity. In particular they will investigate how climate sensitivity varies as the continents move, and tectonics change the height and shape of mountains and ocean floors. As well as allowing a better understanding of the workings of the planet that we (well, Earthlings at least!) live on, it will allow them to identify key time periods in which climate sensitivity was likely to be similar to that of the modern period, and allow new past-climate data to be collected for those time periods.

Knowledge of the value of climate sensitivity, and its associated confidence and uncertainty, can be a starting point for the assessment of the impacts of future climate change, for example the changing risk associated with heatwaves, flooding, droughts, and associated impacts on crops or spread of diseases. This then leads onto the assessment of the costs of climate change, on humanity and on ecosystems and the Earth as a whole, and also leads onto the costs of adapting to climate change (for example the building of flood defences, or migration of populations away from inhospitable regions). This then has to be weighed against the costs (or added to the benefits) of moving to a low-carbon economy and reducing emissions of greenhouse gases such as CO_2. This whole assessment process is handled in the Earth by the Intergovernmental Panel on Climate Change (IPCC), who produce reports every 5 years or so on all aspects of this problem (the Interkingdom Panel on Climate Change in Middle-earth was recently disbanded after a fight broke out in the final plenary session and several Coordinating Lead Authors were sadly beheaded). The reports are summarized in a series of very accessible summaries, freely available online IPCC (2013b). These are aimed at policymakers, because ultimately it is the policymakers who make the big decisions about when and how to act, and how to incentivize action by individuals, industry and commerce. Our future is in their hands.

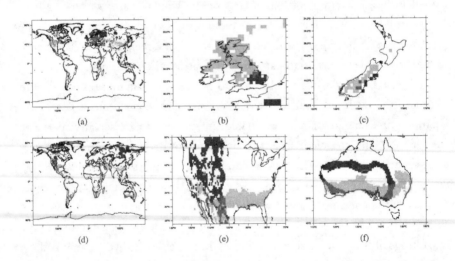

(a)   (b)   (c)

(d)   (e)   (f)

FIGURE 5 (a,b,c)

Darkest gray regions show where on the Earth is most Shire-like in terms of rainfall; lightest gray regions show where on the Earth is most Shire-like in terms of temperature; and mid-gray regions show where on the Earth is most Shire-like in terms of both temperature and rainfall (c,d,e). The same as (a,b,c), but for Mordor instead of The Shire.

6. DISCUSSION

Here I discuss how the simulations of Middle-earth could be improved. The scientists at Bristol do not intend to make these improvements, but they are worth listing because they demonstrate techniques currently being used to better understand the past and future climate of the Earth.

The regional climate of Middle-earth would be better investigated with a higher resolution, nested or regional model. The gridboxes in these regional models are smaller than in a global model, and so more detail is obtained, and small-scale features of the atmosphere and ocean are better represented. In this study I have only used one model, HadCM3L. This brings about uncertainties, because this model is only one of many I could have used, which in general give different results, due to differing representation of key physical processes, in particular those which occur at a spatial scale smaller than one gridbox. There are two approaches commonly used to address this uncertainty: (1) "model inter-comparison projects (MIPs)," and (2) "parameter perturbation studies (PPEs)." MIPs are coordinated studies in which several modelling groups with different models all carry out an identical set of simulations. This allows model–model differences to be investigated. PPEs are similar, but a single model is used, and many simulations are carried out, each with different values of key uncertain parameters in the model. These parameters may be uncertain because there are not enough meteorological or oceanographic observations to properly constrain them, or they may be associated with "scaling-up" small-scale processes to the scale of one gridbox. Both these approaches result in an "ensemble" of results, which allow the uncertainty in our predictions of climate to be assessed.

This work could be developed by assessing the sensitivity of the Middle-earth climate to uncertainties in the applied boundary conditions. In particular, given the undoubted variations in CO_2 due to the nefarious activities of Sauron and Saruman, evaluating the climate sensitivity of Middle-earth would be an interesting exercise. Such activities also manifest themselves in terms of land-use change, and so imposing vegetation distributions from maps found in the libraries of Minas Tirith would also be an important piece of future work. Climate models on Earth are routinely tested and evaluated by comparing results with the extensive global observational network of weather stations (*e.g.* rain gauges and thermometers) and satellite data. Due to time constraints (I am currently overwhelmed by petty adminstrative tasks placed on me by the White Council), I have not aimed to evaluate the simulations of Middle-earth with reference to meteorological records, such as the Red Book of Westmarch (known as "*The Lord of the Rings*" and "*The Hobbit*" by inhabitants of Earth). In any case, care must be taken if this approach were to be used, because the model results I have presented are climatological—*i.e.* averages over a long period of time, as opposed to instantaneous observations of, perhaps, snow-storms in the Misty Mountains. As discussed in Section 2, the model does predict "weather." However, the modeled weather would very quickly diverge from any real weather because the atmosphere is "chaotic," in that very small differences in prediction rapidly grow to cover the whole of Middle-earth—hence the proverb about the flapping of a butterfly's wings in the Shire causing a hurricane in Mordor.

7. CONCLUSIONS

In this paper, I have presented climate model simulations of three time periods/worlds: The Modern Earth, Dinosaur Earth (~65 million years ago), and Middle-earth. The strengths and weaknesses of climate models have been discussed and illustrated through the examples. I have discussed the importance of assessing climate sensitivity in the context of current global warming, and highlighted the key role played by the IPCC, and the responsibilty placed on policy-makers to act upon the information therein. I have introduced a NERC-funded project which will aim to increase our understanding of climate sensitivity in past climates.

The main findings concerning Middle-earth are that:

- The climate of Middle-earth has a similar distribution to that of Western Europe and North Africa.
- Mordor had an inhospitable climate, even ignoring the effects of Sauron—hot and dry with little vegetation.
- Ships sailing for the Undying Lands in the West set off from the Grey Havens due to the prevailing winds in that region.
- Much of Middle-earth would have been covered in dense forest if the landscape had not been altered by dragons, orcs, wizards etc.

- Lincolnshire or Leicestershire in the UK, or near Dunedin in the South Island of New Zealand, have an annual-average climate very similar to that of The Shire.
- Los Angeles and western Texas in the USA, and Alice Springs in Australia, have an annual-average climate very similar to that of Mordor.

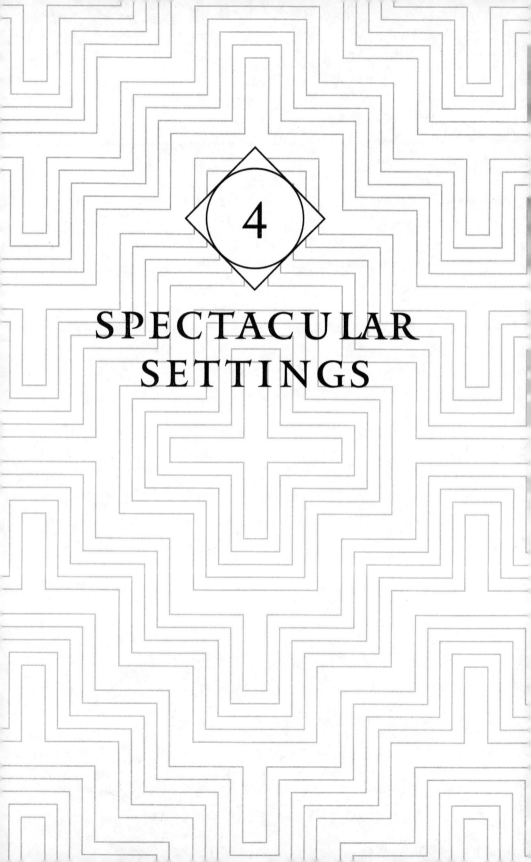

4

SPECTACULAR SETTINGS

PLANTS AND LANDSCAPES IN MIDDLE-EARTH

JEAN-YVES DUBUISSON, paleobotanist,
Sorbonne University
ÉLODIE BOUCHERON-DUBUISSON, botanist,
Sorbonne University

It would be impossible to describe, in a single chapter, all of the remarkable (and often magical) plants that feature in Tolkien's fiction, from the line of white trees created by the Valar and inherited by Aragorn at Minas Tirith to athelas, the medicinal plant used to heal Frodo after he is wounded by one of the Nazgûl, and the various plants cultivated and consumed as pipe-weed (particularly the varieties grown in the Shire, Longbottom Leaf, Old Toby, and Southern Star)—all of these deserve a whole book of their own! It would be more appropriate here, we believe, to examine the various types of vegetation likely to develop in Middle-earth (at least during the Third Age) in order to illustrate the scientific method used by botanists. On our modern Earth, the main types of landscape, other than frozen and hyper-arid deserts, are characterized by their dominant vegetation, which is itself influenced by the local climate—and Middle-earth is no exception to this rule. Tolkien gives us very little information about his world's climates; however, the map of Middle-earth shows an interesting latitudinal zonation, with the great icy expanse of Forodwaith north of the river Arnor, and the vast desert of Harad south of the river Poros. This suggests an extremely cold and harsh climate in the north and a hot and dry one in the south, with an area of temperate climate in between these two zones. As

Middle-earth is bounded to the west by the Great Sea of Belegaer, this temperate climate is presumably mild and oceanic to the west and more continental toward the east, thus corresponding to that of modern Europe.[*]

LINDON, THE SHIRE, AND ERIADOR: A BEAUTIFUL AND BALMY COUNTRYSIDE

The sagas of Bilbo, Frodo, and their companions begin in the Shire, an eminently bucolic region of green hills traversed by rivulets and hedgerows and dotted with patches of woodland reminiscent of the gentle English countryside. Bordered to the west by the Great Sea (Belegaer), the region is probably temperate and mild, with oceanic tendencies (mild winters and summers and inclined to humidity), but also with an increasing tendency to continentalization (cold and wet winters, hot and drier summers) as we get closer to the Misty Mountains.

The pleasant Shire is situated within a larger geographical area called Eriador. Assuming the dominant western winds blowing in from the sea, the whole of Eriador receives plentiful rainfall and is not susceptible to either heat waves or drought. Its forests are typical temperate and deciduous ones; that is, composed mainly of trees that lose their leaves in winter, with few coniferous trees, and dominated by oaks, beeches, and elms, with ash, alder, and willow trees in humid valleys, and large numbers of ferns in the undergrowth—in short, nothing too exotic for a young British reader. Moving toward the east, following along in Bilbo's tracks and later those of Frodo and his companions, the landscape becomes wilder, seeming to alternate forests (such as the Old Forest, which borders Buckland, and the Trollshaws west of Rivendell) and moorland.

THE NORTH AND THE MOUNTAINS: PREDOMINANTLY CONIFEROUS

In modern Europe, with the intensification of the winter freeze, deciduous forests give way in the northern and far eastern regions to mixed forests of

---

[*]   See the chapter "Summer is coming: Climate in Middle-earth," p. 126.

deciduous and coniferous trees, and then to forests composed wholly of coniferous trees called taiga, which are dominated by firs, spruces, and larches.
This gradation occurs both latitudinally and altitudinally, starting in the
foothill zone and rising to alpine meadows (the alpine zone) and ultimately
to the snow zone, where snow persists all year, and including along the way a
zone of mixed forest (the so-called montane zone) below one of forest equivalent to a taiga plain in the sub-alpine zone. In Rivendell, which lies in a deep
valley at the foot of the Misty Mountains, the flora is mixed (deciduous and
coniferous), which corresponds to the montane zone.

In continental Europe, taiga dominates the Scandinavian landscapes, giving
way gradually to deciduous forest as one travels south. This does not seem to
be the case in the north of Eriador, Arnor, the northern part of Arthedain, and
the edges of the kingdom of Angmar, where the majority of the land consists
of cold and barren moors (such as the Ettenmoors, a wild land infested with
trolls). We might suspect that massive deforestation or magical actions (by the
witch-king Angmar?) have hindered the presence of a taiga here. And yet taiga
seems to have been quite agreeable for trolls, which, in Scandinavian legends,
show an affinity for large coniferous forests.

When Bilbo, Frodo, and their companions travel through the Misty Mountains, they navigate coniferous subalpine forests (equivalent to taiga plains)
before reaching the snowy peaks. After leaving Lothlórien and taking the
southern route toward Emyn Muil, the fellowship of the Ring makes their way
through several areas where ridges of pine and fir trees rise alongside deciduous
woodlands heavy with birch trees and undergrowth of blueberries, suggesting
a far more continental climate than that to the west of the Misty Mountains,
given its mixed vegetation, even tending toward a mountainous climate.

THE GREAT FORESTS OF RHOVANION:
WHERE THE OAK IS KING

In modern-day western Europe, lowland forests are populated mainly by oak
trees, with beech trees running second. Beeches do not do well in the cold,
preferring regions with thick cloud-cover; they are also found in montane
zones alongside coniferous trees, while oaks are more tolerant of drier and
sunnier environments. In some regions, however, chestnuts can proliferate
where the soil is sandy and more acidic. Forests heavily populated with oaks

FIGURE I

Climates and vegetation of Middle-earth

Polaire – toundra et desert glacé = Polar – tundra and frozen desert

Froid sec – steppe et lande froides = Dry-cold – steppe and cold moorlands

Montagnard (étage montagnard à nival) – forêts mixte, résineux, pelouse alpine = Mountainous (montane zone to snow zone) – mixed and coniferous forests, alpine meadows

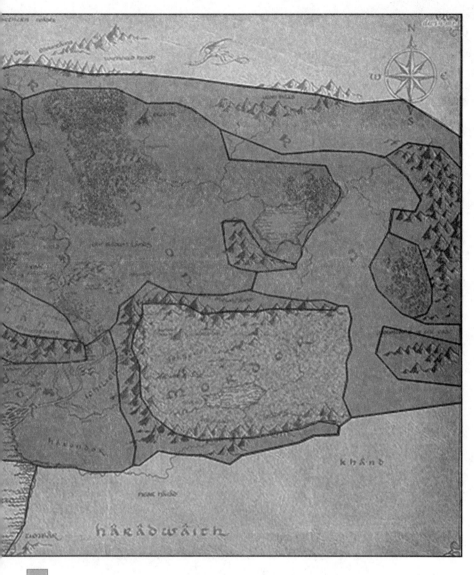

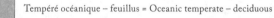 Tempéré océanique – feuillus = Oceanic temperate – deciduous

Tempéré continental – feuillus et résineux = Continental temperate – deciduous and coniferous

Tempéré sec – steppe tempérée = Dry temperate – temperate steppe

Méditerranéen – forêt sclérophylle = Mediterranean – sclerophyll

Chaud sec – semi-désert et désert = Dry-hot – semi-desert and desert

Climat et végétation altérés par la magie = Climate and vegetation altered by magic and beeches generally also include hornbeams and, on their fringes, elms, rowans, spindle trees, dogwood and viburnum (or wayfaring) trees, privet, hawthorn, blackthorn, rosehip, wild plum, and maple trees. The undergrowth and fringes of temperate forests also play host to holly and some vines such as ivy, clematis, and honeysuckle, also known as goatleaf—all three of which are mentioned in Tolkien, but the last of these is the only one used as a family name, that of a Man of Bree named Harry Goatleaf, who, spurred on by Saruman, is involved in the troubles that rock the Shire before Frodo's return.

Humid valleys and the banks of watercourses, ponds, and lakes are largely populated by ash and alder trees and, of course, numerous types of willows. Oak remains the emblematic species, and Thorin Oakenshield uses a shield of oakwood (from which his name derives) to defend himself from Azog the Defiler, the Orc-chieftain of Moria.

The temperate-weather species listed above are the ones found in the woods and forests of Eriador. In Rhovanion, three great forests serve as the havens of the Silvan Elves, or Wood-Elves (in the case of two of them) and many fantastic creatures involved in key events in Tolkien's world. These are Mirkwood, the forested region of Lothlórien, and the Forest of Fangorn. These three vast wooded expanses, like the forests of modern Europe, are dominated by oaks (as well as mellyrn, the mysterious and fantastical trees resembling beeches that are present in Lothlórien and intimately linked to the realm of the Elves Celeborn and Galadriel), often alongside beeches and sometimes elms and, on the mountains, pines, firs, and birches, which are suggestive of a continental tendency in the climate, as well as minority species including the frequently-mentioned rowans.

Fangorn is the ancient southern forest characterized by great diversity of species, including a wide variety of Ents, of whom Treebeard is the eldest. This forest is the remnant of the great forest that covered the hills and plains of the entire western temperate zone during the First Age, including Eriador. When Treebeard brings Merry and Pippin to the Entmoot, the two hobbits notice the resemblance of each creature to specific species, including oaks, beeches, ashes, lindens, chestnuts, rowans, birches, and firs—in other words, Fangorn displays the biodiversity of species of modern European plains and montane zone forests. Treebeard himself is described as resembling an ancient oak or beech. Tolkien also refers frequently to rowan trees, and not only in Fangorn, with the Ent character of Quickbeam. Rowans (or *Sorbus*) include several species of small trees and bushes, the best known of which, as it is frequently planted in parks and gardens, is *Sorbus aucuparia*. These species are no more widespread in temperate deciduous forests, however, than other minority species. Nevertheless, in England, rowan berries are traditionally consumed and used as an ingredient in jellies, jams, and liqueurs. This common British usage of the rowan's fruit may explain why Tolkien assigns such importance to these trees and shrubs.

In Ithilien, in the south of Rhovanion, the climate becomes milder. After a stretch of heath and broom, plants begin to appear that are unknown in the Shire, such as the cypress and cedar trees found in real mountains in Mediterranean regions. The vegetation described as one nears the mouth of the Anduin is typically Mediterranean, with olive trees, myrtle, laurel, and terebinth.

> *Many great trees grew there, planted long ago, falling into untended age amid a riot of careless descendants; and groves and thickets there were of tamarisk and pungent terebinth, of olive and of bay; and there were junipers and myrtles; and thymes that grew in bushes, or with their woody creeping stems mantled in deep tapestries the hidden stones; sages of many kinds putting forth blue flowers, or red, or pale green; and marjorams and new-sprouting parsleys, and many herbs of forms and scents beyond the garden-lore of Sam.*
>
> (*The Two Towers*)

Juniper, though mentioned here, is seen in cooler temperate zones in modern Europe, though it favors sunny meadows (and is the only coniferous tree that grows spontaneously on the plains of Western Europe), where its berries are used, particularly in Great Britain, to make gin (hence the liquor's name). Oregano, thyme, and wild parsley are abundant and eagerly used by Frodo and Sam to flavor their rabbit stew, much to Gollum's disgust.

> "*Gollum!*" *he called softly. "Third time pays for all. I want some herbs." Gollum's head peeped out of the fern, but his looks were neither helpful nor friendly. "A few bay-leaves, some thyme and sage, will do—before the water boils," said Sam.*
>
> "*No!*" *said Gollum. "Sméagol is not pleased. And Sméagol doesn't like smelly leaves. He doesn't eat grasses or roots, no precious, not till he's starving or very sick, poor Sméagol."*
>
> (*The Two Towers*)

In modern-day Mediterranean regions, semi-evergreen forests (which retain their leaves all year) of holm oaks, also called sclerophyll forests, alternate with areas of shrubby scrubland (a forest formation due either to disturbed soil conditions or in which degradation is caused by human activity, particularly pastoralization). The south of Rhovanion adheres to this template, as the holm oaks and boxwoods mentioned illustrate dense forest formations alternating with more open spaces.

THE STEPPES OF ROHAN: GRASS AS FAR AS THE EYE CAN SEE

The kingdom of Rohan, bordered on the south by the White Mountains, is described as a vast steppe (though Tolkien calls it a large grassland) patrolled by King Théoden's cavalry during the War of the Ring.

The presence of a steppe—that is, a wide open expanse—in a temperate region typically covered with deciduous forest is generally related either to drying soil conditions (too much drainage) or simply to the existence of a local climatic peculiarity. We might imagine that the landlocked area between the White Mountains to the south, which would block the damp winds blowing in from the sea (these winds would become drier as they move north due to the föhn effect*), and the Misty Mountains to the north, receives significantly less rain than the northern and central parts of Rhovanion. This would hinder the development of a dense forest but would sustain perennial herbaceous flora. West of Rohan, the region of Isengard is also described as being bare of forests but, in this case, the trees have been cut down in order to supply Saruman's military industry. This massive deforestation and the increasing incursions of orcs into Fangorn has the effect, among others, of angering the Ents, who will eventually besiege the tower of Orthanc, the stronghold of Saruman.

OTHER FLORA

It is not possible to give a complete description of the flora of Mordor, where Frodo and Sam's journey ends, for the land has been profoundly defiled

---

*   See the chapter "Tolkien's landscapes: A geomorphological approach," p. 101.

by the malevolent activities of Sauron and his creatures. In the popular imagination, ancient forests with dark, bramble-filled undergrowth, where numerous fantastical beings dwell in addition to wolves and other carnivores, inspire terror and seem somehow stamped with malignity; the same is true of foggy swamps. With the exception of the part of Mirkwood containing Dol Guldur and its giant spiders, though, Tolkien does his bit (through Fangorn and the Old Forest of Tom Bombadil) to restore the good reputation of this forest environment, which is no less mysterious but commands respect rather than fear. Not so with swamps and bogs, including the Dead Marshes to the north of Mordor, which Frodo and Sam navigate in the company of Gollum. This swamp is a humid expanse of putrid water filled with ghosts of those who died in an ancient battle, where nothing grows but a few stunted trees (willows?) and numerous grasses, sedge, and various types of rushes.

The regions outside of Middle-earth, where Bilbo's adventures and the War of the Ring take place, are only briefly or sketchily described. We might envision a landscape of tundra (sparse and low-growing vegetation) on the edges of Forodwaith, the great frozen desert of the north, and the icebay of Forochel. To the east of Rhovanion, the immense land of Rhûn is peopled by nomads, who might be imagined as reminiscent of the Huns who, coming from the steppes of central Asia, invaded part of Europe in the first millennium A.D. During his vision on Amon Hen, Frodo sees vast plains (steppes?) there, and unexplored forests. In the very southernmost part of Middle-earth, Harad, homeland of the Haradrim, who ally themselves with Sauron during the War of the Ring, is described as a desert region, equivalent to the Sahara and the desert and semi-desert zones of the modern Near and Middle East. The Haradrim ride terrifying mumakil, or oliphaunts, enormous pachyderms that cannot naturally exist in desert environments, but are native to the savannahs and tropical forests we assume to exist in the south of Harad.

Aside from these few outlying regions whose probable flora we must extrapolate, the plant formations of Middle-earth are those of modern Europe, as found from the British Isles to the Ural Mountains in the east and the Mediterranean in the south. Thus, a young Western reader who loses him- or herself in this rich and extraordinary universe with its blend of the mythological and the fantastical is able to walk, like Bilbo, Frodo, and their companions, through natural landscapes that remain familiar.

SUBTERRANEAN WORLDS IN TOLKIEN: AN UNDERGROUND HISTORY

SYLVIE-ANNE DELAIRE, speleologist and veterinarian

Yet [the abyss] has a bottom, beyond light and knowledge. [. . .] Thither I came at last, to the uttermost foundations of stone. [. . .] Far, far below the deepest delving of the Dwarves, the world is gnawed by nameless things. Even Sauron knows them not. They are older than he.

Gandalf, The Two Towers

olkien's subterranean worlds are awe-inspiring and give real depth to the story. Whether they are natural (caverns, goblin caves) or dwarf-built (like the mines of Moria), let us look now at these worlds filled with danger and shrouded in mystery, in an attempt to understand them better.

NATURAL CAVITIES OR CAVES

These are generally formed by the erosion of rocky massifs beneath the action of water, which dissolves calcareous rock (limestone and dolomite, for example). Mechanical fracture also plays a role; tectonics, alternating freezes and thaws, etc. crack, break, and weaken the rock. These natural phenomena lend Tolkien's world a geological and telluric aspect that is timeless but very real, and the dark and mysterious spaces thus created constitute true three-dimensional labyrinths where much action takes place.

In *The Hobbit*, the huge goblin cave beneath the Misty Mountains corresponds to an immense karstic network—subterranean cavities resulting from the erosion of a limestone mountain range. Its arcades and passageways attest to an ancient underground watercourse: surface water seeps down through diaclases (fissures caused by expansion or spreading) or cracks (as in the case of the Front Porch) and erodes the surrounding rock. In this vast underground network, goblins have created a veritable subterranean city called Goblin-town, redeveloping ancient caverns and dried-out tunnels. The main underground river then empties into a larger basin, feeding into a water table below, at the base of the mountain; this is the cave with its lake where Gollum resides. This lake, with its stagnant water as remarked upon by Bilbo, is fed by filtration through the ceiling and remains full due to an impermeable layer of clay situated just below the lake bed. In these spaces where water oozes through the ceiling, magnificent concretions (speleothems) can form: stalactites, stalagmites, columns, fistulas (narrow calcite tubes), and draperies or curtains of flowstone are found in karstic cave networks depending on the calcite levels of the water, the width of the diaclases, etc., as slightly acidic water dissolves the calcareous rock and transports its calcium carbonate. When it encounters a void, it undergoes a drop in pressure that can result in its evaporation and a release of carbon dioxide gas, with the calcium carbonate then re-forming as calcite crystals.

Though it seems far more sophisticated than Goblin-town, the subterranean palace of the Elvenking Thranduil, deep within the mountains of Mirkwood, also appears to be a succession of dried-out natural caves; water has created empty spaces which have then been converted by the Grey Elves into various rooms (reception rooms and rooms to rest in, but also cells where the Dwarves are confined). Seeking to free his friends, Bilbo discovers an underground waterfall in the palace caves; these falls have hollowed out a tunnel used by the Elves to transport merchandise. This watercourse emerges at the foot of the mountain and merges with the Forest River; this is the route by which the barrels in which the Dwarves hide reach Laketown (Esgaroth).

The Trolls' Cave, also featured in *The Hobbit*, is a single cavity in a hillside; these types of caves are created through the erosion of softer rock by storms.

Tolkien's underground spaces often display topographies where almost all of their tunnels have been completely drained of water. The lair of

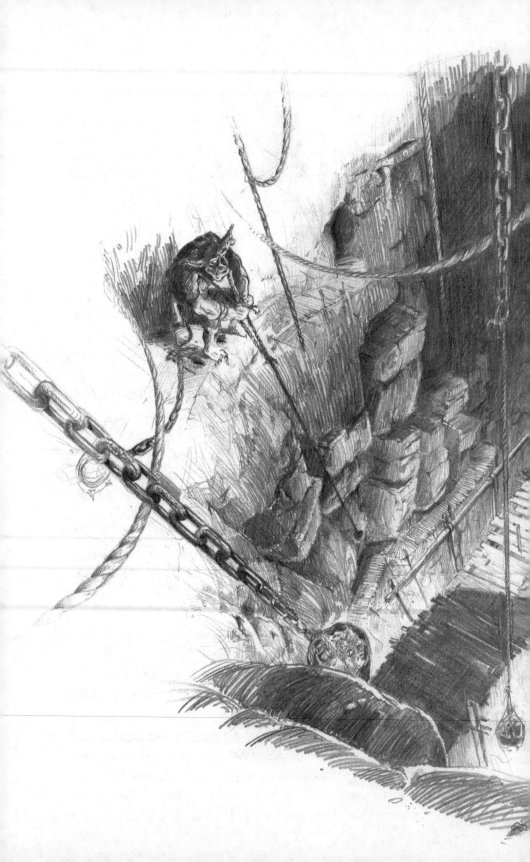

A mithril mine in Moria.

Shelob in *The Lord of the Rings* and the subterranean palace of the elf-king Thingol, Menegroth (the Thousand Caves), in *The Silmarillion* are no exception to this rule. In the real world, these dried-out networks are called "fossil" cave systems, as opposed to "active" cave systems. Watercourses have created spectacular reliefs here, some of them sharp-edged, others smooth and gleaming, such as gours, dam-shaped, carbonate-containing rock formations created when water stagnates in pools. Depending on surface meteorological conditions, below-ground flooding can also occur, further altering subterranean hydrology. Waterfalls loaded with alluvial deposits, powerful underground rivers whose outflow can reach 20 m3/s, and dangerous sumps can form, transforming calm landscapes into violent, raging ones. But there is another hazard that lies in wait for speleologists, whether they are real or imaginary like Tolkien's heroes: the accumulation of carbon dioxide gas underground. The lack of air experienced by Bilbo and the Dwarves in the goblin-tunnels probably corresponds to their passage through a poorly ventilated area where a pocket of carbonic gas has built up, reducing the amount of oxygen in the air.

TOLKIEN'S BATS

We might imagine that these inhospitable underground worlds are devoid of life but, as Gandalf reminds us in *The Hobbit*: "That, of course, is the dangerous part about caves: you don't know how far they go back, sometimes, or where a passage behind may lead to, or what is waiting for you inside." In Tolkien's world, as in reality, caves play host to a number of species, including bats (of the order Chiroptera[*]). During his travels through the Misty Mountains and the Lonely Mountain, Bilbo encounters numerous bats:

> On and on he went, and down and down; and still he heard no sound of anything except the occasional whirr of a bat by his ears, which startled him at first, till it became too frequent to bother about.
>
> *The Hobbit*

[*] For monsters and other imaginary creatures in Tolkien, see Part 6, "A Fantastical Bestiary," pp. 275–371.

Actual physical contact is extremely rare, however, as bats are equipped with echolocation, a sensory system that permits them to "see in the dark" by emitting sound waves from the mouth or nose. The bats of Middle-earth may belong to the family Rhinolophidae, a group typical of cave-dwelling Chiroptera in Europe—Tolkien's home—and relatively sensitive to disturbance. *The Hobbit* characters' various encounters with bats in flight tell us, moreover, that the animals are not in hibernation, suggesting that the surface climate is therefore mild. In reality, bats are more active in summer than in winter, during which season they hibernate, hanging from crevices in the rock, and emerge from this state only periodically, to drink the drops of water that have condensed on their fur, for example.

MORIA:
A MINE OF GEOLOGICAL INFORMATION

Like natural caverns, mines are major theaters of action in Tolkien. There are the mines of the Iron Hills, east of the Lonely Mountain, worked by the Dwarves of the House of Longbeard, also called Durin's Folk; the mines of the Grey Mountains north of Mirkwood, and of course the vast city of Moria (called Khazad-dûm, "Dwarves' Mansion," by the Dwarves) beneath the Misty Mountains. These mines have been hollowed out deep within high mountains; Moria extends beneath three peaks, including Celebdil, "discovered" by Tolkien during a 1911 camping trip at the foot of the Silberhorn, a 12,123-foot-high mountain in Switzerland. This topography, even in an imaginary world, would seem to favor the frequent development of non-flooded tunnels; mines are also possible beneath low-altitude plains, but they require more extensive infrastructures for groundwater pumping, ventilation, and ore extraction.

The mines of Moria (or "Black Chasm" in Sindarin) form a particularly stark and gloomy labyrinth: "[. . .] a dread fell at the mention of that name. Even to the hobbits it was a legend of vague fear [. . .] 'It is a name of ill omen,' said Boromir." (*The Fellowship of the Ring*). Iron, gold, and mithril ores are mined here, the latter an extremely lightweight yet ultra-strong precious metal. Once inside, and despite the feeble light provided by Gandalf, the Fellowship of the Ring advances very slowly and hesitantly, the darkness so deep and dense as to be almost suffocating. Only the dwarf Gimli, accustomed to being underground, adjusts fairly quickly. The heroes—and readers—quickly lose all sense of time here as well; we may correlate the pitch-blackness of Moria with that of real-world iron and magnesium mines, as these ores are very dark by nature. Geologically speaking, then, Moria is likely composed partly of ancient iron deposits deformed by the effects of high pressure and temperatures during the formation of the Misty Mountains, in what geologists refer to as orogenic metamorphism. This type of ore is found in layers, while the veins of mithril correspond to the filling of cracks within the surrounding rock—the same geologists refer to this process as secondary metamorphic recrystallization. Greedily exploiting these lodes of ore, which "lead away north towards Caradhras, and down to darkness," the Dwarves awaken the terrible Balrog, also called Durin's Bane.

HISTORY AND TECHNIQUES OF EXPLOITATION

In Tolkien's world, the Dwarves are historically masters of mine exploitation, somewhat like the kobolds, small beings from Germanic mining folklore. As they make their way through Moria, the Fellowship of the Ring discovers various types of constructions: arcades, winding passages, staircases (that of the West-gate is made up of 200 steps), tunnels, and great vaulted halls bear witness to the different eras and mining techniques employed by the Dwarves.

The serpentine arcades with their sinuous curves, which sometimes intersect with each other, suggest an ancient method of excavation; the Dwarves probably used picks, sledgehammers, and chisels to tunnel through the rock, collecting the ore from "hewing zones" in the extraction area and transporting it in baskets, probably on their backs. This ancient technique was used in real-world Europe into the early nineteenth century; before then, miners—often slaves—broke up the surface to be mined using the "fire-setting" technique, in which the ore seam was heated with a fire and then had cold water thrown on it to weaken the rock. Did the Dwarves also use this technique?

There are no rails or wagons ("tubs" in mining parlance) to be seen in Moria (unlike the mine of Erebor in Peter Jackson's film adaptation of *The Hobbit*). However, some halls are so immense and geometric in form that they attest to a period of mechanization; this is the case for the imposing vaulted room in which the Fellowship of the Ring encounters the Balrog. From the nineteenth century onward in Europe, the mechanization of mines with rails, wagons, and, later on, motorized machinery enabled more symmetry and parallelism in tunnels and arcades, causing them to become wider and more rectilinear with higher ceilings. This progression appears to have happened in Moria at some point as well. Some of these so-called "crosscut" tunnels enable ore to be transported out of the extraction area to outbuildings for processing; it appears, then, that the Dwarves processed their ore directly on-site. Lacking engines (Tolkien's world being fantasy-medieval) or explosives (which seem to have been reserved for Saruman), we might imagine hordes of ponies or Dwarves hauling massive amounts of rock in order to extract such huge volumes of ore. It is also likely that these enormous spaces were hollowed out from top to bottom with the aid of scaffolding, and that they were supported by stone pillars and arches other than the ones the heroes cross when they enter the mine.

The great vaulted chamber inhabited by the Balrog displays huge pillars cut into the rock; some of the stone, then, was left in place by the Dwarves, a technique that provides natural support for vacant spaces. The ratio of the volume of rock removed to that of the rock left in place in the form of pillars is called the "extraction ratio" in the technical language of career miners. The more unstable the rock, the lower this ratio. In Peter Jackson's film adaptation of *The Lord of the Rings*, the chamber is gigantic but its pillars are massive and close together, which implies a low extraction ratio and thus highly unstable rock. In reality, in order to make the site secure, the operators would have carried out a "caving" operation using explosives (known as "depillaring"), or backfilled with barren rock (that is, rock not containing ore). This is not to say that the Dwarves weren't mindful of their own safety, as they constructed enormous lateral staircases in the eastern parts of Moria designed to allow them to access different levels while avoiding inclined planes, which can be extremely dangerous in the event of collapses or rock-slides. The numerous wells (or "shafts") visible on every level also partially illuminate the mine and allow the removal of ore or the evacuation of toxic gases resulting from the breakdown of organic material or from degassing. Finally, the enormous chasm crossed by the Bridge of Khazad-dûm—destroyed by Gandalf during his battle with the Balrog—enabled the Dwarves to protect themselves against possible intruders arriving via the East Gates (or Great Gates).

Moria also contains numerous habitable rooms, such as the Chamber of Mazarbul, where Balin is buried, and the Twenty-first Hall. These chambers were clearly sophisticated living spaces. In our world's below-ground enterprises, living spaces are reserved for workers or miners (break rooms, offices, workshops, and locker rooms on underground work sites), or fitted out for soldiers, resistance fighters, or refugees (catacombs, underground fortifications, redeveloped quarries used as defensive strongholds or hiding places). Tolkien participated in trench warfare in France from June to November 1916 (as a signal officer in the 11th Battalion of the British Expeditionary Force), and some of his descriptions were probably drawn from his life as a soldier.

Finally, though—as we have seen—the mines appear to be natural karstic cave systems, they are often traversed by underground watercourses. In Moria, Gandalf warns his companions against drinking water contaminated by mining operations and, undoubtedly, the presence of malevolent beings: "There are many streams and wells in the Mines, but they should not be touched." The subterranean water in the mine of Erebor is cleaner, on the other hand, for it is the source of the Celduin (or the River Running), which once provided drinking water for The Yale, an ancient city at the foot of The Lonely Mountain, and still does so for Lake-town. This river, intersected by the subterranean constructions of the Dwarves, flows along a rock-cut channel to the Front Gate of Erebor, a development reminiscent of real-world canals that lead to sumps, where water is collected for pumping—the Dwarves weren't just good mining engineers, it seems, but good hydrogeologists too!

In conclusion, Tolkien's underground worlds are much more than simple settings; their detailed descriptions reflect the author's extensive speleological and mining knowledge. Tolkien creates subterranean atmospheres that are truly extraordinary. In addition to the symbolism of below-ground spaces—drawn, undoubtedly, from his studies of mythology—Tolkien gives us an underground rendering that is both realistic and dreamlike. While this British writer, philologist, and passionate taker of imaginary journeys is often compared to a hobbit (Bilbo himself being an author too), there can be no doubt that his in-depth understanding of subterranean worlds is worthy of the Dwarves.

PRECIOUS STONES:
JEWELS OF MIDDLE-EARTH

ERIK GONTHIER, ethnomineralogist, Museum of Paris
CÉCILE MICHAUX, cultural mediator

Diamonds, sapphires, rubies, and emeralds: precious stones are rare and beautiful minerals (also called "gems") that form in the earth's crust. These frequently monocrystalline mineralizations (that is, composed of a single crystal or polyhedron) display sizes, shapes and colors that vary depending on the conditions of pressure and temperature during their formation. Precious stones, coveted by humans since the dawn of time, are taken directly from mines (where they are extracted *in situ*) or indirectly from alluvial deposits, when they are called "placers," and are found in sediments transported by watercourses.

In the popular imagination, precious stones and other gems, aside from their beauty and alluring shimmer, belong symbolically to subterranean worlds, and are often thought to possess various powers: healing, clairvoyance, protection, etc. They are sometimes incorporated into pseudo-scientific theories, according to which the Earth is hollow and contains a hidden—and often evil—world within its depths, equivalent to the Hell of Greek mythology and Christianity, Swartalfheim in Norse mythology, and the Sheol of Judaism. From this same mystical perspective, precious stones, due to their magical character, can also act as a bridge between the visible and invisible worlds.[*]

---

[*] See the chapter "Invisible to Sauron's Eyes?," p. 182.

The Palantiri (Seeing Stones), the Arkenstone of Thrain, Elessar (the Elf-stone), and the Silmarils, to name only a few: in his Middle-earth, Tolkien invented, and sometimes placed at the forefront, many precious stones and other fantastical gems, and in this chapter we will analyze some of them in the light of our modern knowledge. First, let's linger for a moment on the Dwarvish people behind the mining of these stones, whose treatment in Tolkien's fiction was based on numerous Norse and Germanic myths and legends.

MASTERS OF GEMSTONE MINING

Thanks to the characters of Gimli in *The Lord of the Rings* and Thorin Oakenshield and his companions in *The Hobbit*, the great importance of the Dwarves in Tolkien's work is obvious. In the history of Middle-earth, the Dwarves began their mining activity after the awakening of the Elves in the First Age, constructing impressive underground cities such as Moria, where they mined the precious metal mithril. While Tolkien depicts them as fearsome warriors, his portrayal also remains faithful to that of Norse and German mythology, in which dwarves are skilled miners and industrious metalworkers in their subterranean or mountainous realms, creators of weapons for the gods and sometimes endowed with magical powers. In Scandinavian myth, because of the below-ground lives they lead, far from the sun, dwarves have a bad reputation and are even linked to the cult of the dead. Despite their sometimes gloomy outlook and their tough-as-nails temperament, though, Tolkien's Dwarves are neither malevolent nor magical; their civilization is a hard-working and materialistic one, resolutely focused on the riches to be found underground. At once geologists, miners, and gemologists (mineralogists specializing in the study of precious stones and gems), Dwarves are as irresistibly attracted by beautiful minerals as if they had some deep and intimate connection to them—and the line of Thorin Oakenshield displays a marked addiction to a single rare jewel: the Arkenstone.

THE ARKENSTONE,
A MINERALOGICAL CHIMERA

The Arkenstone is a precious stone invented by Tolkien. Discovered by the Dwarvish king Thráin I, its spellbinding beauty ensured the dominance of the House of Durin throughout Middle-earth, until the dragon Smaug drove the Dwarves from their underground kingdom in the year 2770 of the Third Age. The object of the main quest, this jewel is thus at the very heart of the plot in *The Hobbit*, being the centerpiece of the ancestral Dwarvish treasure whose recovery is sought by Thorin and his companions. The Arkenstone is to *The Hobbit* what the One Ring is to *The Lord of the Rings*; it arouses covetousness, and is capable of awakening our avarice, jealousy, and other dark tendencies.[*]

Also called "The Heart of the Mountain," the Arkenstone comes from the depths of Erebor, the Lonely Mountain, in the north-eastern part of Middle-earth. Etymologically speaking, the name of the Arkenstone comes from the Old English *eorcen-stán* or the Old Norse *jarkna-steinn*, both meaning "precious stone." The rarity of the jewel and its "secret" quality are conveyed in the French translation of its name, "Pierre Arcane." As with all of his inventions, then, Tolkien endowed this gem with semantic depth—and neither did he stint on his mineralogical description of it:

> [. . .] *the same white gleam had shone before [Bilbo] and drawn his feet towards it. Slowly it grew to a little globe of pallid light. Now as he came near, it was tinged with a flickering sparkle of many colors at the surface, reflected and splintered from the wavering light of his torch. At last he looked down upon it and he caught his breath. The great jewel shone before his feet of its own inner light, and yet, cut and fashioned by the dwarves, who had dug it from the heart of the mountain long ago, it took all light that fell upon it and changed it into ten thousand sparks of white radiance shot with glints of the rainbow. [. . .] His small hand would not close about it for it was a large and heavy gem [. . .]*
>
> The Hobbit

The Arkenstone is geologically intriguing; it is a white jewel, round and gleaming with its own inner light, and fashioned by the Dwarves so that this light is multiplied. Its power of fascination is neither magical nor malevolent, but rather natural and mineralogical in origin: its extreme purity (the total absence of inclusions in the crystal) evokes that of certain rare diamonds. Its high density is similar to that of chalcopyrite or galena (the sulfides of

---

[*]   See the chapter "*The Lord of the Rings*: A mythology of corruption and dependence," p. 89.

"I will find you, filthy little thief!"

copper and lead, respectively), but its rainbow iridescence is reminiscent of the birefringence (refraction of light-rays) of calcite, for example, a relatively common mineral.

Finally, because Erebor is a volcanic mountain, the Arkenstone must be magmatic in origin, like many precious stones; this gem formed deep in the earth will have resulted from the more or less slow cooling of lava in conditions of extreme high pressure and temperature. With all these characteristics, the Arkenstone is thus of a surprising and unique chemical composition—in other words, a mineralogical chimera!

VARIATIONS ON ELESSAR

Then she lifted from her lap a great stone of a clear green, set in a silver brooch that was wrought in the likeness of an eagle with outspread wings; and as she held it up the gem flashed like the sun shining through the leaves of spring.

—*The Fellowship of the Ring*

Two Elvish stones named Elessar are mentioned by Tolkien, in *Unfinished Tales* and in *The Lord of the Rings*. The extraction, creation (in terms of cutting), and handing-down through Elvish lines of these gems all remain somewhat unclear; however, a stone of this type is given by Arwen, daughter of Elrond, to Aragorn, the last heir to the thrones of Arnor and Gondor, who will reign under the name of Elessar after the War of the Ring.

Let's attempt a bit of analysis: Tolkien's writings clearly describe a large, clear green stone, which may be an emerald, though its deep green color is more characteristic of a rare type of garnet, uvarovite (a chromium-bearing garnet). Mineralogically speaking, emeralds belong to a large group of silicate minerals, and its color is due to the presence of traces of chrome in its chemical composition. Elessar, set in a brooch of silver shaped like an eagle, confers regenerative and protective power upon its wearer. However, the recent film adaptation of *The Lord of the Rings* breaks with Tolkien's books: Peter Jackson shows us a colorless, sparkling stone, similar to a diamond. Did the filmmaker insist on the gem's glittering appearance? Is its lack of color, which seems to reinforce the symbolic quality of eternity, meant to remind us that, in the story, the elves are quasi-immortal?

THE SILMARILS: MYTHOLOGICAL JEWELS

As three great Jewels they were in form. But not until the End [. . .] shall it be known of what substance they were made. Like the crystal of diamonds it appeared, and yet was more strong than adamant, so that no violence could mar it or break it [. . .]. And the inner fire of the Silmarils Fëanor made of the blended light of the Trees of Valinor, which lives in them yet [. . .]. [T]he Silmarils of their own radiance shone like the stars of Varda; and yet, as were they indeed living things, they rejoiced in light and received it and gave it back in hues more marvellous than before. [. . .] [N]o mortal flesh, nor hands unclean, nor anything of evil will might touch them, but it was scorched and withered; and Mandos foretold that the fates of Arda, earth, sea, and air, lay locked within them.

—The Silmarillion

The Silmarils ("radiance of pure light" in Quenya, or High Elvish) are the most precious stones in the Tolkienian universe. These "living" jewels, which appear in both *The Lord of the Rings* and *The Silmarillion*, are vital components in the genesis and evolution of Tolkien's world; *The Silmarillion* tells us that the Elf Fëanor, of the clan of Ñoldor, created them in the First Age from the light of the Two Trees of Valinor (the Silmarils were then in the form of clear globes which were then enchanted by Varda, queen of the Valar). This almost cosmic creation-story enhances the mythological and symbolic dimension of the Silmarils, which display phenomenal powers; in this they are somewhat reminiscent of the Infinity Gems in the Marvel Universe, cosmic science-fiction gems that confer significant power on those wearers able to control them.

Tolkien, preferring to suggest rather than describe, avoids giving us the precise mineralogical composition of the Silmarils. Likewise, he speaks about what a hypothetical observer would feel upon seeing the stones, rather than about the gems themselves. However, he does note that they are harder than diamonds. In geology, hardness is an important property of rocks and the minerals that constitute them. A scale ranging from 1 to 10, still used today, was invented by the German mineralogist Friedrich Mohs (1773–1839). According to this scale, which bears Mohs's name, talc has a scratch hardness of 1 (lowest), and diamond has a scratch hardness of 10 (highest). There are some materials harder than diamond, such as Q-carbon (or quenched carbon), but these are synthetic; that is, created in a laboratory. The fact that no known mineral is harder than diamond thus heightens the supernatural aspect of the Silmarils. However, in Tolkien's description, the mention of an "inner fire" evokes warm colors and shifting tints, the latter of which are found in siliceous minerals (with a base formula of SiO_2), such as red quartz (the color of which is due to inclusions of hematite, an iron oxide) or fire agate (a semiprecious stone whose iridescent tints are reminiscent of flames).

THE THREE ELVEN-RINGS

We are all familiar with the One Ring, which Frodo undertakes a quest to destroy in *The Lord of the Rings*. But there are three lesser known, yet very important Elven rings in Tolkien's world as well, each of them set with a precious stone. These rings, which appear in *Unfinished Tales*, were created by the Elf Celebrimbor ("silver fist" in Sindarin), a great craftsman and jewelsmith and army commander in the Last Alliance of Elves and Men against Sauron. Before meeting a tragic end—Celebrimbor was captured, tortured, and killed by the Dark Lord—he had given his rings to three Elvish figures: Narya, the Ring of Fire, was given to Cirdan ("Shipwright" in Sindarin); Vilya, the Ring of Air, to Gil-galad, High King of the Ñoldor; and Nenya, the Ring of Water, to Galadriel.

The first ring was of gold and set with a ruby. It passed into the hands of Gandalf, who inherited it from Cirdan in around the year 1500 of the Third Age. The ring Vilya, also of gold, was set with a sapphire and eventually came into the possession of Elrond. Finally, the ring Nenya, made of mithril and set with a diamond, remained in the possession of Galadriel. The fact that each ring possesses a proper name elevates them to the status of characters in their own right. In the story, these rings have powers attributed to the gems with which they are set: for example, Narya and its ruby grant the ability to resist oppression and despair. Rubies, composed of aluminum oxide and chromium and red in color, can be evocative of fire. The ring Nenya and its diamond offer protection from evil (and thus from Sauron's influence). The respective hardnesses of rubies (9 on the Mohs scale) and diamonds (10 on the Mohs scale, which makes it one of the hardest substances in the world) undoubtedly inspired their protective powers. Note also that diamonds, a mineral composed wholly of carbon and whose crystalline form (basic geometric structure) is cubic (unlike graphite, which is also composed of carbon, but whose form is hexagonal), can also evoke the purity and integrity which emanate from the personality of Galadriel. The power conferred by the ring Vilya and its sapphire is not clearly stated; it may allow the wearer to manipulate water, as Elrond does when he summons a torrent that sweeps away the Nazgûl. Sapphires, which belong, like rubies, to the corundum family, are composed of aluminum oxide, with a hardness of 9, and a blue color reminiscent of aquatic environments. Whether they confer protection or the ability to control the elements, Tolkien is drawing here on numerous popular beliefs that attribute magical powers to precious stones, and according to which gems represent manifestations of earthly perfection; indeed, the Three Rings, set with precious stones as they are, stand apart from the "simple" metal rings[*] which, in *The Lord of the Rings*, for example, are linked to the forces of evil and make their wearers dependent, and even grasping.

THE RING OF BARAHIR

For this ring was like to twin serpents, whose eyes were emeralds, and their heads met beneath a crown of golden flowers, that the one upheld and the other devoured [. . .]

—*The Silmarillion*

Scientifically speaking, some of the gems imagined by Tolkien cannot be identified—and we can't blame him for it. This is the case with the "Star of the North" (or the "Star of Elendil), a mysterious but extremely powerful white jewel that adorns Elendilmir (the diadem of mithril and gold worn by the kings of Arnor in both *Unfinished Tales* and *The Lord of the Rings*), and, as we have seen, with the Silmarils, the descriptions of which given by Tolkien have to do more with one's feelings upon encountering the gems than with their mineralogical profile.

Some stones, though, are clearly named, as in the case of the emeralds in the ring of Barahir which, though it has no magical powers, possesses a complex history on par with the One Ring. Fashioned in Valinor by Elves of the Ñoldor clan, it was given by the elf-king Finrod Felagund to Barahir, heir to the house of Bëor, in gratitude for the latter's having saved him during a

---

[*]   See the chapter "A Chemical History of the One Ring," p. 190.

fierce battle. After Barahir's death, the ring passed through numerous hands, including those of Isildur and Elrond, finally coming into the possession of Aragorn, who eventually gave it to his wife Arwen at the time of their betrothal in the late Third Age.

Emeralds, as we have seen, are a very rare type of silicate mineral with a hexagonal crystalline structure. Tolkien, a philologist, was undoubtedly aware of the etymology of the word "emerald," which comes from the Latin *smaragdus*, which itself derives from the Persian *zamarat* ("heart of stone"), an image Tolkien borrowed for his Arkenstone ("Heart of the Mountain"). Other emeralds appear in Tolkien's writings as well: in *The Hobbit*, after the Battle of the Five Armies, Bard the Bowman, whose arrow brings down the dragon Smaug, restores to the Elvenking the necklace of his ancestor Girion, composed of five hundred emeralds, "such jewels as he most loved." It is notable that emeralds, within the family of silicate minerals, are a cyclosilicate; that is, a beryl composed of tetrahedra whose apices are formed by four atoms of oxygen linked in the form of a ring—a configuration in perfect harmony with Tolkien's universe!

MEDIEVAL-FANTASTICAL METALLURGY

JEAN-MARC JOUBERT and
JEAN-CLAUDE CRIVELLO, chemists, CNRS

From the Shield of Achilles to Wolverine's adamantium, countless authors have dreamed up fantastical metals and alloys. Whether they have the power to kill or to protect, these elements are endowed with extraordinary, if not magical, properties. In Tolkien, fabled weapons such as the sword Andúril are able to defeat the most fearsome enemies; the metal mithril is precious, ultra-resistant, lightweight, and beautiful all at once; and the Rings of Power, which confer powerful gifts such as immortality and invisibility, form the very foundation of his work. This abundance of metal objects, and their importance to the plot, have led us now to consider Tolkien's fiction from a metallurgical perspective. Tolkien took an undeniably scientific interest in metal objects and the ways in which they were treated and produced—as well as in the symbolism of the forge, frequently to be found in the various myths and legends he studied. Indeed, the terms "metallurgist" and "metallurgy" are employed in *The Silmarillion*. What place does this discipline occupy in his work? Did Tolkien's knowledge go beyond the simple use of the vocabulary and concepts of a metallurgy that seems both medieval and fantastical at the same time?

FORGES: AN ANCIENT HISTORY

From the First Age onward, the forge is of primordial importance in the process of creation in Tolkien's world. Metallurgy appears very early on, used in the forging of weapons and armor to prepare for the first war against Morgoth; the Vala Aülé, a divinity comparable to Hephaestus, Greek god of fire and forges, gives life to the forefathers of the Dwarves, to whom he passes on his knowledge, including that of metallurgy. His best-known creation is Angainor (from *anga*, the Quenya word for "fire"), a forged chain used to bind Morgoth during his first captivity. Aülé also helps the Elves, including the great metalsmith Mahtan, who instructs Fëanor, who in turn secretly produces the Silmarils, jewels which will later lie at the heart of numerous conflicts.

⚔ THE SWORD THAT WAS BROKEN ⚔

Narsil is the name of the sword forged by the Dwarf Telchar of Nogrod during the First Age. Given to the Elves and passing into the possession of the kings of Númenor during the Second Age, it symbolizes the power of the kings of Arnor and Gondor, and Elendil is its inheritor. The sword is broken in two when Elendil falls in battle against Sauron. Isildur, Elendil's son, uses a fragment of Narsil to cut off the finger of Sauron that wears the One Ring. The shards of Narsil are brought by Ohtar, Isildur's squire, to Rivendell after his master falls in the Gladden Fields, marking the end of the Second Age. Elrond predicts that the sword will be reforged, an idea taken up by Gandalf ("Renewed shall be blade that was broken,/The crownless again shall be king," [The Fellowship of the Ring]) though without mentioning Aragorn by name. Indeed, the sword is reforged by the Elves of Rivendell with an "edge hard and keen"; Aragorn then gives it the name of Andúril, "Flame of the West," in reference to Númenor. Galadriel fashions for it a sheath "overlaid with a tracery of flowers and leaves wrought of silver and gold," prophesying: "The blade that is drawn from this sheath shall not be stained or broken even in defeat." The sword is endowed with spectacular powers: death strikes anyone who draws it other than the heir of Elendil, who alone can wield it, and the mere sight of it is terrifying to orcs. For many reasons (above all the recognition of its bearer as a king), this sword can be compared to Excalibur.

Though they share a direct kinship with the Vala Aülé, the Dwarves (unlike the Elves) remain earthbound and landlocked, preferring to work with "iron and copper [. . .] rather than silver or gold" (*The Silmarillion*). They establish the city of Nogrod on the site of a mine in the Blue Mountains, which counts among its native sons Telchar, the smith who forges Narsil, the sword of Elendil (see the text box above) and Angrist, the tempered iron knife used to cut a Silmaril from the crown of Morgoth, its blade so sharp that it had to be worn without a sheath. The Dwarves of Nogrod also invented chain mail.

Elves including the Ñoldor, living secretly in Gondolin, also forged numerous weapons: these included Sting, the dagger found by Bilbo in the Troll's Cave during the Third Age and later given by him to Frodo; Glamdring, the longsword belonging to Gandalf and possessing the same ability to glow blue at the approach of orcs or goblins; and its mate Orcrist, also called "Biter" and "Goblin-cleaver," which was eventually placed on Thorin's tomb. It is important to note that these swords have names and histories as well as lineages, which shows the importance of the fruit of the forge to Tolkien.

The smith Eöl, also known as the Dark Elf, who was trained at the Dwarvish forges of Nogrod, creates a strange black metal, which he named Galvorn ("shining black" in Sindarin), from "iron that fell from heaven as a blazing star" (*The Silmarillion*). Here we can see Tolkien's fondness for meteorites as sacred objects, a theme found in numerous ancient myths and beliefs. But the description given of this metal in *The Silmarillion* as "so malleable that he could make it thin and supple; and yet it remained resistant to all blades and darts" raises questions. Despite the best efforts of imagination, it is difficult to see how a metal could be hard and malleable at the same time. At any rate, Eöl forges twin swords of this mysterious metal, Anguirel and Anglachel, which end up being cursed, as they are inhabited by their creator's malicious soul. The damaged blade of Anglachel is "forged anew for [Túrin] by cunning smiths of Nargothrond, and though ever black its edges shone with pale fire; and he named it Gurthang, Iron of Death" (*The Silmarillion*). Túrin eventually kills himself with this sword, but before making this final, fatal gesture he engages in a dialogue with his sword: "'Hail Gurthang! No lord or loyalty dost thou know, save the hand that wieldeth thee. [. . .] [W]ilt thou slay me swiftly?' And from the blade rang a cold voice in answer: 'Yea, I will drink thy blood gladly [. . .] I will slay thee swiftly.'" For Tolkien, magical objects, especially metal ones, are characters in themselves.

⊶—≺ MITHRIL ≻—⊷

> [. . .] now it is beyond price; for little is left above ground [. . .].
> It could be beaten like copper, and polished like glass; and
> the Dwarves could make of it a metal, light and yet harder
> than tempered steel. Its beauty was like to that of common
> silver [. . .]
>
> —Gandalf, *The Fellowship of the Ring*

Mithril, "grey brilliance" in Sindarin, also called "true-silver" or "silver-steel," is a fantastical ore. Elves and Dwarves nurse an unbounded passion for it (*mithril* is the Elvish term, while the Dwarves' name for it is kept secret). It is all the more sought-after—and costly—after it ceases to be mined with the destruction of Moria in the Third Age; for example, the mail-shirt given to Bilbo by Thorin Oakenshield and passed on to Frodo has a worth "greater than the value of the whole Shire and everything in it," as Gandalf explains in *The Fellowship of the Ring*. Compared with other metals on the periodic table, mithril displays the ductility of copper, the thermal conductivity of silver, the hardness of the best steels, the lightness of titanium or aluminum, the beauty of platinum, and the price of osmium or lutetium; in short, it is a distillation of all these elements combined, with no real equivalent (the American Chemical Society's declaration that mithril is an yttrium-silver alloy notwithstanding).

In Middle-earth, mithril was exploited by the Dwarves in the mine of Moria. After the discovery of the principal seam, their greed led them to dig too deeply, awakening a Balrog that caused their ruin. However, additional veins were found on the western shores of Aman in Valinor, as well as in Númenor. The Elves of Eregion used mithril in the production of ithildin, an alloy with which the inscriptions on the Doors of Durin were created, visible only in moonlight or starlight (the word *ithildin* means "moon-star" in Sindarin). Other well-known mithril objects include Nenya, the Elvish Ring of Power set with a diamond, and the helmets of the Guards of the Citadel in Gondor.

⚬───≺ THE ONE RING ≻───⚬

The Silmarillion describes the creation of the Rings of Power, including the One Ring: "Now the Elves made many rings; but secretly Sauron made One Ring to rule all the others [. . .]. And much of the strength and will of Sauron passed into that One Ring; for the power of the Elven-rings was very great, and that which should govern them must be a thing of surpassing potency [. . .]." The One Ring is made of gold but it cannot be melted; it does not grow hot in normal fire; only the flames deep within Mount Doom, where it was forged, can destroy it. Heat causes an inscription in the Black Speech of Mordor to appear on its surface: "One Ring to rule them all, One Ring to find them,/One Ring to bring them all, and in the darkness bind them." The One Ring possesses various powers: it makes its wearer invisible, enables him to understand any language, and extends his life, but also corrupts his spirit. This ring is "alive"; it is a real character, seeking always to return to its master Sauron. It is for this reason that it falls from the finger of Isildur and is discovered by Déagol, whom his friend Sméagol kills before becoming Gollum, and slips itself onto Frodo's finger at the Prancing Pony inn.

In the Second Age, after the catastrophe that befalls Beleriand and destroys the city of Nogrod, the Dwarves discover mithril (see the text box on page 177) in the mines of Moria beneath the Misty Mountains. This discovery leads to a reconciliation with the Elves of the House of Fëanor, who settle on the plain of Eregion, near the West-gate of Moria, in order to trade with the Dwarves. Sauron, passing himself off as an envoy of the Valar and advisor in metallurgy and magic, is present during the creation of the Rings of Power; he then draws on these smithing techniques to create the One Ring in secret (see the text box above). However, Celebrimbor, head of the guild called the People of the Jewel-smiths in Eregion (*Gwaith-i-Mirdain*, in Sindarin), unmasks him and then forges the three Elvish Rings of Power, which are immune to the Dark Lord's wickedness.

After the important creative advances of the First and Second Ages, a decline in metallurgical activity makes itself felt among the various species and races populating Middle-earth: "[I]n metalwork we cannot rival our fathers, many of whose secrets are lost. We make good armor and keen swords, but we cannot again make mail or blade to match those that were made before the dragon came," the Dwarf Glóin explains to Frodo in *The Fellowship of the Ring*. Tolkien certainly provides far less detail concerning this science in his stories set during the Third Age; we learn only that the Elves reforge the sword Narsil in Rivendell, and that the Dwarves abandon the mines of the Iron Hills to aid their brothers in the Battle of the Five Armies related in *The Hobbit*. Among humans, only the Men of Westernesse are cited for their metallurgical skill by Tom Bombadil, who gives daggers to his visitors: "'Old knives are long enough as swords for hobbit-people,' he said. 'Sharp blades are good to have [. . .]'" (*The Fellowship of the Ring*). Finally, with regard to hobbits, no local forges or weapons are mentioned in the Shire, a region apparently spared all industrial (and industrious) activity. Only the metal of a gate clangs dolefully at the border with Buckland: "[the tunnel] was closed by a gate of thick-set iron bars. [. . .] It shut with a clang, and the lock clicked. The sound was ominous." (*The Fellowship of the Ring*)

METALLURGY AND THE FORCES OF EVIL

In Tolkien, as in much fiction, the forces of Evil are often associated with metallurgy. Metal is symbolic of hardness, of cold and malevolent power: "immeasurably strong, mountain of iron, gate of steel, tower of adamant, he saw it: Barad-dûr, Fortress of Sauron." (*The Fellowship of the Ring*). And the forge, where heat, fire, and molten metal combine, is a sort of hell. The forges of Isengard are truly terrifying: "there Saruman had [. . .] armories, smithies, and great furnaces. Iron wheels revolved there endlessly, and hammers thudded. At night plumes of vapor steamed from the vents, lit from beneath with red light, or blue, or venomous green." (*The Two Towers*). Peter Jackson's film adaptation also plays with this symbolism of industrial "hells." *The Lord of the Rings* was, then, an ecological novel ahead of its time, in which Tolkien pitted the industrialized society of wizards, with its negative environmental impacts (the destruction of the Forest of Fangorn to heat its forges, pollution by black smoke, etc.) against the bucolic and idyllic nature of the Shire.

Thus, fire begets the forge, which produces weaponry and mechanization, which in turn gives rise to evil. The formula functions even more effectively when we add in self-delusion regarding progress ("their greed for knowledge, by which Sauron beguiled them," Tolkien writes in Appendix B to *The Lord of the Rings*); the pride and arrogance of those who possess knowledge of the forge (Fëanor); jealousy ("Melkor was jealous of him, for Aulë was most like himself in thought and in powers; and there was long strife between them" [*The Silmarillion*]); avarice, and the hunger for riches (as with the Elf Maeglin, or the Dwarves of Moria).

METALLURGICAL SCIENCE AND TECHNIQUES

In Tolkien's world, as in the real world, there are classic metals (iron, copper, silver, and gold, employed in our world for thousands of years) and higher-resistance alloys (the bronze doors of Angband and Minas Morgul; the steel walls enclosing the Silmarils)—but Tolkien also features fantastical metals (including mithril and galvorn, as discussed above) and alloys such as ithildin (derived from mithril) and tilkal, the name of which comes, as with many modern alloys, from the initials of the metals that compose it (*tambë, ilsa, latúken, kanu, anga*, and *laurë*, or copper, silver, tin, lead, iron, and gold in Sindarin). If these six metals are added in equal parts to form tilkal, that would make it what modern chemists call a "high-entropy" alloy, meaning an equi-atomic combination created with the objective of obtaining the best properties of each metal constituting it. This type of alloy differs from classical alloys, which are composed of a dominant element whose properties are enhanced by the adding of additional elements (for example, an iron base for steel, an aluminum base for duralumin, a zirconium base for zircaloy, etc.). In this sense, tilkal is Tolkien's most original creation in the domain of metallurgy.

Another of Tolkien's metallurgical winks can be seen in his associating of multiple metals or materials in order to strengthen their symbolism. For example, the noble metals, often reserved for the forces of Good, represent wealth and power; Aragorn's crown is wrought of mithril and gold, and during his reign the gates of Minas Tirith are made of mithril and steel. The common metals, on the other hand, are associated more often with the forces of Evil: Morgoth forges an immense iron crown for himself, and Barad-dûr, the Dark Tower of Sauron, is also built of iron. However, the precious metals are coveted by all, including Sauron, who has a great fondness for mithril. The purity and beauty of gold can thus be combined with spells and curses, as in the case of the One Ring.

Tolkien rarely mentions the techniques used to extract or refine ore. With regard to working and shaping metal, he frequently cites the forge, which appears as an almost magical means of obtaining weapons and jewels (the Silmarils, the Rings of Power, etc.). The author also mentions casting ("[Aulë [. . .] was in his smithy, pouring molten metal into a mold" [*The Silmarillion*), tempering to increase the strength of steel, and damascening, a decorative technique in which a dagger is encrusted with filaments of gold or copper (as with the weapons given to the hobbits by Tom Bombadil in *The Lord of the Rings*). All of this is proof of a certain scholarship on Tolkien's part concerning these techniques, despite his claim of "limited understanding" in a letter mentioning his desire to write an additional volume on various Middle-earth topics, including metallurgy.

MEDIEVAL-FANTASTICAL METALLURGY

The metallurgy of Tolkien, then, is both medieval (in its realistic techniques) and fantastical (in its inclusion of unknown materials with extraordinary properties). Through the frequent use of metal objects (rings, crowns, swords) and their vital importance to the plots, as well as the recurring theme of the forge, the importance of the discipline of metallurgy to Tolkien is made clear.

Finally, the theme of transmission—one of primordial importance for the former professor that Tolkien was—is also very present through metallurgy, creating strong psychological bonds between deities and characters and anchoring itself in the history of the author's legendarium: Aulë instructs Mahtan, who instructs Fëanor, while the Dwarves teach Eöl, who teaches Maeglin, who instructs his apprentices. Thanks to Tolkien, metallurgy, a scientific discipline that cannot be learned alone, is also an art in which the pupil receives instruction from the master in technique and magic.

All of this suggests a particular affection and—why not—a real fascination on Tolkien's part for metals and the creation and working of them. And it is significant that his final work, the fairy tale *Smith of Wootton Major*, recounts the adventures of a kindly metalsmith, Smith Smithson, who is able to roam freely in the Land of Faery while wearing an inlaid silver star on his forehead. At the end of the story, he passes the star on to a young boy and returns to his forge, "to the hammer and tongs," and continues the training of his son, emphasizing again the importance of the transmission of both knowledge and objects down through generations.

INVISIBLE TO SAURON'S EYES?

JEAN-SÉBASTIEN STEYER, paleontologist, CNRS
and Museum of Paris
ROLAND LEHOUCQ, astrophysicist, CEA

Two worlds cohabitate in Tolkien's writings: the "Visible," the tangible and material; and the "Invisible," a hidden world that opens into parallel dimensions, and in which the dead interact with spirits and other supernatural entities both shadowy and benevolent. Some characters, such as the Elves, have access to this world, as Gandalf explains to Frodo: "[. . .] here in Rivendell there live still some [. . .] Elven-wise, lords of the Eldar [. . .]. They do not fear the Ringwraiths, for [they] live at once in both worlds, and against both the Seen and the Unseen they have great power." (*The Fellowship of the Ring*)

In many stories, this invisible world is a way for the author to justify the use of magic. This is the case with the Harry Potter series by J.K. Rowling, published between 1997 and 2007, and its extended universe, where the source of magic lies in a world parallel to that of humans. For Tolkien, however, the Invisible is more complex, and is not limited to a simple magical dimension parallel to reality. In Middle-earth there are many invisible things and beings, as if multiple levels of invisibility exist. Let's examine them through the lens of our modern understanding of the visible world.

DISCREET AS A HOBBIT

"You have nice manners for a thief [. . .]," said the dragon. "[. . .] Who are you, and where do you come from [. . .]?"

"[. . .] I come from under the hill, and under hills and over the hills my paths led. And through the air, I am he that walks unseen."

—The Hobbit

Hobbits are small humanoids with large, hairy feet, capable of moving noiselessly. Though most of them are not really aware of it—and don't see the usefulness of it—this acoustic invisibility is highly renowned throughout Middle-earth, so much so that Bilbo is chosen by Gandalf to be the "official burglar" assisting Thorin, the Dwarf-King, to recover his treasure from the clutches of the dragon Smaug. This ability of hobbits may be due to their elongated, flat feet (their plantar arches are reduced or even absent altogether) or, more precisely, to their highly developed metatarsals (the long bones that make up the instep).* The flattening and elongation of their feet results in a significantly increased contact area of the foot surface with the ground, thus distributing the hobbit's weight over the whole foot rather than just at certain points (the heel, ball, and toes, as in humans). This reduces the pressure—the ratio of weight to contact area—exerted by the foot against the ground, and thus the risk of making anything the foot steps on "crack." With their relatively slight weight (despite their fondness for good food) and their large feet, hobbits are ideally equipped for discretion, and the feet of "he who walks invisibly" are able to avoid snapping twigs in the forest, or making coins clink, even right under Smaug's nose.

FRIENDLY RUNES

Some of the ancient runes and inscriptions in Middle-earth are visible only under certain conditions. This is the case of the letters inlaid into the Doors of Durin, crafted by the great Elf-smith Celebrimbor and composed of ithildin, a material "that mirrors only starlight and moonlight, and sleeps until it is

---

* See the chapter "Why Do Hobbits Have Big Flat Feet?," page 211.

touched by one who speaks words now long forgotten in Middle-earth" (*The Fellowship of the Ring*).

This phenomenon is reminiscent of Chinese and Japanese "magic" bronze mirrors, the earliest examples of which date back to the Han dynasty (206 B.C.–220 A.D.). These appeared at first glance to be ordinary mirrors, but the reflection of an intense beam of light projected on a wall showed the shape etched on the back of the mirror. It took many years for researchers to understand how these mirrors were manufactured and polished; it was a long and complex process involving minute, painstaking modifications of the mirror's surface, invisible to the naked eye, to duplicate perfectly the shape on the back. These modifications were easily revealed by projection in the right lighting, and we might imagine that the Dwarf-smiths developed a similar technique to produce ithildin.

Runes in Tolkien are not only inserted into stone; they are sometimes written on parchment as well, as in the case of the map revealing the secret entrance to Erebor, the Lonely Mountain, formerly inhabited by Dwarves. This map, given by Gandalf to Thorin so that the latter may undertake the quest for his treasure, was drawn by Thror, the Dwarf-king's grandfather. The Elf-lord Elrond detects runes on the map: "Moon-letters are rune-letters, but [. . .] [t]hey can only be seen when the moon shines behind them, and what is more, with the more cunning sort it must be a moon of the same shape and season as the day when they were written. The dwarves invented them and wrote them with silver pens [. . .]" (*The Hobbit*). Unlike the runes described above, which are rather generic, these become visible only under certain conditions: light itself is not enough anymore; the moon must be in exactly the same phase as when the inscriptions were written. Why the writing would be visible only in a particular intensity of light is a scientific mystery, but it may be that they can only be seen in the presence of a certain predetermined wavelength. This is what happens with fluorescence, a property possessed by certain molecules in which they are able to absorb light energy and return it quickly in the form of an emission of light. They can be used, for example, to make ink that can be seen only under ultraviolet light. The use of fluorescent molecules is also common in biomedical research, to mark out certain molecules. These runes are also reminiscent of the "friendly" inks revealed through heating and used in the early days of the secret services. The process here is a chemical one rather than an optical one; the ink oxidizes when heated, so that it is no longer the same color as the surface on which it is written. This is the well-known case with lemon juice, which turns brownish when exposed to a flame.

In our world, it is now possible to make things invisible. In October 2006, researchers at Duke University in North Carolina successfully made an object invisible to microwaves, which is perhaps less spectacular, but impressive nonetheless. Normally, stealth vehicles simply absorb incoming electromagnetic energy or bounce it back in deliberately chosen directions thanks to the specific shape of their fuselage. The researchers at Duke, however, constructed a device capable of forcing light-rays to go around an object, somewhat like how water flows around a pebble. The rays are neither absorbed nor reflected by the object, which therefore becomes invisible. To achieve this, it was necessary to use artificial materials possessing optical properties that were very different from those of standard materials. For example, unlike ordinary substances, they had to refract light at a negative angle; that is, they had to have a negative refractive index. The first of these "metamaterials" were composed of several layers of a fiberglass matrix, between which metal rings were inserted. When subjected to an electromagnetic wave, the metamaterial reacted by creating an internal electromagnetic field that altered the trajectory of the incoming wave and forced it to go around the object. In order for this to work, it was necessary for the size of the structures to be smaller than the wavelength of the incoming ray of light. Manufacturing constraints obliged the American team to create an object a mere millimeter in size, forcing them to use microwaves—but, in principle, what they achieved with microwaves could also be done with visible light, as the phenomena involved are of the same nature. The only hindrance is technological, as the elements to be used must be far smaller, on the order of a few hundred nanometers (billionths of a meter), as well as more complex. In January 2007, a team from the University of Karlsruhe created a lattice composed of holes 100 nanometers in diameter bored into layers of silver fluoride and magnesium on a glass substrate. This material functions at the red end of the visible spectrum, at a wavelength of 780 nanometers. Research is now focusing on the creation of materials that will be invisible in the visible part of the light spectrum.

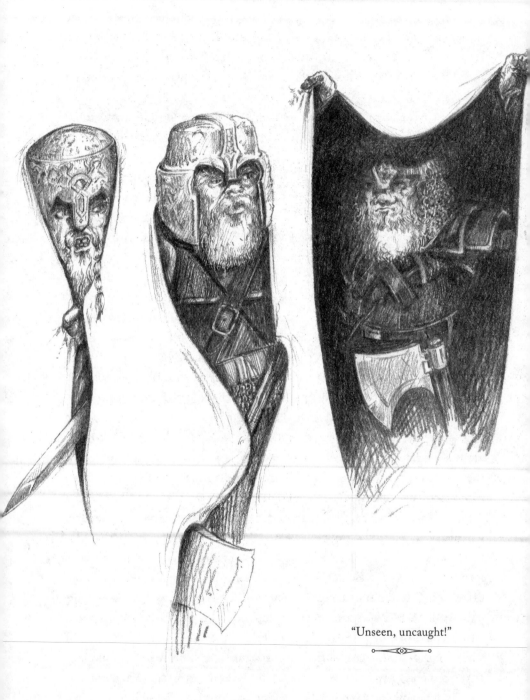

"Unseen, uncaught!"

INVISIBLE BODIES

Tolkien does not limit himself to sonic invisibility and invisible ink. His characters, including Bilbo and his nephew Frodo, are also able to disappear entirely. Everyone dreams of being invisible, to be able to sneak into a movie theater, annoy someone who irritates us, or empty out a bank vault. Once a person is invisible, it seems difficult to preserve a sense of honesty or integrity, as moral constraints disappear too. This is the question raised by Plato in the allegory of the Ring of Gyges: what is it that compels a man who has been made invisible to act morally? H.G. Wells addresses the same theme in his 1897 novel *The Invisible Man*, in which the chemist Griffin discovers an invisibility serum and tests it on himself. Committing thefts and crimes, the character traps himself in a hellish spiral that leads to his death. No longer existing either socially or morally, the invisible individual may also allow himself to give in to his impulses and fantasies, as shown by Milo Manara in his 1986 erotic cartoon *Butterscotch*. The methods employed to induce invisibility vary according to genre. Fantasy favors magic, while science fiction tends to be fonder of physiochemical procedures. For example, Jane Storm, a member of the Fantastic Four (Marvel Comics, 1961), is able to become invisible at will after having been accidentally irradiated by cosmic rays during a space expedition. In the 1975 American TV series *The Invisible Man*, scientist Daniel Westin perfects a "ray machine" capable of making objects invisible, which he uses on himself. Tolkien's characters have recourse to multiple means of invisibility: they may drape themselves in an Elven cloak, or slip the One Ring on their finger.

THE ELVEN CLOAK

When the members of the Fellowship of the Ring leave Lórien, they are given several gifts by the Elf-queen Galadriel, including invisibility cloaks:

> For each they had provided a hood and cloak, made according to his size,
> of the light but warm silken stuff that the Galadhrim wove. It was
> hard to say of what color they were: grey with the hue of twilight under
> the trees they seemed to be; and yet if they were moved, or set in another

light, they were green as shadowed leaves, or brown as fallow fields by
night, dusk-silver as water under the stars.

(The Fellowship of the Ring)

These cloaks have more to do with camouflage and mimicry, a form of contrast invisibility common in many real animals including cephalopods, arthropods, and chameleons. These concealment strategies, retained through natural selection, to go unseen by predators or prey, are possible thanks to pigment-containing cells in the skin (chromatophores and iridophores) that reflect light differently depending on the circumstance. But the Elven cloaks are not "alive"; our heroes are not covering themselves with mimetic organisms when necessary. The cloaks are indisputably made of cloth, as explained by one of the Elf leaders:

> *They are fair garments, and the web is good, for it was made in this*
> *land. They are elvish robes certainly, if that is what you mean. Leaf and*
> *branch, water and stone: they have the hue and beauty of all these things*
> *under the twilight of Lórien that we love; for we put the thought of all*
> *that we love into all that we make. Yet they are garments, not armor,*
> *and they will not turn shaft or blade. But they should serve you well:*
> *they are light to wear, and warm enough or cool enough at need. And*
> *you will find them a great aid in keeping out of the sight of unfriendly*
> *eyes, whether you walk among the stones or the trees.*

(The Fellowship of the Ring)

On a similar note, research teams—often financed by the military—are developing high-tech coatings whose patterns copy the environment and adapt to changes in color. For example, in 2014 a Chinese-American group unveiled the technical details of a "skin" that is both a sensor, "copying" the backdrop behind it, and a displayer of this same design in front. Five layers of different materials make up this skin, their functions analogous to those that have been identified in the chromatophores of cephalopods. For the time being, the coating is only able to display black and white patterns, but it is flexible and responds rapidly to external changes. Also in 2014, a group from France's Directorate General of Armaments (DGA) introduced its own high-tech camouflage, dubbed "Chameleon," which covers a broader spectrum of wavelengths, from the visible to infrared, which is of interest for remaining undetectable at night. While it's doubtful that the Elven cloaks function using this technology, they remain as top-secret as their military cousins.

THE ONE RING

The most efficient way of becoming invisible, though, remains the highly coveted One Ring. But slipping it onto one's finger is not without consequences, as it brings both physical and mental changes, a fact with which Sméagol is quite familiar.* Is it feasible for a ring to possess such powers? In histology (the study of organic tissues), it is possible to make some parts of the body transparent in order to better observe other parts; this technique consists of saturating the subject in various liquids (enzyme-, alcohol-, or dye-based) in order, for example, to make the flesh invisible and to color the cartilage and bones the researcher wishes to observe. This makes it possible to avoid dissection. Scientists at Stanford University have managed to make transparent the brain of a mouse killed by plunging it into a bath of hydrogel, a substance similar to that used to make disposable contact lenses. The molecules of this gel then penetrated the mouse's tissues to form a flexible, transparent frame before drawing out opaque cellular lipids through which light could not pass, making it possible to obtain very detailed snapshots of the molecular architecture of the brain. Does Sauron's ring, like some James Bond techno-gadget, contain substances like this to make its wearer invisible? No, for though it makes anyone who wears it invisible to those around them, it also makes this person *visible* to the eyes of Sauron and his minions, the Nazgûl, or Ring-wraiths. These specters sense and detect the Ring's presence all the more effectively because they are intimately bound to it, as Gandalf explains to Frodo in *The Fellowship of the Ring*: "You were in gravest peril while you wore the Ring, for then you were half in the wraith-world yourself, and they might have seized you. You could see them, and they could see you." Gandalf is correct; in physics, this is called "reversibility of light"—and, according to this principle, a truly invisible person, though they could not be seen, would also be blind! It would be somewhat problematic to rewrite *The Lord of the Rings* according to these conditions . . .

---

* See the chapter "Gollum: Metamorphosis of a Hobbit," p. 226.

A CHEMICAL HISTORY
OF THE ONE RING

STÉPHANE SARRADE, chemist, CEA

Has everything been written about Tolkien's One Ring that there is to write? No! The chemical nature of the ring remains a mystery. The ring is said to be of gold, that noble and rare metal of which wonderful objects are classically made. But is that all? Let's take a closer look at the possible relationships between other materials and the real or magical properties of this malevolent object. Its chemical nature will lead us to conduct a metallogenic and geological analysis of Mordor's metalliferous deposits, the sources of possible gold seams. Can we establish any links between the chemical nature of the One Ring and its strange powers?

ONE RING TO RULE THEM ALL

First of all, what do we know about the One Ring? In *The Hobbit*, it is a simple magical object created by the Dark Lord Sauron, which bestows the power of partial invisibility (the wearer's shadow remains visible in sunlight). In *The Lord of the Rings*, the author tells us that, in his desire to enslave and corrupt living beings, Sauron forged the One Ring to dominate all other rings and thus preserve his tremendous power. The One Ring was born in secret in the depths of Orodruin, Mount Doom, in around the year 1600 of the Second Age. Stolen by Isildur early in the Third Age, the ring was lost in the river Anduin. The hobbit Déagol recovered it by chance, but he was killed by Sméagol. Hidden in the Misty Mountains, Sméagol, who later became Gollum, had the ring stolen from him in turn by Bilbo. Later, in the Shire, Bilbo gives the ring to his nephew Frodo, entrusting him with the mission of preventing it from returning to its creator, which means destroying it in the place where it was forged; that is, in the fires of Mount Doom. This quest, recounted in *The Lord of the Rings*, results in the destruction of the One Ring at the end of the Third Age and the crowning of Aragorn as king. Middle-earth is saved from the destructive power of Sauron, and lasting peace is reestablished in the year 3021.

The One Ring is a malevolent object that manifests the intentions of its creator: to produce a jewel that will concentrate the temptations of Elves, Dwarves, and Men, all of them ready to do anything to possess it and fall prey to it.

THE POWERS OF THE RING: TOWARD AN ENHANCED AND CORRUPTED MAN?

In addition to the gift of invisibility, the One Ring enables its wearers to enter the spirit world, unseen by most mortals, where the Ring-wraiths dwell. But these powers have another effect, subjecting the wearer to a fatal addiction: their wicked tendencies are exacerbated (Sméagol becomes a killer and Bilbo a thief), and the closer they get to the invisible world, the more they lose touch with reality, to the point of losing their humanity (as Sméagol transforms into Gollum). The ultimate risk is to join irrevocably the world of shadows, like the Nazgûl have done.

The Ring's wearers are called, in short, to become new Saurons. To cement the desire for domination, the ring's power also confers supernatural longevity and increased sensory acuity. This brings to mind the effects of some of the drugs and poisons with which science fiction and heroic fantasy are rife. Only a few exceptional beings, lacking any desire for power or possession, are immune to the effects of the ring and are able to resist Sauron's influence, including Tom Bombadil and Gandalf.

Beyond the physical properties of this tangible, pure metal object, the One Ring also seems to behave like an organic being, specifically a parasite: extremely attractive in order to arouse its hosts' interest, it seems to have a will of its own, which it uses to deploy the strategies necessary to rejoin its master. It can change hosts as often as needed, and the more pronounced the negative character traits of a host become, the more effective its corrupting actions. If necessary, the ring can also give its wearer the ability to understand other languages, as is the case when Bilbo is able to understand the language of spiders in *The Hobbit*. To accomplish its objectives, the ring therefore takes control of its host's mind via a cohesive strategy.

WHAT DO WE KNOW ABOUT THE PHYSIOCHEMICAL PROPERTIES OF THE ONE RING?

It is round, smooth, and gleaming (as it is thought to be made of gold); sometimes fiery-hot and sometimes ice-cold. At room temperature it can seem heavy, and at other times lighter, though always retaining the same volume. This may be due to a variable specific gravity or density. When the ring is heated, inscriptions appear in the language of Mordor, summing up Sauron's intentions:

> *One Ring to rule them all*
> *One Ring to find them*
> *One Ring to bring them all,*
> *And in the darkness bind them*
> *In the land of Mordor where the Shadows lie.*

The surface of the ring has obviously been treated in a sophisticated way in order to give its uppermost layer particular optical properties. This implies that this upper layer, ostensibly of gold, has had atoms of another material deposited, or even incorporated, into the auriferous metal lattice. If this new material has been correctly chosen, it will display properties similar to those of gold at ambient temperature (suggesting that it is an element not too far from gold in the periodic table), and its presence will thus be invisible to the naked eye. When the temperature is increased, the differential expansion between the two materials causes heterogeneity in the metal lattice. If these microscopic defects correspond to legible writing, they can macroscopically form letters and phrases. This would mean that Sauron has at his disposal, in Mount Doom, the facilities necessary for a process equivalent to femtosecond laser engraving, used in the real world for additive manufacturing—as is the case here. This seems possible in theory, but its implementation in practice remains complicated. However, if the ring is not made of pure gold but of a mixture containing other metals, the procedure becomes simpler, requiring only a simple electrochemical treatment to cause the desired surface metal heterogeneity!

Don't forget that the One Ring, like all the Rings of Power, is virtually indestructible. Gandalf states that even the hottest of the Dwarves' furnaces could not melt it, nor even the terrible flames of a dragon; only the fires of Mount Doom, where it was created, can destroy it. A simple gold ring would not meet this criterion of inalterability, and so our theory of a sophisticated alloy is confirmed, and leads us to wonder about Sauron's metallurgical skill.

SAURON, A BRILLIANT ENGINEER?

Sauron was originally a Maia, a second-tier divinity in the service of a higher caste, the Valar. He started out as a servant of Aulë, a Vala with the distinguishing characteristic of being a smith, and then became first lieutenant of the fearsome Morgoth. Sauron mastered magic very early, it seems, due to his divine nature, and was an equally quick study regarding metallurgy, the techniques of which he learned by deceiving the Elf-smiths who created the other Rings of Power. As a Maia he is immortal, a master of mental manipulation, and able to change his appearance at will, sometimes even becoming extremely alluring: these are the magical properties we find in the One Ring as well, which, in the end, is merely an extension of Sauron himself.

The combination of metallurgy and magic is known as alchemy. In the Middle Ages, alchemists strove to control matter to the point of being able to transmute it, with the ultimate ambition of transforming lead into gold with the aid of the philosopher's stone, an imaginary catalyst and granter of eternal life. We know today that the chemical elements are composed of protons and neutrons assembled into a core orbited by electrons. In the periodic table, chemical elements are categorized according to the number of protons they possess, which is their atomic number (Z). The number of neutrons for a single element can vary, while Z remains constant. According to the number of neutrons it possesses, a chemical element has variants called isotopes. Some of these are radioactive, and therefore unstable; these will shed particles over time in order to return to the state of being a stable isotope, the most common version of the element. In the periodic table, gold (Au) has an atomic number (Z) of 79. Its neighbors to the right 80 (mercury), 81 (thallium), and 82 (lead) protons, while those to its left have 78 (platinum) and 77 (iridium). By bombarding platinum or mercury with neutrons, it is possible to obtain gold (a phenomenon called nuclear fission). This transmutation was first achieved in 1941 in a nuclear reactor, which is simply a source of quick-moving neutrons, meaning that the gold thus obtained is radioactive. Remember, finally, that the stable isotope of gold is 197Au (79 protons and 118 neutrons), and that the isotopes 195Au, 196Au, and 199Au also exist. The last two of these are radioactive and have a half-life of two to three days, transmuting spontaneously into the mercury isotopes 198Hg and 199Hg.

Going by this information, then, Mordor may be an area of significant nuclear activity, which is compatible with the level of desolation described by Tolkien. The subsurface of Mount Doom would be rich in veins of gold or in ultrabasic rocks containing gold inclusions. Since Sauron does not appear to possess a particle accelerator to carry out his transmutations, it seems that his radioactive source must be natural; that is, geological. This phenomenon exists in the real-world region of Oklo, in Gabon, where a natural nuclear reactor was discovered in 1972, where uranium ore embedded in sandstone and supported by granite becomes naturally radioactive in the presence of water (a rapid neutron moderator). A natural nuclear reactor of this type in Mordor would explain both the specific characteristics of the Mount Doom subsurface and the fact that the One Ring could only be created, and can only be destroyed, in this specific place, due to the presence of such an immense continuous energy source (the power of the Oklo site is between 500 and 1000 MW). This would make Sauron an alchemist possessed of significant atomic engineering knowledge and equipped with impeccable radiation safety gear—his heavy armor must be made of lead! Chemically speaking, the energy in a place like this would enable the transmutation of mercury or platinum (naturally contained in the auriferous rocks of Mount Doom) into gold, but also into the iridium and platinum that also make up the ring. The presence of these elements, known for their hardness, would explain the One Ring's increased mechanical resistance.

The ring is formed, then, of a material obtained via the neutronic bombardment of auriferous veins or source rocks, resulting in an improbable alloy composed of radioactive isotopes of gold, mercury, platinum, and iridium. This highly unstable alloy would explain the ring's sudden temperature increase

Periodic Table of the Elements

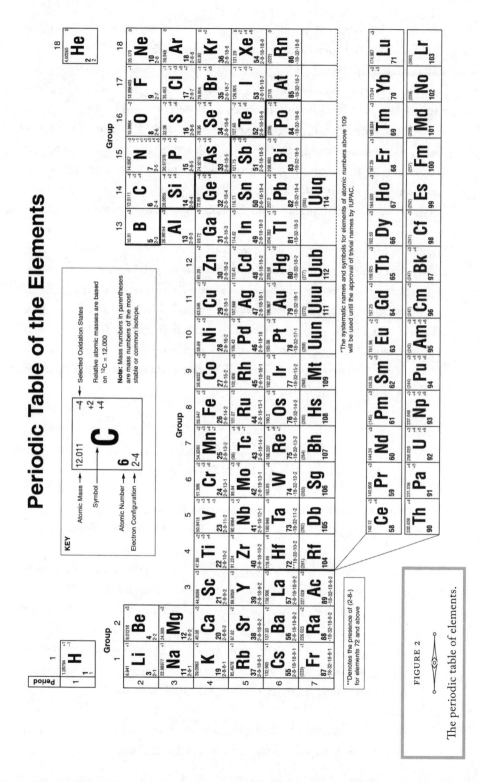

when it comes in contact with certain sources, as well as its change in density. The powerful radioactive emissions of fission products would explain the mutation of poor Sméagol into the dreadful Gollum—remember that the other wearers of the ring, Sauron and Bilbo, were protected by armor of lead and mithril, respectively. Finally, the ring's radioactivity could also explain its power of invisibility. To demonstrate why, consider the Vavilov-Cherenkov effect, which is well known to physicists. It describes the luminous flash that is produced when a charged particle passes through a dielectric (or insulated) medium at a speed greater than that of light; this effect spreads like a shock wave and causes the bright blue illumination surrounding the core of a nuclear reactor. Let's imagine that a similar effect is produced when a wearer slips the ring onto his finger: upon contact with the cells of the epidermis, the ring would find a source of energy to draw on in order to emit high-speed particles. This effect could alter the ring's properties of reflection or diffraction of light. A guided and astute modification of the light could then diminish or even obliterate the image of the ring's wearer! But it is absolutely the wearer who would provide a large part of the energy necessary for this phenomenon, which would also explain Frodo's extreme fatigue and even the physical deterioration of its other wearers.

SAURON, AN EXPERT IN BIOTECHNOLOGY?

Nematomorphs are parasitic worms that establish colonies in insects belonging to the Orthoptera order (crickets, locusts, and grasshoppers). When the worm passes from the larval stage to the adult stage, they become larger than their hosts and must return to an aquatic environment to reproduce. These creatures are so-called "manipulator" parasites because, in order to continue their reproductive cycle, they induce their hosts to "commit suicide" by throwing themselves into water. To do this, they produce mimetic neuroactive molecules, structurally comparable to those produced by the insects themselves, and secrete them directly into the central nervous systems of their hosts, indirectly taking control of the host organism! What does this have to do with Sauron's Ring?

The act of wearing the ring involves actual contact between the piece of jewelry and the skin. Imagine nanoparticles able to pass from the ring through the wearer's epidermis, and then to travel through the bloodstream to reach the central nervous system. These nanoparticles, which would be radioactive because they come from the ring, would act like drugs, altering or stimulating the host's mind on various levels. They would also be capable of prolonging the host's life, but at the price of altering his cells, as occurs with radiation therapy. These nanoparticles, then, would resemble the neuroactive molecules secreted by manipulator parasites, and, since Sauron created them, that would make him an expert on biotechnology as well!

In conclusion, Sauron's ring cannot be made of pure gold. It must be composed of an unstable alloy made up of radioactive isotopes obtained via a natural nuclear reactor situated beneath Mount Doom. This high-energy site would make it possible both to create and destroy the ring via an ultimate atomic transmutation or transformation. Upon contact with a wearer, the ring would draw out energy in order to emit high-speed particles leading to an invisibility effect, but the wearer would also be infected by dangerous neuroactive nanomaterials that would permanently alter his behavior and metabolism. A master of war, a great smith, and an experienced alchemist, Sauron, it seems, is an expert in radioactivity and biotechnology as well—a decidedly multidisciplinary supervillain!

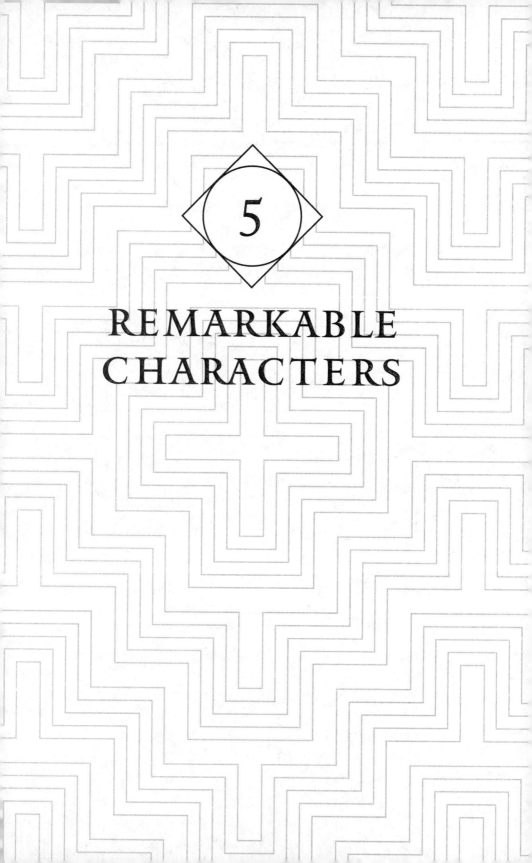

5

REMARKABLE CHARACTERS

TALES OF A YOUNG DOCTOR . . .
IN MIDDLE-EARTH

LUC PERINO, clinical physician and essayist

Doctors, nurses, obstetricians, psychologists, and other medical practitioners: none of them have ever met with much success in Middle-earth. It's as if health care providers must forever be excluded from fantasy, under the pretext that a heroes' body needing to be tended would diminish them somehow, or spoil the fun. Heroes certainly have points of great fragility—therein lies their charm, after all—but these weaknesses are rarely biomedical in nature; they tend rather to be emotional or spiritual. There are two additional reasons that might explain this medical vacuum in Middle-earth: the first has to do with the author, and the second with his characters.

THE KING'S HAND,
AND THE AUTHOR'S

Tolkien must have enjoyed enviable good health in order to produce, over the span of eight decades, the torrent of neurotransmitters and the blood-pressure levels necessary to make such a tumultuous brain function.

At the same time, it is clear that he drew inspiration from the medieval era, rich in knightly exploits but lacking in health-related ones; saturated with mysticism but devoid of clinical sense. If the author had included a few doctors in his stories, it would undoubtedly have been in order to make fun

The Herb-master of Minas Tirith.

of their incompetence, or their cowardice in the face of the plague and other epidemics far more murderous than orcs.

Medical medievalism is everywhere. The Black Breath of the Nazgûl is clearly meant to evoke pneumonic plague, which killed 35% of the population in a three-year epidemic, a fact incomprehensible today, at a time when—though the comparison has its flaws—AIDS has killed 0.5% of the population in thirty years. This is why the Nazgûl arouse such paralyzing terror even before they strike. And medieval doctors, as ignorant about physical illness as they were about psychological trauma, have no real reason to figure into this tale of heroes.

The only treatments mentioned in the medieval era involved herbalism and, in Gondor, the Houses of Healing in Minas Tirith boasted an Herb-master. The action of medieval roots and leaves on minor physical complaints was doubled by a strong and mystical placebo effect, as they came from the "simple-gardens" (or medicinal gardens) of 9th-century monasteries. All one needs to do to see the parallels is to reread the passages in *The Lord of the Rings* in which Aragorn uses athelas (or kingsfoil) to heal Frodo, Éowyn, Faramir, and Merry. The talents of this versatile hero stem more from incantation, shamanism, or exorcism than they do from apothecary science; the athelas-leaf itself is less important than the hand of Aragorn. Tolkien's medieval inspiration is confirmed by the passage in which the elderly healer Ioreth speaks of the healing power of a king's hands, "And so the rightful king could ever be known." This evokes the ritual of "touching the scrofulous," in which the monarch touches the scrofulous lesions of tuberculosis patients, more specifically those affected with chronic tuberculous cervical lymphadenitis, in order to heal them.

Passages like these, and the ones describing the Houses of Healing where Éowyn is treated both for the war-wounds she has sustained and for her lovesickness, were excised from the film adaptation, which is a shame, because they would have provided a rest from battle scenes. Clearly, not even shamanism and psychiatry are thought worthy of a place in fantasy!

Only wartime field-surgery (a thriving industry in the Middle Ages, obviously) appears here and there, but it amounts basically to the classic extraction of blades and arrows by Elrond or Gandalf.

VARIABLE HEALTH DEPENDING
ON THE CHARACTER

Who would ever be drawn to practice medicine among people who would never call on them, or listen to their advice if they did? Hobbits can't even imagine having need of a doctor, or a nurse, or any healer whatsoever; they feel protected, it seems, by their sense of communalism or clannishness. Certainly they feel pain, and do not disregard illness, but any requests for assistance are reserved for catastrophic situations when wizards become their only recourse. Though there are evil wizards as well, the hobbits depend on their intuitive faculties to find a good one, and anyone who has chosen a malevolent wizard through lack of discernment is thought to have gotten the treatment they deserved. In short, medically speaking, there is no middle ground between abstention and magic, and the risk of being permanently marginalized discourages young hobbits from undertaking medical studies.

Among the Dwarves, if there were once a doctor or two, these are merely the stuff of tall tales after an evening of drinking, told to poke fun at those who would have dared consult them.

Men, implicitly more fragile than hobbits or Dwarves, do not count many model patients among their numbers. The statistics are difficult to compile, but I would divide them into three categories in terms of their behavior in the face of illness: 80% scorn treatment, 15% turn to magic, and a mere 5% seek medical treatment.

Orcs cannot get sick because they are sicknesses themselves. Their population is a semiology of pustules, cancers, secretions, sores, pruritis, suppurations, hemorrhages, and deformities. They are fragile, and die from the slightest sword-wound. Moreover, the aim is not to treat disease objects, but to treat patients!

The Elves, immune to disease and senescence, deserve a clinical reflection on immortality, which will follow shortly.

A clinical physician who accidentally wandered into Middle-earth would find himself able to practice only half of the medical craft; that of diagnosis, for treatment is doubly absent, in terms of both believers in it and curative therapies and drugs themselves. Let's look now at a few clinical cases among Tolkien's characters, without worrying about their care.

THE PROBLEMATIC IMMORTALITY OF THE ELVES

Senescence, the group of physiological processes by which a subject ages, does not exist among the Elves. The same is virtually true for real-world medicine, which does not recognize senescence as such, which sometimes imparts a grotesque quality to death management by physicians. In medically-accepted thinking, people do not die because the specific life expectancy of *Homo sapiens* has a maximum limit; they die as a result of disease. The incontestable successes of modern medicine in preventing premature death have been followed by the demand for (and provision of) treatments that are alleged to delay belated deaths as well. This paradox, though unsustainable and undeclared, is tacitly understood by both doctors and patients, leading to a addictive game of dialectical and clinical hide-and-seek comparable to what an Elven physician might engage in. We know that Elves are immortal, but can choose to become mortal. The choice is one that they agonize over, as would any of us. Opting for immortality immediately causes all other life choices to evaporate, and provokes a sense of ennui infinitely more painful than simple depression. The job of an Elven doctor, then, would be to steer his patients toward the choice of a mortal life—all the more so because the principal source of happiness for Elves is the company of Men and Dwarves, two mortal species that will presumably disappear forever someday. Imagine the endless collective despair of an immortal people that has lost its primary reason for living!

The percentage of Elves he can convince to die one day would undoubtedly be the standard by which any doctor of theirs is judged. Unlike today's real-world transhumanists, who monetize false survival and whose only contribution is a painful obsession with living longer, the Elven physician is an expert who prescribes the no-cost and comfortable self-medication of mortality. Tolkien was one such expert himself, encouraging Arwen to choose mortality out of love for Aragorn.

THE QUESTIONABLE THERAPIES OF RIVENDELL

This brings us to the excellent relations that exist between Elves and Men, particularly in Rivendell, that haven of peace also called Imladris, where a stay of any duration is an enchanting delight. The half-elven Elrond is the

Lord of Rivendell. He is at once healer and surgeon; it is he who extracts the blade from Frodo's shoulder. Clinical observation suggests that his talents stem not only from magic, but also from a very real drug that is both anesthetic and psychotropic.

Two or three times in the story, Gandalf offers his wounded or exhausted companions a mouthful of a precious drink that he keeps in a leather flask, and which, he specifies, was given to him by Elrond. This is *miruvor*, which Gandalf evocatively calls the "Cordial of Imladris," a potion that is once again suggestive of the Middle Ages with its 'theriac', an electuary that was more than a match for Tolkien's fantasy, being composed of—among other ingredients—viper's flesh, squill, birthwort, bitumen of Judea, the odorous secretions of the sexual glands of beavers, and more than a hundred other substances. But its main component was opium, and it was undoubtedly to this that it owed its reputation for healing all manner of ills!

The great pleasure of stays in Rivendell, the many delights tasted there, and the warmth and conviviality of the place are due not solely to Elrond's hospitality, but also to the enjoyable properties of the opium or other psychedelic drugs playfully hinted at by the author.

A PSYCHOSOMATIC CARICATURE

Frodo, the central character in the story, is a clinical caricature. This young hobbit, guided by Bilbo and fascinated by Gandalf, is plagued by every possible doubt. He acts like a hero and a coward by turns, a loyal friend and a traitor, and is easily influenced. It's amusing to note that Tolkien gave Aragorn the extraordinary, superhero-like talent of being able to heal through the power of suggestion, and made Frodo an antihero whose suggestibility results in his being constantly ill.

Frodo faints; he almost drowns several times; he allows himself to be hypnotized; he suffers for long periods from his many wounds. Most of his symptoms and illnesses appear to be psychosomatic in nature.

Let's weigh our words carefully, so as not to anger Frodo's devoted fans. He is a hobbit, which means that he is not a hypochondriac; that is, he doesn't believe he has an illness that no doctor can diagnose or treat. Neither, as a hobbit, can he be a nosophobe (someone with an irrational fear of contracting any disease) or a normopath (the term used to designate patients who submissively agree with every real or imagined pathology with which a doctor diagnoses them). Nor is he a hysteric who exaggerates his symptoms, or a histrionic who fakes them. Rather, he simply demonstrates isolated psychosomatic ailments triggered by the weightiness of his mission, his own doubts, his low self-confidence, and his lack of bodily introspection, and, when provided with an adequate explanation, his symptoms disappear—a phenomenon exclusive to this type of disorder.

It is for this reason that Tolkien caused him to be accompanied by Sam, his psychosomatic opposite, a character whose pragmatism is fundamentally biological, and who, brimming with rustic vitality, is immune to any mental or physical pathology.

GOLLUM THE SCHIZOPHRENIC

It seems too simple to view Gollum as a simple representation of the conflict between good and evil, for this trait is shown by every character who possesses or even comes near the One Ring. To the clinical physician, Gollum is far more than this—partly because a little more medicine is sorely needed in Tolkien's sadly-lacking saga!

Did Gollum, when still Sméagol, kill his friend who possessed the One Ring simply out of covetousness or jealousy? That, to me, seems astonishing for a hobbit. Doesn't it make more sense for him to have committed this criminal act while under the control of voices commanding him to do it, voices he couldn't stop himself from hearing?

Hearing compelling voices; feeling as if one is being constantly watched, as if one's mind is being manipulated by someone else—all these are constant and major symptoms of schizophrenia, a disease with varying manifestations that also includes many other, less disabling symptoms.

Sméagol never stopped hearing these voices once he became Gollum. They have continued to speak to him, to cause him to commit several attempted murders that failed due to his weak, slight body. Today, Gollum would be prescribed antipsychotic drugs that would make him obese and apathetic. A schizophrenic can never bring himself to admit that the voices they hear are only voices. Gollum's periods of extreme friendliness and submissiveness are not due to the influence of the forces of Good; they are simply what are called intercritical periods. During these periods, Frodo and Sam do not feel threatened by him, even though Sam has always had better intuition in terms of diagnosis. Like all schizophrenics, even during his intercritical periods Gollum still demonstrates a visibly troubled personality, unlike those suffering from bipolar disorder, a few clinical cases of which we will now examine.

MENTAL MANIPULATION AND BIPOLARITY

It is sometimes said that man is a bipolar monkey—an amusing way of implying that the development of conscious awareness has brought with it a variety of anxieties and mood disorders from which no *Homo sapiens* is wholly exempt. Bipolar disorder, formerly known as manic-depressive psychosis, is the pathological form of this human trait.

As is always the case with this type of illness, there is a dispute between those who limit the diagnosis to its serious forms, composed of psychotic manic and depressive episodes, and those who, influenced by the pharmaceutical industry, tend to broaden the diagnosis to include all human beings. Tolkien had nothing to do with this debate, but there are some bipolar patients in his work, simply because it would be impossible for there not to be.

Like many powerful people in this world, Sauron is undoubtedly bipolar, with mania his predominant and perhaps only state. Abandoning the quest for the One Ring is, for him, the only situation capable of triggering a major depressive and suicidal phase.

Bipolar individuals are deft mental manipulators. Sauron is an expert at it, able to manipulate many kings—directly, as in the case of Denethor, and indirectly, as with Théoden, through Grima (Wormtongue). This mental manipulation is clearly evoked by the poisonous words that penetrate their ears: "Think you that Wormtongue had poison only for Théoden's ears?" This manipulation is frequently aimed at other, weaker sufferers of bipolar disorder. Théoden, after the typical spell of catatonia from which he is stirred by Gandalf, reveals himself to be bipolar as well, with his episodes of heroism and depression, while Denethor ends by sinking into suicidal madness and attempting to drag his son Faramir down with him.

Tolkien, a younger contemporary of Freud, could not have been aware of the future of psychoanalysis and its deceptions, including the Machiavellian planting of false memories through manipulation—so I will not go so far as to call Wormtongue a psychoanalyst, for fear that everything I have just said will be labeled pure fantasy!

WHY DO HOBBITS HAVE BIG HAIRY FEET?

JEAN-SÉBASTIEN STEYER, paleontologist,
CNRS and Museum of Paris

What is a hobbit? I suppose hobbits need some description nowadays, since they have become rare and shy of the Big People, as they call us. They are (or were) a little people, about half our height, and smaller than the bearded Dwarves. Hobbits have no beards. There is little or no magic about them, except the ordinary everyday sort which helps them to disappear quietly and quickly when large stupid folk like you and me come blundering along, making a noise like elephants which they can hear a mile off. They are inclined to be fat in the stomach; they dress in bright colors (chiefly green and yellow); wear no shoes, because their feet grow natural leathery soles and thick warm brown hair like the stuff on their heads (which is curly); have long clever brown fingers, good-natured faces, and laugh deep fruity laughs [. . .]

—The Hobbit

hough small in size, hobbits are large in terms of the role they play in Tolkien's writing. They are mentioned only briefly in *The Silmarillion* and *Unfinished Tales*, but they are at the heart of the plot in Tolkien's two best-known works: Bilbo "the Burglar" in *The Hobbit* is caught up in a daring quest and helps the Dwarves to recover their treasure, while his nephew Frodo, "the Ring-bearer," saves Middle-earth by destroying Sauron's One Ring in the fires of Mount Doom in *The Lord of the Rings*.

Also called "Little Folk" (by the humans of Bree), or "Halflings," or even "Periannath" (by the Elves), hobbits are curious beings: short in stature (between 60 and 120 centimeters tall, with an average of 107 cm, according to the prologue to *The Lord of the Rings*) but possessed of great bravery, they are sometimes fearful bourgeois village-dwellers, sometimes valiant adventurers. Fond of good living and deeply attached to the land, they can also become indefatigable and courageous explorers. Mischievous and lazy, they are also loyal and lovable friends. In short, hobbits are living paradoxes! Even their appeal and their morphology seem to defy common sense: their way of dressing is rather urban despite their countrified lifestyle, and they possess adult features on bodies as small as a child's. But what sets hobbits apart from the other "races" (it would make more sense to talk about "species") in Middle-earth—Elves, Dwarves, goblins, and humans, for example)—are their large, hairy feet, which they care for meticulously.

Why did Tolkien give these characters such animalistic, even monstrous traits? What adaptive benefits would these hairy appendages confer? Our modern knowledge of biology, comparative anatomy, and biomechanics (the analysis of the body's and the skeleton's possible movements) can shed some light on these questions.

OUTSIZED METATARSALS

First, let's look again at the size of these feet, the enormity of which is accentuated by the wearing of the short trousers so popular in the Shire. No shoes for hobbits; their bare feet are a source of pride, as attested to by the names of some very old families such as the Proudfoots (or is it Proudfeet?).

As bipedal humanoids (that is, human-shaped and walking upright), hobbits constitute a biological species relatively similar to humans: "It is plain indeed that in spite of later estrangement Hobbits are relatives of ours: far nearer to us than Elves, or even than Dwarves" (*The Fellowship of the Ring*). This presumed phylogenetic proximity* (also mentioned in Tolkien's correspondence) enables some interesting comparisons: compared to the feet of *Homo sapiens*, for example, those of hobbits are proportionally broader and, more importantly, longer.

---

* See the chapter "The Evolution of the Peoples of Middle-earth: a Phylogenetic Approach to Humanoids in Tolkien," p. 263.

FIGURE I

Comparative study of a human foot and that of a hobbit.

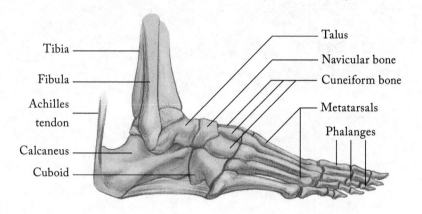

Tibia

Fibula

Achilles
tendon

Calcaneus

Cuboid

Talus

Navicular bone

Cuneiform bone

Metatarsals

Phalanges

HUMAN FOOT

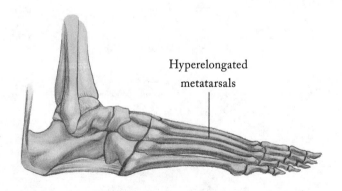

Hyperelongated
metatarsals

HOBBIT FOOT

The hobbit's despair

Moreover, the position of the heel on the foot—that is, the position of the calcaneus in relation to the other bones—is roughly similar in both species. This calcaneus, and the phalanges, also have similar proportions. The most striking difference between the feet of hobbits and humans lies in the widening, and especially the elongation, of the metatarsals (the bones of the sole of the foot), as shown in Figure 1.

Mammals that possess very long metatarsals (in proportion to the length of their feet) include unguligrades; that is, herbivores (and some omnivores) that walk on hooves: cows, horses, goats, boars, tapirs, and almost 200 other living species. Note that this term arbitrarily encompasses species that are not closely related to one another, phylogenetically speaking, but simply share the same manner of locomotion—the suffix "grade" is used to designate a grouping of this sort. Are hobbits unguligrades, as is the case with other fictional species such as centaurs (mythological half-man, half-horse creatures) and satyrs (beings with the ears, lower body, and tail of a goat)? The answer is "no," obviously; hobbits do not walk on hooves but on the soles of their feet. They are plantigrades.

Plantigrade mammals possessing highly elongated metatarsals include squirrels and rabbits (or lagomorphs), which encompass nearly 300 species! The comparison to rabbits recurs relatively frequently in Tolkien, who describes hobbits as small beings that are somewhat fearful in nature and make their homes underground. Hobbits and rabbits, then, seem to share the same kind of habitat; indeed, the very first line of *The Hobbit* reads: "In a hole in the ground there lived a hobbit." The word *hobbit* itself is similar to the word *rabbit*. And the comparison is unavoidable when Bilbo and his companions, *en route* to Mirkwood (also known as the Forest of Great Fear), are captured by three mountain trolls: "'P'raps there are more like him round about, and we might make a pie,' said Bert. 'Here you, are there any more of your sort a-sneakin' in these here woods, yer nassty little rabbit,' said he looking at the hobbit's furry feet [. . .]."

Why such large feet, with such elongated metatarsals? Could it be to help hobbits hollow out such lovely (and luxurious) underground homes, so reminiscent of some of the passive (low-energy) houses that are so trendy right now? Apparently not, because real-world burrowing species have long claws in addition to their large feet, and this is not the case for hobbits, who have flat nails (a characteristic of most primates).

Wouldn't these long metatarsals be useful for a better grip on the ground? Apparently this isn't the case either, for the Shire is not, geomorphologically speaking,[*] a windblown gully, but rather a large, calm plain scattered with hills, prairies, and woods. And hobbits, due to their small size, already have a low center of gravity.

Would their big feet help them to travel more easily on sandy or shifting terrain, increasing lift and making them less likely to sink? We might imagine that hobbits' flat feet act like snowshoes, allowing those creatures equipped with them to avoid sinking into sand (as with dromedaries) or snow (in the case of the artificial ones used by hikers). Let's dismiss the question of sand, because the Shire, with its green valleys and bucolic prairies, bears no resemblance to a desert. With regard to snow, do hobbits' large feet perhaps suggest ancestors that came from cold climates, that appeared during an ancient ice age or descended from some snowy peaks like the Misty Mountains or the Grey Mountains? No. Hobbits never historically lived in these high-altitude areas, but rather the valleys of the river Anduin.

On the other hand, these impressive soles, in direct contact with the earth, possess no arch, which would distribute the pressure from the ground more evenly, allowing hobbits to walk silently and even stealthily, an ability which, coupled with their small size, equips them perfectly for their reputed roles as "burglars."

Another possibility: were these large, flipper-like feet inherited, perhaps, from a water-dwelling ancestor? This is unlikely for at least two reasons: one, hobbits do not have interdigital webbing, which remains characteristic of organisms adapted to aquatic environments. Two, hobbits traditionally dislike water, and most of them don't know how to swim—let's not forget that Sam Gamgee is saved from drowning by Frodo at the end of *The Fellowship of the Ring*, and that Frodo's own parents drowned when he was a child.

We could investigate other possible functions of these enormous feet, and come up with any number of theories—but at the end of the day, why must we always associate an organ with a function? To ask another question, are hobbit's oversized appendages realistic? The matter becomes even more of a mystery given that the earliest representations of Bilbo, drawn by Tolkien himself, show a hobbit with feet that are certainly hairy, but that are more modest in size, unlike later depictions by Alan Lee and those in the various film adaptations, whether Ralph Bakshi's 1978 animated version or the more recent films by Peter Jackson. Hobbits' feet, it seems, have grown larger over time . . . perhaps to make them more unique, more different?

---

[*]   See the chapter "Landscapes in Tolkien: A Geomorphological Approach," p. 101.

Another problem is that hobbits' hands are of "normal" size. Genetically speaking, the same sets of architect genes (or "homeogenes") are involved in the morphogenesis of the upper and lower limbs—so morphogenetically realistic hobbits should also have large hands, with long metacarpals (the bones of the palm). They *should*, perhaps, yes—but hobbits with werewolf-hands wouldn't be as appealing (remember that *The Hobbit* was originally a story intended for children). For modesty's sake, we won't mention genital size, which also seems to be controlled by the same set of architect genes . . .

Speaking of sex, isn't it possible that those big, hairy feet are, quite simply, a sexual or intraspecific characteristic? This is a valid theory, as foot-size varies among hobbit families; the Proudfoots are known throughout the Shire for passing their large feet down from father to son. On a similar note, do male hobbits generally have larger feet than female ones? In any case, hobbits' feet seem to grow faster than other parts of their bodies, a process specialists call "positive allometry." But does the growth curve of hobbit feet swerve sharply upward before, during, or after puberty? Can their feet continue to grow indefinitely? The answer, of course, is no; very large feet could quickly become heavy and cumbersome. Beyond a certain point, they would require a significant energy expenditure with each step, as well as very large strides, necessitating highly developed muscles so as not to trip. Biomechanically realistic hobbits, then, would have far more muscular calves and thighs.

"MAY THE HAIR ON HIS TOES NEVER FALL OUT!"

What about the hair on hobbits' feet? This hirsuteness of the tops, toe-knuckles, ankles, and perimeter of the feet may play a protective role (remember that hobbits go barefoot, no matter the terrain navigated during their perilous adventures), as well as regulating heat: if the hair is dense, it insulates the skin from the cold, thus reducing the risk of getting sick in the winter. If the hair is sparse, however, it allows for a more effective release of body heat, as is the case for elephants. As the seasons change, then, the hobbits' foot-hair will act as both an air-conditioner and a radiator. And unlike those of some humans, hobbits' feet never smell. Is this perfectly-distributed hairiness stimulated by testosterone, the male hormone which, remember, is also secreted by women in the adrenal glands? To answer this question, we would have to see whether

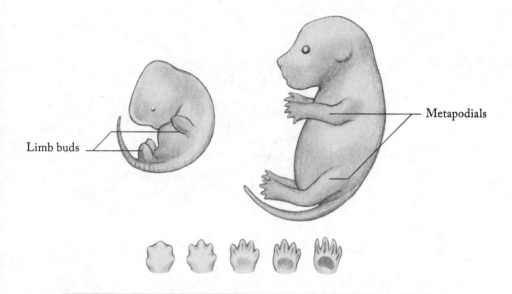

Limb buds

Metapodials

FIGURE 2

Morphogenesis of the hand and foot is regulated by the same type of genes, called "architect" genes. Fingers and toes form at the same time, through the proliferation of cells in the skeleton as well as apoptosis (programmed cell death) in the interdigital areas.

females of the species have equally hairy feet. At any rate, this hirsuteness of foot (emphasized by the fact that hobbits, unlike Dwarves, are beardless) is apparently a source of great pride; consider how Thorin begins his speech in *The Hobbit*: "Gandalf, dwarves and Mr. Baggins! We are met together in the house of our friend and fellow conspirator, this most excellent and audacious hobbit—may the hair on his toes never fall out!"

Finally, to conclude with a bit of paleontology, *Homo floresiensis* is a surprising human fossil discovered in 2003 on the island of Florès in Indonesia.[*] This curious species lived around 12,000 to 13,000 years ago, at the same time as *Homo sapiens*. With its small stature and its proportionally longer and flatter feet, it was quickly nicknamed "Hobbit." Science may often inspire fiction, but the reverse is also true!

---

[*]   See the chapter "When a Hobbit Upsets Paleoanthropologists," p. 218.

WHEN A HOBBIT UPSETS PALEOANTHROPOLOGISTS

FRANÇOIS MARCHAL, paleoanthropologist, CNRS

I n a hole in the ground there lived a hobbit. Not a nasty, dirty, wet hole, filled with the ends of worms and an oozy smell, nor yet a dry, bare, sandy hole with nothing in it to sit down on or to eat. This was a large space situated at the fork of two small streams, ensuring a supply of fresh water and the raw materials needed to create solid stone tools. Around 500 meters above sea-level and open to the north, it also provided its occupants with welcome coolness and fresh air. Are we in the Shire, home to hobbits, with its temperate climate? No. Welcome to the island of Florès in the Indonesian archipelago, characterized by a far hotter tropical climate. And the hobbits? We'll get to that—but these aren't the hobbits born in J.R.R. Tolkien's imagination; rather, this is the paleoanthropologists' "Hobbit," a human that lived in the cave of Liang Bua around 50,000–100,000 years ago. And it is no less extraordinary than Tolkien's creation.

THE PALEOANTHROPOLOGISTS' HOBBIT?

In September 2003, a team led by Mike Morwood, a New Zealand-born archaeologist with the University of New England in Australia, had been digging for two years in the sediments of the Liang Bua cave in Indonesia when they unearthed the partial skeleton of a human, as well as the scattered bones of several other individuals, none of which resembled anything previously known. Just over a year later, on October 28, 2004, the skeleton was presented in an article in the journal *Nature* under its official name, LB1: LB for Liang Bua, and 1 for the inventory number. It wasn't enough. Like all extraordinary fossils, it needed a nickname—after all, everyone knows A.L.288-1 much better as "Lucy." At the time of the article's publication, Tolkien fever was at its height, revived by Peter Jackson's cinematic adaptation of *The Lord of the Rings* trilogy, which had hit screens in 2001, 2002, and 2003. Mike Morwood, a compatriot of Jackson's, didn't have to think for very long before coming up with a moniker for this human skeleton which, among other remarkable characteristics, stood only around one meter (3.2 feet) tall. Skeleton LB1, rechristened "Hobbit," quickly gained in fame and attracted media attention the world over. The tremendous visibility of the discovery was quickly matched, however, by the controversy that swirled around it. Fifteen years after the fact, this controversy has yet to die down, an indicator of the scope of the scientific questions raised by the fossil.

How can we explain such upset? The 2004 article begins by describing the anatomical characteristics of skeleton LB1. Its small size, the source of its nickname, is a particularly remarkable feature of this fossil. Yet the individual was a full-grown adult, possessing its permanent teeth. The very well-preserved skull shows that its cranial capacity was around 380 cm^3, equivalent to that of a chimpanzee. To find this cranial capacity in our line, we must go back to the australopithecenes; that is, at least 2.5 million years (see Figure 1, "An overview of human evolution"). And yet, according to the dating, individual LB1 lived only around 50,000–100,000 years ago!

Besides its height and cranial capacity, the fossil's unsettling quality stems from its unique combination of human and other, more primitive characteristics; we'll come back to this. In view of so many unique characteristics, LB1's discoverers decided to designate it as a new species, *Homo floresiensis*. Their original theory regarding the evolutionary origin of this small human species was that Florès man was a descendant of *Homo erectus*, previously present on the island of Java, which evolved after the arrival of this species on the island of Florès through insular dwarfism. This common evolutionary phenomenon involves species of relatively large size that find themselves isolated on islands; their bodily size diminishes rapidly (on a biological scale) in response, particularly, to limited access to available resources, which become less and less abundant the smaller the island is, and to changes in predation stress if predators do not exist on the island, which is often the case. This phenomenon has been observed on Mediterranean islands containing dwarf deer and elephants—and in the Liang Bua cave, where, in addition to the Hobbit, archaeologists unearthed remains of *Stegodon*, a dwarf elephant that was probably hunted by *Homo floresiensis*.

THE HOBBIT:
A PATHOLOGICAL MODERN HUMAN?

A lively debate arose immediately, spurred mainly by *H. floresiensis*'s small cranial capacity. In 2003, an example of the *Homo* genus dating from around 50,000–100,000 years ago and with a cranial capacity of barely 400 cm^3 did not fit in at all with the contemporary perception of human evolution. Certainly, the definition of the genus *Homo* had evolved over the course of the history of paleoanthropology, but the possibility of a human being, at least in the zoological sense, with the cranial capacity of a chimpanzee had never been considered. In fact, the then-current definition of the genus *Homo* specified a minimum cranial capacity of 600 cm^3.

Researchers were quick to put forth an alternative hypothesis to the one proposed by LB1's discoverers: the Hobbit skeleton, they theorized, could be that of a *Homo sapiens* afflicted with a deformity, in this case microcephaly. An intense debate then began, in the form of dueling scientific articles, between partisans of the "pathological" hypothesis and those in the "insular dwarfism" camp. The former pointed out the various pathologies (cretinism, Laron syndrome, Down syndrome) that could cause microcephaly, with the latter systematically dismantling their opponents' arguments.

Was DNA analysis at the start of the 21st century advanced enough to settle the debate? After all, it had been done for Neanderthal man, whose discovery predated that of Florès man. Unfortunately, the hot and humid climate of Florès island is not favorable for the preservation of DNA. Despite the remarkable progress made in recent years, the DNA of Florès man is likely to remain inaccessible. In any case, most specialists today believe that the Hobbit is in fact representative of an extinct, short-statured species and not a deformed individual. If this is the case, the implications for our understanding of human evolution are myriad and profound.

HAVING A HOBBIT AMONG OUR CLOSE RELATIVES CHANGES (ALMOST) EVERYTHING

First of all, the definition of the genus *Homo* would have to be reexamined, and the "cerebral" criterion eliminated. The fossil register shows that the average brain size increases in our evolutionary line over time, most notably after the emergence of the genus *Homo*. Its oldest representatives, *Homo habilis*, have cranial capacities greater than 600 cm³, or the average of modern African great apes. Our honor is safe! That would no longer be the case with a hobbit definitively accepted as a representative of our genus. Brain size would no longer define the human species, and would no longer stand as an objective difference separating humanity from the animal kingdom.

Other criteria would have to be modified as well. For example, representatives of the genus *Homo* all share the same bodily structure; they are relatively tall, with long lower limbs. But the Hobbit is not only short; its bodily proportions are also unlike ours. In particular, it has lower limbs that are proportionally very short in relation to other parts of its body. It also has large feet in relation to its thighs and legs, which is not unreminiscent of Tolkien's hobbits.* This type of bodily proportions is a characteristic also found in ancient hominids (predating the genus *Homo*) as well as in modern great apes. As with cranial capacity, then, bodily proportions would no longer be a criterion differentiating humans in the zoological sense from the rest of the human evolutionary line.

A second important point concerning human evolution called into question by the discovery of *Homo floresiensis* is the link between brain volume and cognitive ability. Of course, we know that intelligence is not determined solely by brain size; indeed, this characteristic varies considerably within a single species. For example, in our own species, it can range from 1,000 to almost 2,000 cm³, and individuals with small brains are not necessarily less intelligent than those with large ones. However, we can say with certainty that humans are the most intelligent primates, and that their brain is, on average, significantly larger than those of other primates, whether this value is relative or absolute. The fossil register shows that brain size has been increasing for at least two million years and that, at the same time, the increasing complexity of dressed stone tools is

---

*   See the chapter "Why Do Hobbits Have Big Hairy Feet?," p. 209.

a manifestation of the increasing cognitive abilities of our ancestors. It all seems to follow a logical pattern . . . until we come to Florès man, who suggests that it is possible, with a brain the size of a chimpanzee's, to create relatively sophisticated dressed-stone tools, and to hunt *Stegodon*. These complex behaviors attest to cognitive abilities much greater than those of any non-human primate.

THE HOBBIT'S ANCESTORS

A third important point raised by our Hobbit has to do with the incompatibility of the fossil's specific anatomical characteristics with its dating. No truly satisfying evolutionary hypothesis has yet been developed to reconcile these two factors. If it had been found in Africa and dated to two or three million years ago, this fossil, while still just as remarkable, wouldn't have posed any problems—or, at least, it wouldn't have caused such complete upheaval. With its mixture of modern characteristics (meaning those shared with our species, or at least with other representatives of the genus *Homo*) and archaic ones (those shared with ante-*Homo* forms of our evolutionary line), it would represent a so-called transitional fossil, much like *Australopithecus sediba*, for example, discovered in South Africa some years ago. But how can we place the Hobbit fossil in a logical evolutionary framework, or (and this is really the same thing) modify the evolutionary framework we use, to make room for this fossil in it?

The first theory, suggested by the discoverers of the fossil, is that the Hobbit is a descendant of the Asiatic *Homo erectus* that evolved via insular dwarfism. If true, the dwarfism was extremely acute. But how does this theory explain the reappearance of a whole variety of primitive characteristics that had not been present for a long time in the Asiatic *Homo erectus* (the bodily proportions; a pelvis more like that of the australopithecines than representatives of the genus *Homo*, etc.)? It is an unlikely scenario, so much so that the discoverers themselves have, since then, rejected it in favor of a more ancient origin for the Hobbit. This would make *H. floresiensis* a direct descendant of species such as *Homo habilis*, or even the australopithecines. In this case, the primitive characteristics would have been inherited from these ancestors and, since they were smaller in stature, the phenomenon of insular dwarfism would have been less extreme.

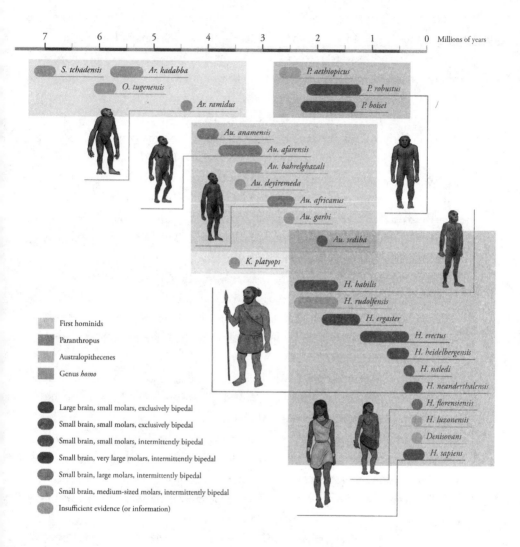

7   6   5   4   3   2   1   0   Millions of years

S. tchadensis   Ar. kadabba
O. tugenensis
Ar. ramidus

P. aethiopicus
P. robustus
P. boisei

Au. anamensis
Au. afarensis
Au. bahrelghazali
Au. deyiremeda
Au. africanus
Au. garhi
Au. sediba
K. platyops

H. habilis
H. rudolfensis
H. ergaster
H. erectus
H. heidelbergensis
H. naledi
H. neanderthalensis
H. florensiensis
H. luzonensis
Denisovans
H. sapiens

First hominids
Paranthropus
Australopithecenes
Genus *homo*

Large brain, small molars, exclusively bipedal
Small brain, small molars, exclusively bipedal
Small brain, small molars, intermittently bipedal
Small brain, very large molars, intermittently bipedal
Small brain, large molars, intermittently bipedal
Small brain, medium-sized molars, intermittently bipedal
Insufficient evidence (or information)

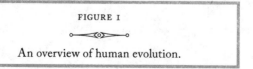

FIGURE I

An overview of human evolution.

It remains to be explained, then, how more highly evolved characteristics (reduced tooth size, reduced cranial bone thickness, greatly retracted and vertical face, etc.) appeared independently of what happened in the evolutionary line leading to our species. While the reemergence of numerous archaic characteristics seems unlikely, the parallel development of many "evolved" characteristics in the Hobbit's line at the same time as in our own seems equally improbable. In the end, none of these theories is wholly satisfactory in terms of parsimony—that is, involving the fewest possible evolutionary events. It is easy to understand why some paleoanthropologists favor a "pathology"-type hypothesis.

The fourth point called into question by the Hobbit of Florès island concerns the earliest human migrations out of Africa. For the Hobbits to be direct descendants of the Asiatic *Homo erectus* would mean that Florès was populated via the maritime route around a million years ago, which is already quite difficult to imagine (fossil records of coastal navigation go back only 100,000 years). However, at least this scenario doesn't question the fact that the first settlements outside Africa happened after the emergence of the genus *Homo*, characterized by a slightly larger brain, taller stature, and a bodily structure generally similar to our own. On the other hand, if we consider the theory of earlier origins, it seems even more difficult to imagine an island settlement predating the previous million years. Worse, this would mean that we would need to envision the first settlements outside Africa being made either by the very earliest representatives of the genus *Homo*, who had retained certain australopithecine characteristics, or by the australopithecines themselves. And while australopithecine fossils are abundant in Africa, none have ever been found in Asia, even though excavations on that continent have been going on for more than fifty years longer than those conducted in Africa.

Ultimately, there are still two main groups of theories in competition to explain the presence of the Hobbit on the island of Florès. Those who consider *Homo floresiensis* a pathological (or diseased) *Homo sapiens* have the major advantage of not threatening to topple the beautiful house of cards patiently constructed by paleoanthropologists over the past century and a half. However, they are at a loss to explain the variety of characteristics observed in the fossil remains of these beings as a consequence of one pathology or another. Those theories suggesting that Hobbits are a new species in the human evolutionary tree, while they may seem more convincing, lead to profound uncertainty with regard to our understanding of the evolution of man.

In Tolkien's work, hobbits are probably the only creatures wholly imagined by the author, without relying on existing mythology, but they are unquestionably the most ordinary as well. By making them tranquil, home-loving, unremarkable beings, he made them very similar to us, despite an unusual physical characteristic or two. Because their very ordinary ordinariness makes them so easy to identify with, they are the perfect vector by which the reader can enter Tolkien's marvelous universe. Though the Hobbits of paleoanthropologists share certain anatomical traits with the fictional beings to whom they owe their nickname, they are, unlike those fictional hobbits, extraordinary creatures in the strictest sense of the word, no matter which theory you subscribe to.

Because of them, or rather, thanks to them, we will now have to rewrite part of our evolutionary history. After years of study and analysis to come, we will have a better grasp of who they were and how they fit into our past, which we can only understand more deeply tomorrow. The history of man's evolution is a saga every bit as amazing as the one invented by Tolkien, and rich with just as many beings. And among all these beings, in this marvelous history as in the fictional one, it may well be that, here too, hobbits have an important role to play.

GOLLUM:
THE METAMORPHOSIS OF A HOBBIT

JEAN-SÉBASTIEN STEYER, paleontologist,
CNRS and Museum of Paris

At the bottom of the tunnel lay a cold lake far from the light, and on an island of rock in the water lived Gollum. He was a loathsome little creature: he paddled a small boat with his large flat feet, peering with pale luminous eyes and catching blind fish with his long fingers, and eating them raw.

—The Fellowship of the Ring

his is the story of a hobbit gone bad. Sméagol was his name once, a name its bearer has taken pains to bury deep down inside himself. Born into a good family, the young Sméagol was the most inquisitive of Stoors, a breed of hobbits who dwelled mostly in the Anduin valley. He was clever and lively, an amateur naturalist and explorer curious about everything, who loved to dig and poke around in nature: "He was interested in roots and beginnings; he dived into deep pools; he burrowed under trees and growing plants; he tunnelled into green mounds [. . .]" (*The Fellowship of the Ring*).

Alas, Sméagol goes over to the dark side one cursed day in the year 2463 of the Third Age when, during a fishing expedition on the river Anduin, his cousin Déagol discovers the One Ring by chance. Sméagol wants it—he asks to be given it for his birthday—but his cousin, who has already given him a gift, refuses. So, gripped by a sudden, murderous rage, consumed with hate and the desire for possession, Sméagol strangles Déagol, and the fishing trip becomes a crime.

A long descent to hell follows. The thief and murderer is banished by his people and wanders at night, avoiding the sun and taking refuge deep in the Misty Mountains, in a squalid cave that he turns into his domain. Here, the hobbit transforms into a monster, and Sméagol into Gollum.

Of all the characters invented by Tolkien, Gollum is indisputably one of the most complex, and the most important; no one lacks an opinion about him. A minor character in *The Hobbit* (1937), Gollum becomes one of the main protagonists of *The Lord of the Rings*, published seventeen years later in 1954–55. Gollum merely provides an opportunity for a game of riddles at first, a way for the author to put his favorite hero to the test, lost in the goblin-infested mountains. Later his status evolves, and he goes from shadow-creature dogging the footsteps of the Fellowship to guide and servant, ending as the key participant in the conclusion of the quest: at the denouement of *The Lord of the Rings*, Tolkien elevates his *monster* to the position of *savior* of Middle-earth, for it is Gollum who ends by destroying the One Ring, and himself in the process.

A shadow amid shadows, Gollum wavers throughout the story between good and evil. Sometimes corrupt and treacherous, sometimes loyal and faithful, the psychology of this tortured character impels us to question the relationship between our thoughts and our actions, as well as our own relationship with the power symbolized by the Ring.* While Gollum's psychology has been discussed at some length, few biologists have examined his anatomy and morphology—yet, from the kindly hobbit he once was to the vile, slimy creature he becomes, Gollum is, from a naturalist perspective, a completely singular case!

A MUTANT IN SPITE OF HIMSELF

The transformation of Sméagol into Gollum could be considered a case of accelerated polymorphic mutation. This drastic and rapid change of form recalls the famous metamorphoses of insects and amphibians. Gollum's agility is also often compared to that of animals: "With a jump like a grasshopper or a frog, Gollum bounded forward into the darkness" (*The Two Towers*). These

---

*   See the chapter "Tales of a Young Doctor . . . in Middle-earth," p. 201.

"My precious."

species undergo metamorphosis naturally during their development, passing from the larval stage (tadpole, grub, or caterpillar) to the adult stage (frog or winged insect). These very real metamorphoses have undoubtedly inspired a host of imaginary ones: the frog that turns into Prince Charming, the man who transforms into a werewolf, etc. These radical changes in form and psychology add to the fantastical qualities of fictional stories. And Gollum is not the only character to undergo metamorphosis in Tolkien; Beorn, for example, is a powerful Skin-changer that turns into a bear, the last member of his species, who helps Thorn and his companions to find their treasure in *The Hobbit*. If Tolkien was inspired by the character of Grendel* in creating Gollum, his transformation itself is more reminiscent of Dr. Jekyll, invented by Robert Louis Stevenson in his 1886 novel *The Strange Case of Dr. Jekyll and Mr. Hyde*, affecting both the psychology and morphology of this pathetic character.

However, Gollum's metamorphosis has nothing to do with either his growth or his personal development; his is not a natural transformation, but one caused by the One Ring, the embodiment of Evil in Tolkien's work. And, as for all representations of Evil, the author has taken care to make it complex; in addition to altering one's morphology and corroding the spirit, the ring also confers some "benefits": invisibility, greatly increased strength, resistance to pain, longevity . . .

Speaking of longevity, note that Gollum, already greatly compromised by the Ring, certainly dies prematurely, but at the age of 579 years, while Bilbo, born more than a thousand years earlier, hardly looks different at all. Does the ring affect Sméagol more strongly because he wears it more often? Unfortunately, Tolkien doesn't give us much information about how much time the Ring spends on its various wearers' fingers. However, he does say:

> Gollum used to wear it at first, till it tired him; and then he kept it in a pouch next his skin, till it galled him; and now usually he hid it in a hole in the rock on his island [. . .]. And still sometimes he put it on, when he could not bear to be parted from it any longer [. . .]
>
> (The Hobbit)

Bilbo and Frodo, on the other hand, wear the ring only very rarely. It seems, then, that the ring's effect is dependent on its proximity to, or even physical

---

* The humanoid monster featured in the Anglo-Saxon epic poem *Beowulf*.

contact with, its wearer, as if it were a mutagenic agent. The metamorphosis it triggers, which is irreversible and affects both the possessor's physiological and psychological makeup, is therefore a mutation.

This idea of Sméagol's mutation—or transmutation—into Gollum suggests that Tolkien was an evolutionist, a possibility reinforced by the remarks that occasionally appear in his writings; for example: "There are strange things living in the pools and lakes in the hearts of mountains: fish whose fathers swam in, goodness only knows how many years ago, and never swam out again, while their eyes grew bigger and bigger and bigger from trying to see in the blackness" (*The Hobbit*). If we take into account the markedly gradualist (or adaptationist) nature of these remarks (corresponding to the context of Tolkien's time), we can see that Sméagol does not evolve into Gollum, but rather *regresses*, as suggested by his mode of locomotion, which starts out bipedal (like any self-respecting hobbit) and ends up four-legged, and even as a crawl: "Gollum moved quickly, with his head and neck thrust forward, often using his hands as well as his feet." Gollum has become, then, a degenerate hobbit, half-elderly half-infantile, and readapted to aquatic environments. His "large, flat feet," so characteristic of hobbits, now act as paddles for swimming, a rare thing among the Little Folk, who are generally unable to swim. Poor Gollum alone, then, is symbolic of nearly 380 million years of evolutionary regression, going back to the very first tetrapods, the crawling aquatic vertebrates of the ancient past.

A PROTEAN MONSTER

Let's now attempt to compare Gollum to some species that currently exist, or existed in the past. Apart from his terrible metamorphosis, which is nothing but an evolutionary regression, Gollum is often compared to an amphibian: "[A]nd then with marvellous agility a froglike figure climbed out of the water and up the bank" (*The Two Towers*). In the Rankin-Bass animated films of 1977 and 1980, Gollum looks like nothing so much as a humanoid frog. Still, it is difficult to slot the beast Sméagol becomes into any one existing animal category, all the more so because Tolkien uses different comparisons and anatomical descriptions throughout the books. Aragorn describes him as a creature "slyer than a fox, and as slippery as a fish," while Faramir compares him to a kingfisher and a tailless squirrel, which is more in keeping with Legolas's remark that "he had learned the trick of clinging to boughs with his feet as well as with his hands." Sam Gamgee, who hates Gollum, compares him to "a nasty crawling spider on a wall," and, finally, Tolkien himself speaks of "some large prowling thing of insect-kind."

So, it is no easy matter to label the monster based on these descriptions! Yet, his anatomical characteristics speak for themselves: his "thin limbs," his "big round pale eyes," his mainly piscivorous diet, his large head, his "eyes like telescopes," and his flat hands with long fingers give this excellent swimmer an undeniably amphibious quality, one further enhanced by the wet and slimy appearance frequently mentioned by Tolkien, which suggests water-permeable, permanently moist skin. Is Gollum, like a frog, capable of breathing through his skin? Perhaps, but when he is captured, Aragorn notes that he "stank" and was "covered with green slime," which would hinder—at least temporarily—exchanges of gas between the skin and the external environment. Does Gollum perhaps have mucous glands in his epidermis? Though his skin does not seem to be dotted with bumps or warts as is common in toads and frogs, the presence of such glands still emphasizes the idea of genetic anomalies in the skin, accentuating the mutagenic power of the Ring, which acts directly on its wearer's genome. It is interesting to note that the filmmaker Peter Jackson, who adapted part of Tolkien's oeuvre for the screen, gives Gollum extremely broad fingertips, which also denotes very round, flat terminal phalanges, similar to the webbed digits of some frogs! Is Gollum, then, simply an ordinary hobbit who has undergone a sort of amphibian mutation like the evil mutant Toad in *X-Men*? Or is he a victim, because of his "Precious," of digital hippocratism (or clubbing), a rare disease that causes hypertrophy of the terminal phalanges accompanied by a deformation of the nails and sometimes even cyanosis (a blue coloration of the skin)? It seems clear that Gollum's is a pathological case.

With regard to his color, Tolkien often describes a "black" or "shadowy" shape, but he never really specifies whether this is due to Gollum's clothing, his skin, or his temperament. Cave-dwelling animals can sometimes develop extremely pale skin due to the absence of surface pigmentation, as is the case with the proteus (Proteus anguinus), a cave salamander, and with the small crustacean *Niphargus*, which lives in European cave networks and cannot survive exposure to light (20,000 lux can kill it in a few days, and bright sun can generate up to 100,000 lux). Is this why Gollum prefers to travel at night? Are his huge, protuberant eyes sensitive to light? Whatever the case, after centuries spent in the shadowy dimness of the Misty Mountains, the monster's other senses—hearing and smell—have become highly developed. This hyperacuity, combined with the malevolent power of the Ring ("He became sharp-eyed and keen-eared for all that was hurtful") and his razor-sharp teeth make Gollum, all in all, a dangerous predator.

SUPERHEROES OF FANTASY

The former hobbit, then, has become far-from-desirable company. In addition to his fragmented identity (in both psychological and phylogenetic terms), he displays what specialists call a mosaic of traits: in addition to his protean characteristics, which can be classified as neotenic (that is, retaining juvenile traits), he also possesses hyperadult characteristics such as a thin and haggard face, a hunched, prominent spine, a "long, scrawny neck," "yellow teeth" reduced in number to six (while healthy hobbits possess thirty teeth, rather than thirty-two, like humans*), and "thin lank hair" (this last trait brings up some interesting questions about the side-effects of the Ring in terms of the secretion of male hormones responsible for hair loss). All of these characteristics point to Gollum being prematurely aged or even an ultra-resistant zombie: tireless, unbreakable, and resilient, he is able to travel kilometers without eating, to stand up to torture in Mordor, to survive extreme cold, and to swim in frigid underground water without risking hypothermia.

In addition to this "enhanced" physiology, Gollum's "Precious" also grants him exaggerated strength; his grip is "soft but horribly strong" according to Sam, who engages in a physical struggle with him. In this Gollum is reminiscent of the Nyctalope, the first literary superhero, who first appeared in 1911 (well before the American comic-strip heroes of Marvel Comics, for example) in Jean de la Hire's novel *Le Mystère des XV* (later translated into English as *The Nyctalope on Mars*). Can Gollum be considered a fantasy superhero? Don't forget that this character does indeed deliver justice despite himself, for he destroys the One Ring, while Frodo, even at the end of his quest, is unable to resist temptation. Hideous hybrid though he might be, Gollum will always be the avenger of Tolkien's world. Gandalf was right: "My heart tells me that he has some part to play yet, for good or ill, before the end." A requiem for Gollum!

---

*   According to Bilbo's riddle "Thirty white horses on a red hill,/First they champ,/Then they stamp,/Then they stand still."

THE EYESIGHT
OF ELVES

ROLAND LEHOUCQ, astrophysicist, CEA
JEAN-SÉBASTIEN STEYER, paleontologist,
CNRS and Museum of Paris

Gandalf walked in front, and with him went Aragorn, who knew this land even in the dark. The others were in file behind, and Legolas whose eyes were keen was the rearguard.

The Fellowship of the Ring

Which of our five senses is the most useful, especially in inhospitable lands filled with beings who are not necessarily well-intentioned, like the ones navigated by Tolkien's heroes? Sight, undoubtedly, for it is the quickest provider of information about one's surroundings. Vision relies on the eye, an organ that adapts rapidly to very different situations; it can accommodate—that is, focus on—an object a mere few dozen centimeters from itself, and then adjust in an instant to focus on another object a great distance away. It adapts to significant variations in light intensity and recognizes a range of wavelengths, known as the visible spectrum, from violet to red, from 400 to 800 nanometers (billionths of a meter) in length. Before discussing the eyesight of elves, let's go over the basics of vision.

HOW DOES VISION WORK?

The human eye is an approximately spherical organ around 25 millimeters in diameter, enclosed by a tough, fibrous, white membrane called the sclera. The sclera is transparent in the front of the eye, forming the cornea. Light penetrating the eye passes through the cornea and then the vitreous humor, a liquid containing a circular muscle, the iris, pierced by a hole, the pupil. The iris acts like an aperture, regulating the diameter of the pupil in order to control the amount of light that enters the eye. In bright light, the pupil's diameter can shrink down to two millimeters, while in darkness it can expand to eight millimeters. Behind the iris, the lens behaves like the converging lens of a camera. In a normal eye, and when the lens is at rest, images of distant objects form on the retina; in this case, the eye does not "accommodate," to use the clinical term. On the other hand, when the eye looks at a nearby object, its lens contracts, which reduces the focal distance and allows images to form on the retina (which itself cannot move); this means that the eye is now accommodating. By contracting, the lens can adjust to a focal length ranging from 15 to 24 millimeters.

The retina is formed of two types of photoreceptor cells called cones and rods, which have complementary functions. Rods are extremely sensitive to light, but cannot distinguish colors, while cones react only to light beyond a certain brightness, but are capable of distinguishing colors. There are three types of cones, each sensitive to a specific range of wavelength: red, green, or blue. Cones are inactive in low light, and this is when the rods take over, at the cost of losing color vision—all cats really *are* gray at night.

A small area of the retina, the fovea, located on the pupillary axis, contains only cones and plays an important role in vision; this is the part of the retina with the highest resolution and color perception. The vital overall function of the retina is to enable the brain to pinpoint potentially interesting areas in order to orient the fovea in the direction of these areas for optimal processing. This is the reason our eyes are in permanent rapid motion, as we continually re-center the image of the important area on the fovea, which is what is happening right now, as you're reading these lines!

The last stop on our tour of the ocular anatomy is the choroid, an absorbent membrane that covers the eyeball behind the retina to keep light from getting out. The obvious complexity of the human eye makes it seem to be a "perfect" organ, and has long been cited by opponents of evolutionary theory, who insist:

how could the intermediary, and thus imperfect, stages of the eye's evolution have served any purpose whatsoever? And yet natural selection has indeed played a role; our eyes are not perfect—and other types of eyes have appeared throughout the course of the evolution of species.

HIGH-RESOLUTION VISION

Though the eyes and eyesight of elves operate according to the same principles as those of humans, several passages in *The Lord of the Rings* show that their capabilities are far superior. For example, in *The Two Towers*, when Legolas, Gimli, and Aragorn are in Rohan, on the trail of the orcs who have taken Merry and Pippin, the following exchange occurs:

> *"Riders!" cried Aragorn, springing to his feet. "Many riders on swift steeds are coming towards us!"*
>
> *"Yes," said Legolas, "there are one hundred and five. Yellow is their hair, and bright are their spears. Their leader is very tall."*
>
> *Aragorn smiled. "Keen are the eyes of the Elves," he said.*
>
> *"Nay! The riders are little more than five leagues distant," said Legolas.*

Five leagues? Remember that the league was originally conceived as the distance a man could travel on foot in an hour. This depends on the terrain, of course, but we can estimate five leagues as being around four kilometers, which means that the riders are some twenty kilometers (12.4 miles) away. Can this be?

We would point out, first of all, that an observer's sightline cannot stretch beyond the horizon. In an unobstructed area like the sea, for example, the distance between an adult and his horizon is just under five kilometers (roughly three miles). To bypass this consequence of the earth's roundness, all you have to do is raise your elevation; this is what lookouts do when they climb to the top of a ship's mast, and it can extend their visual radius to the horizon beyond twenty kilometers. To see the group of riders five leagues away, Legolas must be at least thirty meters above the average level of the plain. Since the group has just spent the night at the top of a hill that is undoubtedly much higher than that, the Elf's vision can extend over a good sixty kilometers.

Next, the light rays emitted by the source must propagate in the atmosphere, which absorbs part of this light. The rays are also subject to all sorts of turbulence and disruption, due to differences in temperature or atmospheric movement, for example. This will make the source appear faint and deformed; so much so, on occasion, as to become imperceptible. This phenomenon is more pronounced underwater, where nothing more than a few meters away can be seen; outlines are blurred, and the light is weak. A similar phenomenon occurs in the atmosphere, with the slight difference that, because air is a thousand times less dense than water, interference thickness can be measured in kilometers. In order to see a long way, the atmosphere must also be perfectly clear.

Finally, the eye must be capable of picking out a detail seen from a very small apparent angle, which depends on the ratio between the real size and the distance of the object. If the eye does not have good angular resolution, it will be impossible to distinguish the finest details. The use of a spyglass or binoculars helps to increase the angle from which the source is seen. The quality of angular resolution is limited by two factors: diffraction of light, and the size of the cells of the retina (see Figure 1). Diffraction is an angular "dispersion" of light that occurs when this light encounters an obstacle whose size is comparable to that of the light's wavelength.[*] Penetrating the eye through the pupil (which has an average diameter of 3 mm), light undergoes diffraction at an angle of around 13 thousandths of a degree. The image of a perfectly punctiform (dot-shaped) light source will appear on the retina, located 1.6 cm from the iris, as a spot a few micrometers in diameter, called an Airy disk. The cells of the fovea are separated by a distance of around two micrometers, smaller than the size of the image-disk. For two punctiform and angularly close light sources to be seen separately, the centers of their Airy disks must be separated by a distance greater than the distance between two neighboring cells. Otherwise, their image-disks will be superimposed on one another and the eye will perceive a single source, as shown in Figure 1.

In order for Legolas's eye to capture such fine details, it is necessary both for his pupil to be larger than that of humans (to limit diffraction) and for his retinal cells to be smaller (to separate two Airy disks). The disadvantage of a pupil with a large diameter is that it causes glare, which is unfavorable to night

---

[*]   More precisely, the angle of diffraction is proportional to the quotient of the wavelength of the incident light and the dimension of the object. In order for diffraction not to hinder the observation of an object, light with a wavelength smaller than the size of the object must be used.

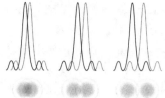

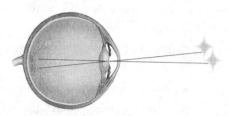

Appearance of Airy disk resulting from the diffraction of light from a punctiform source via a circular hole

Not separate

Slightly separate

Separate

For two sources to be seen separately, their Airy disks must not overlap.

For the eye to separate two punctiform sources, their image-disks must be projected onto two different retinal cells so that they are seen separately.

FIGURE 1

Diffraction of light and its effect on visual acuity.

vision. This suggests that the Elf has a way to avoid this inconvenience; this might be, for example, a more pronounced brow arch or supraorbital ridge, which act like a visor for nocturnal birds.

To estimate the density of Elves' retinal cells, let's take a cue from birds. For them, the eye is an organ even more important than it is for humans. In fact, birds' eyes are proportionally much larger than human eyes. For example, the eyes of a starling represent 15% of the mass of its head, while in humans this percentage doesn't even reach 2%. In addition, the fovea of a bird's retina contains significantly more cones than that of a human's; for example, a sparrow's fovea contains 400,000 cones per square millimeter, twice the number in our own eyes! And the falcon, which hunts its prey from the sky, has a density of cones in its fovea five times greater than ours. More generally, raptors have the sharpest vision among birds. They are able to distinguish details at distances two to three times farther than what is possible for us; a common kestrel, which hunts at low altitude and eats mostly insects, can spot a 2-millimeter long object from a height of 18 meters, and the eagle, hunting small mammals at high altitude, can spot a 15-cm long object from 1,500 meters in the air. The eyes of raptors have an angular resolution three to five times better than our own, and an extraordinary ability to estimate distances.

A remark made by Gandalf as he travels with King Théoden's army between Edoras and Helm's Deep, seems to suggest that Legolas's vision is even sharper than that of a raptor: "'You have the keen eyes of your fair kindred, Legolas,' he said; 'and they can tell a sparrow from a finch a league off. Tell me, can you see anything away yonder towards Isengard?'" For Legolas to possess angular resolution seemingly one hundred times sharper than that of a human (if you can tell a sparrow from a finch from forty meters away!), his eyes would have to be *truly* different from those of humans, both in terms of pupil size—one hundred times larger—and the density of retinal cells in the fovea—ten thousand times greater! This seems incompatible with descriptions of the Elf given elsewhere: a tall, slim humanoid, but one lacking an outsized head or eyes! He undoubtedly has extraordinary visual capabilities, but not as great as some might claim . . .

A final point: in humans, the parts of the brain dedicated to processing visual information greatly prioritize the data provided by the fovea. Equipping the optical system with very high-resolution sensors would be too costly in terms of the neural resources required to process the information. In Legolas's case, it's almost as if his brain is devoted almost wholly to managing visual information! Leave a little space free for everything else . . .

NIGHT VISION

In addition to their extraordinary powers of resolution, Elves' eyes are perfectly adapted to seeing at night, as Uglúk, the leader of the orc-company that captures Merry and Pippin, tells his troops: "There's only one thing those maggots can do: they can see like gimlets in the dark. But these Whiteskins have better night-eyes than most Men, from all I've heard [. . .]."

To see only faintly luminous objects, the eyes must receive as much of their light as possible. The obvious solution is to increase the size of the area collecting this light—that is, to have a pupil with a large diameter. This is how our eyes adapt to seeing in low light, called scotopic vision: by dilating the pupils. The other solution is to heighten the sensitivity of the detector; in other words, to make it capable of reacting to a smaller number of incoming photons. The rods in our retinas are so sensitive that they react to the impact of a single photon; hard to do much better than that. Their density is greatest on the edges of the retina, and decreases inward until there are no rods in the fovea itself. This is why, to see a faint star at night, it is better to use your peripheral vision, so that the starlight is perceived by the rod-rich outer edges of the eye.

Increasing the eye's overall scotopic sensitivity comes down, then, to increasing the number of rods. Since the size of the eye is fixed, this could only be done to the detriment of the number of cones, and thus our color vision. In other words, a very sensitive eye does not see colors as well. To see well at night, a person would have to have eyes with large pupils and retinas full of rods, like those of owls. Nocturnal animals share a very interesting ability in this area: they possess a special membrane called the *tapetum lucidum*, or choroidal tapestry. This is a layer of reflective cells behind the retina which act like mirrors, reflecting up to 90% of the light that touches them back to the retinal cells. This efficient recycling of light significantly increases the eye's sensitivity in low lighting and, in near-darkness, the *tapetum lucidum* gives these animals' eyes the same "phosphorescent" glow as those of cats. A night-sighted Elf, already equipped with a large pupil for high angular resolution, must also have a large quantity of rods, or even a *tapetum lucidum*, to recycle incoming light. If light hits this Elf's eyes at night, the reflection of its *tapetum lucidum* should be clearly visible. This is also the case with the cave-dwelling Gollum,[*] of whom it is said that he peers through the darkness "with his eyes palely shining."

Central characters in Tolkien's work, the Elves are fascinating beings who, in addition to sharp vision, have highly sensitive hearing, as we learn when the companions are searching for the West-gate of Moria in *The Fellowship of the Ring*: "Legolas was pressed against the rock, as if listening." This sensory acuity is explained by their large pointy ears, but there is no corresponding morphological clue to their super-vision. In any case, the ability to see a long way into the distance is an important one, especially when one is immortal . . .[†]

---

[*]   See the chapter "Gollum: The Metamorphosis of a Hobbit," p. 226.

[†]   See the chapter "Tales of a Young Doctor . . . in Middle-earth," p. 201.

ARE DWARVES HYENAS?

SIDNEY DELGADO, biologist, Sorbonne University
VIRGINIE DELGADO BRÉÜS, doctor in French linguistics

f all the creatures that people Tolkien's legendarium, which are distinguished by their hardy, brave and courageous nature, shaded by a certain taste for gold and a hair-trigger temper? Dwarves! These beings, with physiques much like those of men, are proud, deliberate, unyielding, and obstinate, resistant to domination in any form, quick to make both friends and adversaries.

Tolkien's oeuvre is characterized throughout by a fairly precise description of the "race" of Dwarves. In this chapter we will analyze, from a biological perspective, the evolution of this people who lives hidden beneath the earth—for, while Tolkien developed a highly complex mythology to explain the origins of his world and its inhabitants, remember that this world is also our own, in a fictional past, with the end of the Third Age occurring around 6,000 years before our time. The inhabitants of Middle-earth, then, are subjected to the same laws of evolution as the species at the origin of modern biodiversity were.

Dwarves are described as short-statured humanoids that live in caves. They work in mines and underground cities and are accustomed to dwelling in darkness and tunnels for long periods. They stand between 1.2 and 1.5 meters (3.9–4.9 feet) in height and are sturdy, tough, and extremely strong for their size. They can carry extremely heavy loads and are hardworking, and have a high tolerance for pain, long voyages, and even hunger—which does not prevent them from having very hearty appetites.

Dwarves working the mines of Erebor.

Dwarves have an average life expectancy of 250 years, but some live as long as 400 years. The education of Dwarves continues until they reach the age of 30, and another ten years are required to toughen them up. They do not marry until at least the age of 90, and have their first child at around 100.

BEARDED LADIES?

This is undoubtedly the Dwarves strangest characteristic: their women are rarely seen, to the point that they have become mythical to the other inhabitants of Middle-earth. They are almost never mentioned in the books, which casts some doubt on their very existence. However, there is a precise description given in the appendices to *The Return of the King*:

> *They seldom walk abroad except at great need, They are in voice and appearance, and in garb if they must go on a journey, so like to the dwarf-men that the eyes and ears of other peoples cannot tell them apart. This has given rise to the foolish opinion among Men that there are no dwarf-women, and that the Dwarves 'grow out of stone'.*

From this we can deduce that female dwarves are, in fact, bearded ladies! And this is no haphazard supposition, for their hairiness is confirmed in *The History of Middle-earth*:

> *For the Naugrim have beards from the beginning of their lives, male and female alike; nor indeed can their womenkind be discerned by those of other race, be it in feature or in gait or in voice, nor in any wise save this: that they go not to war, and seldom save at direst need issue from their deep bowers and halls.*

Dwarves, then, are an unusual species, whose creation Tolkien recounts in *The Silmarillion*:

> *It is told that in their beginning the Dwarves were made by Aulë in the darkness of Middle-earth; for so greatly did Aulë desire the coming of the Children, to have learners to whom he could teach his lore and his crafts, that he was unwilling to await the fulfilment of the designs of*

Ilúvatar. And Aulë made the Dwarves even as they still are, because the forms of the Children who were to come were unclear to his mind, and because the power of Melkor was yet over the Earth; and he wished therefore that they should be strong and unyielding. But fearing that the other Valar might blame his work, he wrought in secret: and he made first the Seven Fathers of the Dwarves in a hall under the mountains in Middle-earth.

But we will not content ourselves merely with this explanation, and a scientific analysis may provide clues to the mystery of the Dwarves.

WHEN THE DWARF APPEARS . . .

Remember that a new species emerges when a group finds itself isolated from its ancestral population. After this isolation, species originally capable of reproducing continue to evolve and adapt locally, and if the new environment is very different, their morphologies diverge. We will suppose that the ancestors of the Dwarves were a group of Men who found themselves isolated at some point in their history, and adapted to a new environment, to the point that a reproductive barrier arose between the two species.

According to the American biologist Rudolf Raff, "You can't understand evolution without understanding the evolution of embryonic development." One of the Dwarves' most distinctive characteristics is their height. If we start from the hypothesis that the Dwarves' ancestors had a similar height to Men, we must then look for the evolutionary mechanisms behind this reduction in size—and they can be found in the work of the American paleontologist Stephen J. Gould. In *Ontogeny and Phylogeny*, which he considered one of his most impactful books, he develops the idea that evolution (phylogenesis) should be imagined as a process of modification of embryonic developments (ontogenesis) rather than a simple accumulation of differences between adults. One chapter is devoted to a particular mechanism of embryological change: heterochrony of development; that is, changes in the duration and speed of the embryonic development of an organism in the course of evolution. These ontogenetic shifts explain the differences between an ancestor and its descendant, and the changes caused by heterochrony of development include changes in size.

A CAVE IS AN ISLAND

The type of heterochrony that interests us here is dwarfism, a phenomenon that occurs relatively frequently in the evolution of living beings, particularly in island environments; this is the case, for example, with the dwarf elephants, measuring 90 cm (35 inches) at the withers, that populated the islands of the Mediterranean Sea during the Pleistocene epoch!* It was probably the discovery of their skulls that gave rise to the legend of the Cyclops in ancient Greece, due to the highly noticeable presence of the nasal cavity in the center of the skull. Surprisingly, while large animals grow smaller in island environments, smaller animals such as rodents tend to increase in size in the absence of predation-related stresses.

Island ecosystems, with their limited resources, are a key factor in the development of dwarfism. But we must consider these ecosystems in the broader sense of their definition, as other types of places can impose similar limitations. For example, the caves and tunnels in which Dwarves work are environments with scarce nutritional resources, in which smaller individuals can move around more easily. In the early days of their evolution, the Dwarves were probably forced to abandon another way of life to seek treasures beneath the earth's surface. Food must have been in short supply at first and, before becoming rich thanks to their mining operations, they must have gone through a difficult time during which individuals capable of fasting for relatively long periods survived more frequently. The hardiest, smallest, strongest Dwarves were the ones to survive this period of famine, and transmitted their characteristics to their descendants. The Dwarves would also have developed their large appetites at this time, as a way of storing reserves of fat for their bodies to use in harder times.

WORMS AND DWARVES

As we have said, Dwarves can live for up to 400 years, a record among vertebrates, who never possess such longevity (the longest known lifespan is that of the Seychelles giant tortoise, which can live for up to 250 years). The question of lifespan is an important one in biology. Since the 1990s, researchers have

---

*   See the chapter "When a Hobbit Upsets Paleoanthropologists," p. 218.

been studying aging in a small transparent worm, *Caenorhabditis elegans*, the flagship model of modern biology since the 1970s. Mutant examples of this worm with an extended lifespan have been found, confirming an old suspicion: some "builder" genes vital to our biological construction and function become, with age, "destroyer" genes that cause death, with the effects of the latter delayed by "protector" genes, as described by the biologist Jean Claude Amesein in his book *Dans la lumière et les ombres* (*In Light and Shadow*). All of these genes are linked to crucial processes, such as the storage and use of energy, DNA repair, embryonic development, and fertility.

The oppositional relationship between lifespan and fertility had already been demonstrated through an experiment. The lifespan of a *Caenorhabditis elegans* worm with its germ cells removed (spermatozoids and oocytes, originally) increased by two times, and even six times for some individuals! A point of interest is that some of these long-lived mutants became less fertile, which is also the case with Tolkien's Dwarves, though he claims that this is due to the small number of women. Other mutant worms with extended lifespans are more resistant to environmental stresses; their bodies are more robust, as is also true of the Dwarves. Is it possible, then, that the Dwarves of Middle-earth are the product of mutations that have given them stronger constitutions, prolonged their lives, and reduced their fertility?

WOMEN ARE HYENAS FOR DWARVES

In addition to size, a typical Dwarven characteristic is highly developed facial hair. Beards are an integral part of their culture, and they take great care of them. How can we explain this unusual hairiness, even among female Dwarves? Are there equivalent examples in the real-life animal kingdom?

More than 2,000 years ago, in his *History of Animals*, Aristotle examined the anatomy of *Crocuta crocuta*, the spotted hyena, an animal then widely believed to be hermaphroditic. The species was indeed divided into males and females, he revealed, but the genital organs of females were astonishingly similar to those of males, which explained the confusion. Since then, numerous biologists have tried to explain this mimesis between the sexes, unique among mammals. The answer lies in a closer look at their social behavior.

Females are dominant in spotted hyena societies, a unique case. They are heavier and more aggressive than the males, and eat first when a carcass is available. In addition, they respect a complex social hierarchy with well-defined ranks that may be passed down from generation to generation. Natural selection has favored the most aggressive females, capable of defending their status, their young, and their food.

The aggressiveness of female hyenas has probably been increased by another feature of their way of life. To protect her litter, a mother hyena isolates herself and hides her babies in a den. During nursing, she places herself in the narrow entrance to the den and the cubs fight, sometimes to the death, to access her teats. In these conditions, the most combative cubs (male or female) have better odds of surviving, as the researchers Wolfgang Goymann, M.L. East, and Heribert Hofer have noted.

Other studies conducted by Stephanie M. Dloniak's team have shown that, during their embryonic development, hyena fetuses float in amniotic fluid rich in androgenic—that is, masculine—hormones. The more elevated the levels of these male hormones are, and the more aggressive the hyenas, the higher the hierarchical status of these hyenas becomes.

It was long thought that the pseudo-penis (clitoris) of female hyenas was a direct result of this embryonic masculinization. However, this trait is merely the secondary consequence of selection pressure favoring aggressive individuals. And the story gets even more complicated: the formation of the female pseudo-penis is no longer controlled by male hormones, even if this was the case at one time, according to the researcher Dr. Gerald R. Cunha.

How can we explain the existence of this organ, which represents such an enormous burden of evolutionary stress? At birth, cubs must exit the mother's body via a relatively narrow passage, and are commonly deprived of oxygen; in fact, 60% of hyena cubs die at birth, and 8% of females die giving birth to their first litter. One reason for this has to do with the fact that female cubs are subjected to violent aggression until the age of three months, particularly on the part of other females also seeking to assert their domination. Difficulty in distinguishing females from males reduces the risk of such aggressions, as shown by the primatologists Richard Wrangham and Martin N. Muller. And it is indeed true, at least for human observers, that the sex of a hyena cub cannot be determined until after the age of three months.

Another theory suggests that these increased levels of male hormones are the reason for deformations of the genital organs in both females and males, the latter of whom have penises of increased length. It may be that these changes have triggered a mutual evolutionary path, so that the genital organs of males and females remain compatible!

Let's return to the Dwarves. We won't speculate on the form of their genital organs here, especially because Tolkien remains silent on the subject. But let's imagine that they underwent selection pressures identical to those of the hyenas; it would explain many of their characteristics. Did the necessity of surviving in a hostile environment, deprived of food, favor the most aggressive and physically robust Dwarves? If baby Dwarves subjected to higher levels of male hormones than others have a greater chance of surviving, then it makes sense that natural selection would have led to the appearance of bearded female dwarves. In humans, excessive testosterone levels in women can also cause facial hair to develop, a phenomenon called hirsutism, which turned many women into sideshow attractions until the 20th century.

As we have seen, there are numerous parallels between Dwarves and hyenas. We might wonder, too, if female Dwarves are dominant, like female hyenas. Wouldn't that be the ultimate secret of the Dwarves? Does power in fact lie in the hands of their women?

WOMEN: A RARITY

Let's conclude by examining one last Dwarven oddity: the gender ratio, or the number of males in relation to the number of females, which is skewed in favor of men. This imbalance may be caused by the environment, which sometimes favors one gender less than the other (as in the case of war, which kills mostly men), or by a genetic problem. In nature, we almost always find a one-to-one ratio. However, contrary to what we might think, this is not the optimal ratio— that is, the one that enables a population to grow as rapidly as possible. Farmers have known for centuries that one bull can impregnate a great many cows! But this one-to-one ratio is most common, because it is the most stable: when one gender becomes rare, it is automatically at an advantage, with a greater chance of having descendants, and thus it tends to reestablish equilibrium, which cancels out this advantage. This is known as the Fisherian sex ratio hypothesis, as explained by the biologist William Donald Hamilton.

However, while male and female descendants don't have the same advantages in terms of selective value for parents, natural selection favors a more significant investment in the more advantageous sex. This sexual competition then causes gender ratios that are biased in favor of one of the two sexes. For example, in many polygamous vertebrates, large, healthy males mate with more than average frequency, while smaller, weaker males sometimes do not mate at all. On the other hand, virtually all females mate, regardless of their physical condition. The females with the best physical condition produce healthier young, and these babies tend to become larger adults. Consequently, females in good condition should produce a larger proportion of males, which has been confirmed more than once, particularly in the case of the opossum.

The same mechanism appears to be at work among humans. Dominant women tend to produce statistically more boys than other women, and this may provide us with an initial explanation for the overabundance of males among the Dwarves in Tolkien's world.

We also see a gender ratio strongly biased toward males in . . . the spotted hyena, particularly among females that have a high status in the hierarchy. We might also presume that, in a particularly harsh environment like that of the Dwarves, the individuals with the highest resistance would be males, and it would then become more advantageous for a female to produce males, who would have a better chance of reaching reproductive age.

The Dwarves are a people renowned for having accumulated enormous wealth, attracting the envy of many, including Smaug the dragon. Yet, biology shows that, before they lived in such opulence, the Dwarves survived a tormented past, difficult years that shaped both their morphology and their behavior. As in the rest of the animal kingdom, we find in the Dwarves characteristics that are reflective of past evolution and ancient environmental conditions. It should be remembered, however, that natural selection is not all-powerful, and that some biological characteristics can result not from adaptations to the environment, but rather from mechanical or embryonic stresses, or quite simply from chance.

Stephen J. Gould defended non-adaptive explanations against the "Panglossian paradigm"—that is, the attitude of biologists who, like Doctor Pangloss from Voltaire's *Candide*, believe that all is for the best in this best of all possible worlds. What would the Dwarves' bearded women think of that?

COULD AN ENT REALLY EXIST?

BRUNO CORBARA, ecologist,
Université Clermont Auvergne

"I arose and began to ascend a tree which stood near. As I raised myself by its limbs, it gave a low, yet shrill scream [. . .]." These words might have come from the mouth of Merry or Pippin, telling a gaggle of marveling hobbit-children about their first encounter with the Ent Treebeard, as it occurs in *The Lord of the Rings*, but it is nothing of the kind. These lines are taken from the Danish writer Ludvig Holberg's 1741 book *Niels Klim's Underground Travels*. In this passage, the hero, who has just landed on a planet contained inside our own hollow Earth, is stunned to encounter its strange inhabitants: trees endowed with reason.

THE LIFE OF A DENDRANTHROPE

Man-trees and tree-men existed in science-fiction and fantasy literature long before Tolkien; however, the Ents of Middle-earth are undeniably the dendranthropes (from the Greek *dendros*, "tree," and *anthropos*, "human being") that have made the greatest impact on our imagination—all the more so because, in the film adaptation of *The Lord of the Rings* by Peter Jackson, the character of Treebeard is, if I might interject a personal opinion here, particularly wonderful.

The possibility that a creature can be a plant *and* a thinking, talking being is a striking one. But the most confusing question for a biologist is how it can

The Ents attack Isengard.

be both a plant and an animal at the same time. After all, being living proof of the fact, we know that the transition from animality to reason—which occurred with the process of hominization—is entirely possible. But what about belonging jointly to the animal and plant kingdoms? How could such *ent*ities exist?

The History of Middle-earth tells us that the Ents have always acted as protectors of the forests. The male Ents continued this work alone when the Entwives abandoned the forests in favor of orchards and gardens (we do not know what became of the female Ents after Saruman destroyed their splendid gardens). The Ents then ceased to reproduce, and there were no more young Ents, or "Entings." By the end of the Third Age, when the events of *The Lord of the Rings* take place, there are only a small number of Ents surviving within Rohan, in the Forest of Fangorn. It is mainly through the episode here in which the two young hobbits encounter Treebeard that we learn about the appearance and behavior of the Ents.

Ents' appearances vary depending on the species with which they identify: oak, rowan, etc. If we go by the description given in the book, they are tall (more than four meters, or around thirteen feet, in Treebeard's case) and humanoid in form, which means that they have a head, a neck so thick as to be nonexistent, a trunk (obviously), arms, and legs. They have bark-like skin, beards, and hair like bushy twigs. Peter Jackson's film makes Treebeard taller than he is in the book, and his resemblance to a tree, particularly in terms of his abundant branches and plentiful foliage, is meticulously executed.

When they meet the old Ent, Merry and Pippin are impressed by his deep and penetrating eyes as well as his voice. Ents can see, speak (and thus hear), and have their own language, Entish. Generally slow in their movements and extremely calm in nature, they can become fearsomely efficient in combat: after three days of deliberation—incredibly speedy for Ents—several dozen of them launch an attack on Isengard and destroy Saruman's forces there.

So, the Ents possess both animal attributes—they move, walk, and communicate; in short, they "behave"—and those of plants, meaning that an Ent, when perfectly still, is indistinguishable from a tree. Additionally, Tolkien tells us more than once in *The Lord of the Rings*, an Ent can *turn into* a tree. This is apparently the standard fate of very ancient individuals. We are also told that, after the disappearance of the Entwives, many Ents, in their loneliness, made the decision to become simple trees. Conversely, according to Treebeard, a tree can transform into an Ent: "Most of the trees are just trees, of course;

but many are half awake. Some are quite wide awake, and a few are, well, ah, well getting *Entish*. That is going on all the time."

Tolkien also mentions beings with an in-between status, called Huorns. Considered Ents that have reverted to a primitive state (are we to take this as meaning similar to trees?) but which can awaken; indeed, some Huorns take part successfully in the decisive battle against Sauron.

At the end of the day, are Ents animals, plants, one and the other by turns, or both at the same time? In the context of Tolkien's world, we can accept their in-between status, but what would a biologist say about such creatures?

Firstly, Ents, as living creatures, are necessarily the result of a long and continuous evolutionary process. This may not be readily apparent in Tolkien's world, where the Ents—"the earthborn, old as mountains," according to the poem—seem be the fruit of an act of creation rather than biological evolution, but nevertheless we will stick to an evolutionist perspective here.

In this context, how can we explain the possibility of a living thing being both plant and animal at once? And, for that matter, what exactly do we mean by *plant* and *animal*?

PLANTS VS. ANIMALS, ANIMAL-PLANTS, AND PLANT-ANIMALS

Plants, which we will equate here with tracheophytes (meaning "vascular plants"), are multicellular beings that possess two significant characteristics: they are autotrophic, and they are sessile. "Autotrophic" because they are self-sustaining in terms of nourishment: via photosynthesis, and thus the light energy of the sun, plants are capable of producing the basic molecules of life, in particular sugars, by utilizing water from the ground and carbon dioxide from the air. Plant cells are equipped with organelles called chloroplasts, where these chemical reactions take place. "Sessile" means that plants live attached to a substrate; they do not move, but find all their resources *in situ*. Though some plants, watched via time-lapse photography, appear to move, this is in fact due to their growth and is in no way a behavior.

Animals, or metazoans, are also multicellular beings, but the vast majority of them (there are exceptions, especially among marine species, which we'll get back to later) have two characteristics that distinguish them from plants: they are heterotrophic (taking nourishment from preexisting organic material) and mobile. To feed themselves, animals depend on plants, either directly, as in the case of herbivores, or indirectly, as with carnivores. Their mobility enables them to access these plant or animal food resources.

The characteristics of plants and animals—autotrophy vs. heterotrophy and immobility vs. mobility—have consequences for every aspect of their respective lives. To give one example, in the presence of predators, animals can flee. They are capable of this because they have sensory organs (to detect the threat), muscles (to move), and a nervous system (to process the information and make the muscles work). In the same situation, plants, which cannot flee, have developed numerous "on-the-spot" defenses, including spines and toxic substances they produce. A plant's most basic defense mechanism is that it can be partly eaten without truly endangering its survival. This is impossible for most animals, as they have highly specialized organs whose functions they cannot live without.

Plants and animals, then, have taken two separate evolutionary paths. The divergence of the two lines occurred several hundred million years ago, in the time of their single-celled ancestors. In terms of the classification of living beings, we can certainly clearly differentiate between plants and animals. But certain organisms present characteristics of both lines, as do the Ents. Some of these are plants that live like animals in some ways (we will call them "animal-plants"), and some are animals that display some plant-like characteristics ("plant-animals").

While plants are not mobile in the ethological sense of the term, some of them are capable of motility (the ability to move independently). This is the case, for example, with the carnivorous plants whose modified leaves can close quickly to capture insects and digest them. Among these "animal-plants," movement goes along with a certain form of heterotrophy, which does not involve carbon in this case (they still have chlorophyllian leaves), but rather nitrogen. Carnivorous plants, which generally grow in nitrogen-poor soils, have, in the course of their evolution, developed this original way of obtaining nitrogen. These predatory plants, which are capable of moving, have—unsurprisingly—inspired many fiction writers, some of whom have peopled their imaginary jungles with "anthropophagic (man-eating) trees."

As for plant-animals, the most notable examples are undoubtedly corals. Mobile during one stage of their life cycle (the "medusa stage"), they spend most of their lives stationary. Each polyp (individual living unit) secretes a skeleton, the polypary, composed of calcareous elements, with a reef being formed by the piling-up of these skeletons over thousands of generations. The ecological success of tropical corals owes a great deal to a mutualistic relationship between polyps and zooxanthellae, single-celled algae that live in some of their cells. Thanks to their photosynthetic activity, these algae produce sugars, part of which are redistributed to the coral polyps. Animals that are stationary in the manner of plants, corals benefit, then, like plants, but indirectly, from the sun's energy. In return, the polyps provide the zooxanthelles with nitrogen from captured prey.

Mobile plant-animals are rare, but they do exist. The most extraordinary example is that of the sea slug *Elysia chlorotica*. This mollusk is virtually indistinguishable from the filamentous algae from which it feeds, except for its movements. Astonishingly, this slug, commonly known as the eastern emerald elysia, can go for several months without food, as it . . . manufactures sugars via photosynthesis! Its cells actually play host to chloroplasts obtained through a mechanism called kleptoplasty. These chloroplasts originate in the cells of the algae on which the slug feeds, and are "captured" in the digestive tract of the sea slug and settle in other cells of its body.

ENTS: ANIMAL(-PLANTS) OR PLANT(-ANIMALS)?

To sum up, a plant is a plant and an animal is an animal. It is impossible to be both at once, even though some plants *behave* like animals and some animals remain *planted*. Given everything we have just discussed, are Ents more likely to be animal-plants, or plant-animals?

Let's imagine that Ents are plants. This means that they are autotrophic, and the problem of their nourishment is thus largely resolved, as they are able to manufacture glucids in their leaves via photosynthesis. Given that their vegetal "legs" are poorly suited to the function of absorption that characterizes normal roots (which require a whole network of fine, highly fragile rootlets), they can procure nutrients through their leaves, as do many epiphytic plants (that is, plants that grow on other plants), for example tank bromeliads. Now, what would they need to become animal-plants capable of moving, walking, perceiving their environment, and communicating verbally (we'll leave aside the idea of thought here)? As we have seen, they would need muscles, sensory organs, and a nervous system, all highly complex structures that it is difficult to imagine being easily constructed from plant cells.

Now let's imagine that Ents are animals. They possess everything needed for "behavior": muscles, sensory organs, and a nervous system. They are also heterotrophic. And we know that they have a mouth, because they speak—and we also know that many animals (most land vertebrates) generate sounds with the entrance to their digestive system. So, it is highly probable that Ents eat with their mouths. As to the nature of their diet, given their environment, we can safely conclude that they eat leaves, fruits, and seeds.

Of these two scenarios, the second seems far more probable. From this viewpoint, Ents would be animals whose external appearance is that of a plant. Natural terrestrial environments, particularly tropical ones, teem with animals—stick-insects (phasmids) and leaf-insects (Phyllidae), for example—which, thanks to their resemblance to plants, or at least parts of plants, avoid the attention of predators. Likewise, the Ents, whose role is to guard the forests, are undoubtedly, with their bark-like skin and their limbs that resemble branches and leaves, beings that mimic trees.

What remains now is the question of how Ents change into trees, and vice versa. How is such a metamorphosis possible if Ents are animals? It has to be because the beings created by an Ent's transformation are not real trees, but tree-mimicking animals that live in a stationary manner, like coral. In this state of complete immobility, how do these "sleeping" Ents feed themselves? They must, like all sessile beings, be autotrophic. The model of the sea slug *Elysia chlorotica* offers an elegant solution. Would we find chloroplasts (obtained through kleptoplasty) in the animal cells of the pseudo-leaves of these Ents? If so, this would make the Ents plant-animal chimeras peopling Middle-earth, sometimes giant photosynthetic phasmids; sometimes land-dwelling branching coral. Did Tolkien have images like this in mind when he created the Ents? It's understandable if you doubt it. Whatever the case, of all the creatures to spring from the mind of the Master of fantasy, they are inarguably among the most . . . *ent*eresting.

SARUMAN'S GMOS
(GENETICALLY MODIFIED ORCS)

SIDNEY DELGADO, biologist, Sorbonne University
VIRGINIE DELGADO BRÉÜS, doctor in French linguistics

 nitially, Saruman was only introduced by Tolkien to explain Gandalf's absence from the rendezvous he had arranged with Frodo in the autumn, for the latter's departure from the Shire. However, he went on to become a central character in *The Lord of the Rings*, the symbol of corruption by one's thirst for power.

THE SHREWD LEADER OF THE ISTARI

Saruman is, according to *The Silmarillion*, the head of the group of five wizards dispatched to Earth by the Valar to aid the Men and Elves in defeating Sauron. In an ironic twist, he ends up allying himself with the one he is supposed to help destroy. This powerful mage is attracted by the Rings of Power, which he has studied for a long time, and in particular the most powerful of them all, the One Ring. Faced with the rising power of Sauron's forces, Saruman decides to submit to the Dark Lord, planning to take the ring for himself at the opportune moment. As he explains to Gandalf before taking him prisoner:

> *There is no hope left in Elves or dying Númenor. This then is one*
> *choice before you, before us. We may join with that Power. It would*

be wise, Gandalf. There is hope that way. Its victory is at hand; and
there will be rich reward for those that aided it. [. . .] And why not,
Gandalf? Why not? The Ruling Ring? If we could command that,
then the Power would pass to us.

(The Fellowship of the Ring)

Joining with Mordor, Saruman begins assembling an army of orcs and wargs in his stronghold of Isengard, in the shadow of his tower of Orthanc, with the goal of invading neighboring Rohan, an invasion which would enable Sauron to conquer Middle-earth. But Saruman also means to use his army to scour the land for the One Ring, with the intention of keeping it for himself rather than returning it to Sauron. In *The Two Towers*, Saruman's orcs, along with those of Sauron, capture the two hobbits Merry and Pippin, but a conflict erupts between the factions of orcs when those that serve Saruman want to take the prisoners back to Isengard in order to search them, which enables Merry and Pippin to escape.

Saruman's alliance with the forces of Mordor is therefore not a true submission, but a strategic choice attesting to his great cunning. Moreover, Saruman's name derives from *searu*, from the Old English (a language studied by Tolkien the philologist), meaning "strategem." In *The Silmarillion*, the mage is also called Curunír in Sindarin, which translates to "man of skill."

Using his magic to further his sinister plan, Saruman undertakes to transform the Uruk-hai, an already powerful race of orcs, in order to make them even stronger and, more importantly, able to tolerate the sun. We also learn in *The Lord of the Rings* that Saruman has engineered orcs intended to serve as his spies, and which are also unusual in that their physiognomy is closer to that of humans than other orcs. However, the books do not offer any precise description of his "manipulations," and Tolkien prefers to let his characters theorize, without confirmation, about human-orc crossbreeding. For example, in *The Two Towers*, the Ent Treebeard muses: "I wonder what he has done? Are they Men he has ruined, or has he blended the races of Orcs and Men? That would be a black evil!"

Everything suggests that these manipulations are the result of some magic. However, in light of current scientific knowledge, we can envision how Saruman might have been able to create his creatures using genetic technology—and associating Saruman with modern technology is not as absurd as it might seem . . .

SARUMAN'S MIND OF WHEELS

Somewhat paradoxically, even though Saruman is the archetypical wizard, a character normally far removed from modern technology, he is also described as symbolic of the negative effects of industrialization. Tolkien had a deep attachment to nature. In many ways, *The Lord of the Rings* is a celebration of unspoiled nature, of flora, its beauty, its contemplation. The Elves, quasi-perfect beings, live in total harmony with nature. Tolkien's biographers are unanimous in stating that he detested the effects of industrialization on the rural landscapes of England. It is said that he was devastated to recognize almost nothing of the area near Birmingham where he spent his childhood when he revisited these places when he was in his forties, during the 1930s. In his books, evil is associated with the destruction of nature, and Saruman is the very one ravaging it, cutting down trees and building factories in their place. Treebeard says of Saruman: "He has a mind of metal and wheels; and he does not care for growing things, except as far as they serve him for the moment." And Gandalf tells us that "[. . .] the valley below seems far away. I looked on it and saw that, whereas it had once been green and fair, it was now filled with pits and forges."

The Shire of the hobbits is described at the beginning of *The Lord of the Rings* as an idyllic rural landscape, far from the hustle and bustle of the world, but at the end of the trilogy, when Frodo and his friends return home without knowing that Saruman sought refuge there, they discover a scene of desolation: "An avenue of trees had stood there. They were all gone. And looking with dismay up the road towards Bag End they saw a tall chimney of brick in the distance. It was pouring out black smoke into the evening air." (*The Return of the King*)

A MAD SCIENTIST AHEAD OF HIS TIME?

Saruman bears a great resemblance to the archetypal "mad scientist" who manipulates science to achieve his personal ends, without worrying about the negative consequences of his actions. In literature and films, these characters are often marked by arrogance, the desire to play God, the pursuit of their experiments without ethical consideration, a lack of respect for nature, and a solitary existence except, perhaps, for the presence of an assistant. Saruman fits this profile perfectly.

Fritz Lang created the prototypical mad scientist in his 1927 film *Metropolis*, with the character of Rotwang. In the film, a huge city is divided into a skyscraper city, where the privileged elite dwell, and an underground city, where the workers toil in extremely harsh conditions. This state of affairs is threatened by a romance, and Rotwang is enlisted to restore order. To do this, he builds a robot in the likeness of the young worker with whom the hero is in love (George Lucas, incidentally, drew inspiration for the character of C-3PO in *Star Wars* from this film), and this female robot sows chaos and discord which almost destroy the city. We can see some of Rotwang's characteristics in Saruman: a malevolent mind, an obsession for goals set without consideration for the lives of others, and a fascination with machines.

Mad scientists are also to be found among geneticists, who manipulate genes in defiance of the laws of nature, as in Aldous Huxley's futuristic 1932 novel *Brave New World*. In this science-fiction classic, the director of the Central London Hatchery and Conditioning Centre is irresistibly reminiscent of Saruman: tall, thin, arrogant, with a slight edge of menace to his voice, speaking dispassionately of the brutal treatments to which embryos are subjected, and boasting enthusiastically about the large numbers of clones produced.

SARUMAN'S GENETIC MANIPULATIONS

How has Saruman produced his modified orcs? Firstly, as the members of the Fellowship of the Ring surmise, through orc-human crossbreeding.

When animals belonging to different species are crossbred, biological mechanisms prevent fertilization and embryonic development. However, with species that are genetically similar enough, interspecies hybrids can be obtained, which display characteristics midway between the two parent species. Additionally, the result will be different depending on whether a male from one species is crossed with a female from another, or vice versa. For example, the crossbreeding of a male horse with a female donkey will yield a hinny, while crossing a male donkey with a mare produces a mule. Likewise, among big cats, crossing a male lion and a female tiger creates a liger, while crossing a female lion and a male tiger gives a tigon. Numerous hybridizations are possible in the animal world, then—on the condition that the animals are in the same genus (*Panthera*, *Equus*, *Ursus*, etc.), which ensures adequate genetic proximity. Hybrids are usually sterile, though a small percentage can reproduce. The female tigon, for example, has a 100% fertility rate, but can only reproduce with a tiger.

In *The Lord of the Rings*, crossbreeding orcs and humans is therefore possible, assuming a genetic similarity, as suggested by the orcs' humanoid appearance. Moreover, the direction of the crossbreeding, as well as the race of orc utilized (for we have seen that Uruk-hai are different from basic orcs) leads to a race of beings able to be exploited by Saruman. Uruk-hai warriors have the strength of orcs and the human ability to tolerate sunlight, while the physiognomy of Saruman's spies is closer to that of humans, which enables them to pass unnoticed.

But one problem still remains for Saruman: that of producing enough half-orcs to create an army numbering in the thousands, even with part of this army being made up of the swarthy Men of Dunland. Interspecies hybridization necessarily involves sexual reproduction, though the wizard could gain some time by effecting *in vitro* fertilization and then implanting the hybrid embryos in orc or human females. This would enable him to produce at least *some* soldiers and spies, such as the ones that make up the party that captures Merry and Pippin. But to produce an army like the one that attacks Helm's Deep, Saruman would need to produce embryos on an industrial scale. It is impossible not to be reminded here of the production of huge numbers of clones in *Brave New World*.

The fortress of Isengard, then, is nothing less than a genetics laboratory in which orc embryos are mass-produced to suit Saruman's needs. Tolkien remains wisely silent on the wizard's methods, even though magic plays a large role in this sort of situation. In his film adaptation, Peter Jackson provides his own vision of the birth of the Uruk-hai, with the leader of Saruman's orcs emerging from a kind of viscous mud, suggesting supernatural origins.

Saruman may also employ other techniques used in genetic engineering for his manipulations: for example, he may inject human stem cells into an orc embryo, which would give the eventual fetus human characteristics; this would make them chimeric organisms. These techniques have been used since 1969 to create chimeric quail-chicken embryos, enabling important advances in our knowledge of cellular biology.

GMO ORCS

To give his orcs the desired characteristics (strength, tolerance of sunlight, a human appearance, etc.), Saruman may also be creating genetically modified organisms, or GMOs. To accomplish this, all he would need to do is inject a DNA fragment containing the relevant gene(s) into an embryo. To ensure the integration of this DNA sequence, it is paired with viral DNA that has the ability to insert itself into a host genome. This is how the first genetically modified animal (a mouse) was created in 1982. Since then, procedures have been improved thanks to genetic scissors technology (CRISPR/Cas9). Cows can now be created that produce milk that is casein-rich or has lower lactose levels. Recently, transgenic goats have been created that can produce in their milk the protein that constitutes spider-silk, for use in manufacturing surgical sutures, fishing line, and even clothing! Pigs have been made resistant to swine fever by means of gene manipulation, and salmon have been genetically modified so that they grow faster. Stranger still, some animals (fish, mice, rabbits, and even sheep) have been made fluorescent.

Bioethical laws prohibit the sale or marketing of these transgenic animals, so it will be some time before we know what a fluorescent chicken looks like when it is cooked! Saruman, on the other hand, is bound by no such limitations, and is free to produce all the genetically modified orcs he wishes.

One of the genes likely to interest the wizard is myostatin, mutations of which can result in increased muscle mass. We know of species of cows (the Belgian Blue) and dogs (bully whippets) that possess significant muscle mass due to mutations in this gene. Bully whippets, being extremely fast, are sought after as racing dogs. Other hormones might also be useful in the genetic doping of orcs, including erythropoietin (EPO, used widely by cyclists), vascular epithelial growth factor (a protein that causes the formation of blood vessels), and IGF-1 (which has an anabolic effect on muscle mass). Today these genes are monitored by the World Anti-Doping Agency, which is currently researching methods to detect the presence of additional DNA in the bodies of athletes.

The extreme sensitivity to light that afflicts orcs is reminiscent of the symptoms of *Xeroderma pigmentosum*, a genetic disease caused by mutations in DNA enzyme-repair genes. Children with this illnesses are often referred to as "moon children" because they cannot tolerate exposure to the sun's UV rays. Saruman may have been tempted to introduce functional versions of the genes involved in this disorder to create his Uruk-hai warriors, which are able to bear sunlight.

What is true of orcs is also true of wargs, the giant wolves they ride. In his films, Peter Jackson portrays wargs as resembling a mixture of wolves and hyenas, only much larger. Here again, we might imagine that Saruman has unhesitatingly modified the genetic makeup of smaller creatures (probably other canid species) to create fearsome war-machines. And he shares one last thing in common with many other mad scientists: he is killed by his slave, Grima (Wormtongue).

THE EVOLUTION OF THE PEOPLES OF MIDDLE-EARTH: A PHYLOGENETIC APPROACH TO HUMANOIDS IN TOLKIEN

JEAN-PHILIPPE COLIN, associate professor of
Life Sciences and Earth Sciences

When you open *The Lord of the Rings* for the first time, the adventure doesn't begin right away. *The Fellowship of the Ring* begins with a detailed map, followed by an encyclopedia-esque text called "Concerning Hobbits," in which the author introduces the setting of the story to come, and the beings that populate it. Tolkien, like any self-respecting academic, brings us into his world in a serious and methodical manner, as if we were opening an ancient book by the cozy fireside in a hobbit hole, a cup of tea in hand. All the magic of the tale is there, in the earnestness and realism instilled by Tolkien in his complex universe; the author is appealing to our curiosity, the better to lay the foundation for the entertainment he is about to provide. His chapter on hobbits brings up some interesting questions about the kinship relations among the species, or "races," that populate his fiction.

> *It is plain indeed that in spite of later estrangement Hobbits are relatives of ours: far nearer to us than Elves, or even than Dwarves. [. . .] But what exactly our relationship is can no longer be discovered.*

It is noteworthy that Tolkien compares hobbits, key characters in his work, to humans. This relationship (still unresolved, if we go by the last sentence of the quote above), imposes limitations on our understanding of the subject, as if a sort of prescience is already explaining Tolkien's work—he is deliberately blurring the line between fiction and reality, it seems. Let's take him at his word, and try to clarify, scientifically, the ties of kinship between the species of Middle-earth!

NO ONE CAN ACHIEVE THE IMPOSSIBLE . . .

There is a scientific (that is, testable) method that can be used to establish the kinship relations between populations (called "taxons" by specialists): phylogenetics. Using Tolkien's assertion that the relationship between humans and hobbits cannot be known as our starting point, it makes sense for us to consider all the species of Middle-earth: hobbits, yes, but also Men, Elves, Dwarves, orcs, and all other analogous species.

Phylogenetics is an interesting method, because it takes into account both modern and fossil taxa, without necessarily considering a fossil as the ancestor of a recent example. It was introduced in the 1950s by the German biologist Willi Hennig in his book *Basic outline of a theory of phylogenetic systematics*. In the book, Hennig, an entomologist specializing in dipterans (a group of insects including flies and mosquitos), proposed a fairly simple and transparent method called cladistics. Greeted with little fanfare at first, the book was translated into English in 1966 and helped to change the principles of classification worldwide. But the overthrow of the traditional classification system inherited from Linnaeus during the Enlightenment was no easy pill to swallow, and remains a work in progress even today. The emergence of DNA sequencing technology has also contributed to the change, adding new traits for comparison. And finally, thanks to the calculating power of computers, ever-more sophisticated algorithms have been introduced to process larger and larger quantities of data.

Phylogenetic analysis results in the creation of a tree—that is, a graphical network linking taxa according to their common derived characteristics (or synapomorphies). The nodes (or branch points) of the tree correspond to what specialists call "ancestral morphotypes," for, unlike a genealogical tree, a phylogenetic tree does not link ancestors and descendants in a direct relationship

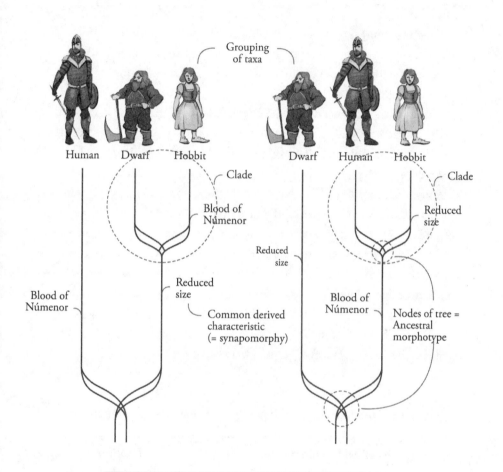

Grouping
of taxa

Human Dwarf Hobbit   Dwarf Human Hobbit

Clade   Clade

Blood of   Reduced
Númenor   size

Reduced   Reduced
size   size

Blood of
Númenor   Blood of   Nodes of tree =
Reduced   Númenor   Ancestral
size   morphotype

Common derived
characteristic
(= synapomorphy)

FIGURE I

○━━◇━━○

Two phylogenetic trees showing different theories of
kinship among Dwarves, hobbits, and humans.

(see Figure 1). Figure 1 shows two examples of phylogenetic trees, showing different kinship theories. The more intuitive theory (on the left) places Dwarves and hobbits in the same clade, with the reduced size of these two races thus being a characteristic acquired once by a common ancestor of both of these taxa, and specific to this clade. The hypothesis proposed by Tolkien himself, on the right, places humans and hobbits in the same clade; in this case, small stature is an evolutionary convergence appearing multiple times in the course of evolutionary history.

Firstly, then, we must define the characteristics of the group to be studied. These characteristics can be morphoanatomical (based on the taxas' skeletons, for example) or molecular (for example, based on DNA sequences). To study the phylogenetics of the peoples of Middle-earth, the ideal would be to examine representatives of each taxon from every angle, to measure them, and possibly to dissect and sequence them, all in order to document as many traits as possible. However, for lack of anything better, I have simply defined some characteristics based on extrapolations made from Tolkien's descriptions, the illustrations of Alan Lee, John Howe, and other artists, and Peter Jackson's cinematic adaptations.

TAXA OF MIDDLE-EARTH

We have analyzed ten taxa:

Humans. Humans probably represent the majority of the population in Middle-earth, despite their recent appearance in the timeline. Various ethnic groups are described in Tolkien's work; for example, the Haradrim with their oliphaunts, and the Northmen, to whom the Horse-lords, or Rohirrim, belong. As diverse as they are, the men and women of Tolkien's world seem wholly comparable, from a biological perspective, to the real humans that you and I are, and so we will not linger further on a description of them.

Dwarves. Dwarves are a species of fierce warriors who live reclusively within certain mountain ranges in Middle-earth, from which they mine colossal wealth. Besides their small stature, Dwarves are marked by a particularly startling characteristic: the absence of sexual dimorphism—that is, their females resemble the males, even down to their beards! Dwarvish women are also extremely discreet, making up no more than one third of the population and rarely leaving their underground cities. Biologically speaking, dwarfism can have several origins, but the most likely theory here is undoubtedly that of hormonal dysfunction, since this is associated with a loss of sexual dimorphism.

Hobbits. Represented most notably by Frodo and Bilbo, hobbits are Tolkien's heroes.[*] Sometimes called "halflings," their physical attributes are not, however, those of a typical hero: large, hairy feet and short stature are inarguably hobbits' most characteristic traits.

Elves. These beings are extremely ancient, with a history extending back over several millennia. However, in recent times their influence has decreased, as many are leaving Middle-earth to return to their ancestral realm of Valinor. The principal biological characteristic of the Elves is their indefinite lifespan; they are affected by neither illness nor aging. They also possess highly developed senses, pointed ears, and pale skin, and are—almost without exception—beardless.

Ents. Ents are forest-giants whose bodies appear to be composed of plant tissue: bark-like skin, branches and leaves for hair. But their bipedalism, articulate language (Entish), and the existence of female Ents—now lost—make them a full-fledged people in their own right. Like the Elves, they can reach extremely advanced ages, which involves interesting evolutionary innovations, for example an extremely high-performing immune system, which protects them from infection, and anti-aging mechanisms, including an effective antioxidant system, which would be crucial for the body to fight off free radicals. These tiny molecules, which are unstable and highly reactive, are continuously formed as oxygen is utilized in metabolic processes, or when the immune system is active. Their reactivity causes damage at the molecular level, which is thought to be a major cause of physical aging. Another characteristic of Ents is the presumed presence of the enzyme telomerase, which would be active in all their cells, whereas in humans, this enzyme is active only in reproductive cells and certain cancerous cells. To understand the important role played by this enzyme, it's important to know that each time they divide, our cells must replicate all of their DNA, which is organized into chromosomes. However, a small end section of our DNA is never replicated, and with each cycle, the tips of our chromosomes, called telomeres, become a little bit shorter. As if they were genetic hourglasses, the depletion of telomeres leads to cellular senescence

---

[*]   See the chapter "Why Do Hobbits Have Big Hairy Feet?," p. 209.

and, on a body-wide scale, to age-related illnesses. Telomerase eliminates this problem by copying the missing terminal section at each replication.

Orcs, goblins, and Uruk-hai. Orcs constitute the majority of the troops serving the forces of Evil. Sensitive to light, they dwell mainly in Mordor, where the volcanic Mount Doom generates a permanent state of shadowiness, limiting its inhabitants' exposure to the sun. Peter Jackson portrays an interesting characteristic of the orcs in biological terms; their blood is black, probably due to a specific type of hemoglobin. The blood of most vertebrates is bright red due to the presence of an iron atom within their hemoglobin, a protein contained in red globules which transports oxygen. It is possible that this iron atom is replaced in orcs by another element able to bind to oxygen; certain mollusks and arthropods, for example, show copper within a protein, called hemocyanin, which means that their blood is blue. These general descriptions of orcs,[*] however, should not cause us to lose sight of their great diversity, with several "wild" populations known to exist in the Misty Mountains, for example. As Gandalf says in Chapter 7 of The Hobbit: "Before you could get round Mirkwood in the North you would be right among the slopes of the Grey Mountains, and they are simply stiff with goblins, hobgoblins, and orcs of the worst description." To illustrate this diversity, we have included in our analysis those smaller orcs sometimes called "Snaga," a Black Speech word meaning "slave." Finally, certain large orcs, chosen by Saruman, have been made wholly resistant to sunlight and can fight both day and night; these are the orcs known as Uruk-hai.

Trolls and Olog-hai. Like orcs, trolls are intolerant of light, to the point that they turn to stone if they are exposed to the sun's rays. This tendency toward mineralization is undoubtedly linked to their extreme strength and robustness, and we may suppose that the deeper layers of their skin are largely ossified. In many vertebrates, bony skin plates can act as veritable armor, as in the case of the Placoderms, a Paleozoic-era armored fish that lived between 440 and 358 million years ago. This natural toughness, combined with colossal strength and a certain servility, constitutes a military asset exploited by the orcs, which have "domesticated" trolls. Here again, numerous races coexist: cave trolls, hill trolls, and the fearsome Olog-hai, which are able to tolerate light, and which force their way past the gates of Minas Tirith, the capital of Gondor, in Peter Jackson's cinematic adaptation of The Return of the King.

---

[*]   See the chapter "Saruman's GMOs (Genetically Modified Orcs)," p. 257.

CHARACTERISTICS AND THEIR POLARIZATION

Once characteristics have been defined, they are compared between taxa in order to identify different character states. One method of determining whether a character state is ancestral (plesiomorphic) or derived (apomorphic) consists of using a reference group (external group or extra-group) that is considered to present ancestral character states. The "hairy feet" trait, for example, shows two states in the sample: the "hairy" state, observed in hobbits, and the "naked" state seen in humans, Elves, and others. If I use chimpanzees (*Pan troglodytes*) as an extra-group, then the fact of possessing hairy feet is an ancestral characteristic, and naked feet are a derived characteristic. Choosing an extra-group is therefore a delicate process, as it constitutes a premise based on previously established knowledge. In our current case the extra-group is hypothetical for lack of a good real-world candidate, a non-human primate in Tolkien's universe.

When the sample contains multiple taxa and characteristics (or character states), phylogenetic analysis is often computerized. There are a number of software platforms for this (PAUP, Hennig, etc.), with most of these used to calculate the sparsest tree(s); that is, showing a minimum number of (character) states. This software, called "sparse" software, is rather like chess software, which calculates all possible moves in order to pinpoint the best one; the software we have used here is R-3.3.2.* The kinship between humans and hobbits shown in Figure 1, proposed by Tolkien, was manually established, while Figure 2 shows the results of computer analysis.

ONE TREE TO GROUP THEM ALL, AND IN EVOLUTION BIND THEM!

The tree obtained using R-3.3.2 software (Figure 2) is interesting on multiple levels. At its center, those creatures that serve the forces of Evil (goblins, orcs, trolls, and their derivatives) are assembled into a natural group, or clade, called "Malphotophobes"; this group is characterized in particular by their sensitivity to light. The prefix "mal" refers here to the malevolent nature of these beings,

---

*   Available online at no cost, for readers who may be curious.

while "photophobe" refers to their intolerance of light. This sensitivity, which restricts their way of life, is subject to evolutionary regression which, in the case of the Uruk-hai and Olog-hai, facilitates the invasion of Middle-earth. Another noteworthy result of the software analysis is that Elves are linked to Ents, due particularly to their considerable lifespans. This group has been dubbed the "Perennosylvans" for their long lives and the fact that they dwell in wooded ("sylvan") areas. Surprisingly, this clade is itself linked to the Mal-photophobes because of two shared derived characteristics: pointed ears and hairless faces! Finally, on the other side of the tree, hobbits stand as a sister taxon of humans, which confirms the quote cited at the beginning of this chapter: hobbits are indeed related more closely to us than they are to Dwarves, despite their similarly short statures.

CAN ONE SPECIES BE MORE HIGHLY EVOLVED THAN ANOTHER? FRIENDS OF THE DWARVES, JUST SAY "NO" TO ELVOCENTRISM!

In the French language, "evolution" is often synonymous with progress, but in a scientific context it is important to understand that no one organism is more evolved than another. In reality, every organism currently alive on Earth is the result of a very long evolutionary history that goes back to the origins of life on our planet. A microbe, then, is just as advanced as a human, in the sense that its evolutionary history is just as long. Complexity is a difficult concept to define, but because the earliest forms of life were extremely simple, they could *only* evolve toward being more "complex" versions. In some cases, however, evolutionary simplification is known to occur, with the loss of organs, such as limbs in snakes. So, with all due respect to Elrond and his airs and graces, the Elves are no more highly evolved than the Dwarves they scorn, or the trees among which they dwell . . .

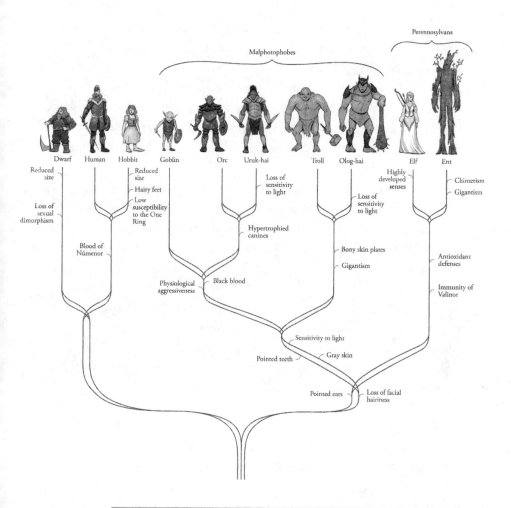

Perennosylvans

Malphotophobes

Dwarf   Human   Hobbit   Goblin   Orc   Uruk-hai   Troll   Olog-hai   Elf   Ent

Reduced size

Reduced size

Hairy feet

Low susceptibility to the One Ring

Loss of sexual dimorphism

Blood of Númenor

Loss of sensitivity to light

Hypertrophied canines

Highly developed senses

Chimerism

Gigantism

Loss of sensitivity to light

Bony skin plates

Gigantism

Antioxidant defenses

Immunity of Valinor

Physiological aggressiveness

Black blood

Sensitivity to light

Pointed teeth

Gray skin

Pointed ears

Loss of facial hairiness

FIGURE 2

Phylogenetic tree showing kinship relations
between the various peoples of Middle-earth.

THE PROBLEM OF EVOLUTIONARY CONVERGENCE

It sometimes happens that two taxa are distant from one another in evolutionary terms, and yet share a common characteristic. One frequently-mentioned example is that of bats' wings, which are structurally similar to those of birds; another is the hydrodynamic silhouette of the dolphin (a mammal) or the ichthyosaurus (a fossil reptile), both of which greatly resemble that of the shark. These similarities are not inherited from a common ancestor that possessed wings or a hydrodynamic outline, however, but have in fact appeared multiple times, independently from one another, throughout the course of evolutionary history, in response to identical selective stresses. If a dolphin displays the same profile as a shark, it is because this profile is an effective solution to the problem of aquatic predation. These evolutionary convergences are real traps for evolutionary scientists, for they can lead us to connect taxa that are actually quite far removed from one another. Of the convergences we are seeing here, we would cite the reduced size of Dwarves and hobbits, which appeared twice—independently—within these two populations, being an evolutionary advantage for life below-ground. For this reason it is extremely tempting to place these two taxa in the same clade, as they are in Figure 1; however, Tolkien tells us specifically that hobbits are far more closely related to us than they are to Dwarves.

WHAT EVIDENCE IS THERE FOR THE HUMAN-HOBBIT RELATIONSHIP?

At first glance, there is no external (or morphoanatomical) characteristic that appears to support the human-hobbit biological kinship so dear to Tolkien. However, we can imagine that there are common traits at the molecular level; that is, DNA sequences or genetic markers shared by both taxa.

These days, there are several companies that are able to sequence your DNA very easily; in return for fifty dollars, a saliva sample, and a couple of weeks' processing time, they can provide the key to discovering your geographic and familial origins. Imagine for a moment that we possessed samples of saliva from Frodo (a hobbit), Aragorn (a human), and Gimli (a Dwarf): what would we learn?

The most likely result, if we go by Tolkien's writings, would be a great similarity between humans and hobbits, which is represented on the phylogenetic tree as the "Blood of Númenor" characteristic. In fact, according to *The Silmarillion*, before settling in the Westlands, or Eriador, where much of the action of *The Lord of the Rings* takes place, humans lived on a large island called Númenor for part of the Second Age; at that time they enjoyed longer lifespans, a fact borne witness to by Aragorn, a representative of this line of Men who is aged 87 at the time of the Quest of the Ring, and whose reign as king lasts for 120 years before his death at the age of 210! Hobbits, as we know them, did not yet exist during this period, unlike Dwarves, which already inhabited the Westlands. Therefore, if hobbits are indeed related to humans, they are also descendants of ancestors who lived on Númenor, from whom they inherited certain genetic characteristics before diverging in their turn. Elrond describes this divergence as a decline during the meeting of the council bearing his name in *The Two Towers*: "And ever since that day the race of Númenor has decayed, and the span of their years has lessened."

It's important to note that genetic homogeneity is reinforced within a group when it is reduced to a small population and reproduces via only a limited number of individuals in isolation. This endogamy, or inbreeding, whether it occurs for geographical reasons (the island of Númenor) or social ones (Aragorn's royal lineage), limits the genetic intermingling associated with sexual reproduction, which is often deleterious, but convenient when it comes to genealogy or, on a broader scale, phylogeny. This example is illustrative of the fact that phylogenetic classification is not merely about grouping taxons; it can also be used to trace their geographical history.

CREATION OR EVOLUTION?
THE PARADOX OF *THE SILMARILLION*

Tolkien explains the origins of his world and its various peoples in *The Silmarillion*, a quasi-mythological tale that provides a detailed chronology of the emergence of each group: during the Years of the Trees, the Elves were created by Eru Ilúvatar, the powerful deity that had also created the Ainur, including Sauron. Next the Dwarves appeared, followed by humans, whose emergence marked the dawn of the First Age. This account may seem rather creationist at first, but it should be noted that Tolkien prefers to use the terms "awakening" and "birth" to "creation." Moreover, Tolkien displays significant knowledge of evolutionary science.[*] And so, *The Silmarillion* should be considered a book of genesis, an extremely ancient tome that would have been stocked by Middle-earth's libraries for generations. In his desire to explain his world, Tolkien proposes a sort of cosmogony, with its associated myths. Despite some dubious origins, lost in the mists of time, each race has a complex history that includes major migrations that go a long way toward explaining the diversity found within the different clades.

IN CONCLUSION

While similar to ours, Tolkien's world is enriched by the great diversity of peoples that inhabit it—as is often the case in other worlds in the heroic fantasy genre. However, the question of the origins of these species and their biological relationship to humans is rarely discussed in a realistic or scientific manner, as we have just done. Is this due to some sort of taboo, aimed at preserving the reader's vision of the imaginary world? Elves, Dwarves, and orcs are stalwarts of the genre whose roots lie in Germanic and Scandinavian folklore, and whose characteristics differ for each author. This diversity of "humanoids," sometimes criticized as overly Manichean with its noble Elves and vile orcs, nevertheless remains a formidable tool for the creation of rich characters. But some universes, such as that of the comic strip Wollodrïn, created by David Chauvel and Jérôme Lereculey, or the one in the video game *World of Warcraft* (Blizzard Entertainment), do not hesitate to portray "kind" orcs while still respecting the codes put in place by Tolkien. Finally, the often antagonistic cohabitation of these species provides an interesting tableau of the way things may have been when, for several hundred thousand years, multiple species of the genus *Homo* lived alongside one another, until *Homo sapiens* attained dominance . . .

---

[*]   See the chapter "Gollum: the Metamorphosis of a Hobbit," p. 226.

6

A FANTASTICAL
BESTIARY

MYTHOTYPING ORIGINS

LUC VIVÈS and CHRISTINE ARGOT,
researchers and educators, Museum of Paris

 olkien's work combines the natural history of fantastical beings with a supernatural human history in a land that would eventually become England. This double duality gives rise to a perfectly-imagined picture of the cohabitation and evolution of half-natural, half-mythical flora and fauna. In the pages of Tolkien's writings, the earliest forms of human genealogy rub elbows with the mutant figures of imaginary specimens both extinct and surviving to the present day, by virtue of the lucky survival of written and oral versions of their individual fates. In this way, a number of creatures have propagated, branched out, and declined (literary and artistically, that is) into new images that either resemble the originals, or blur them into unrecognizability. These fantastical beings, taken all together, establish Tolkienian prose as a discourse on history both natural and supernatural; a reflection on, and questioning of, the biological dimension of mythical beings that is equally curious about the amazing properties of natural, real-world creatures. Tolkien's work is unique in that it forms the first interface between cycles, sagas, and epics on one hand, and nomenclatures, dictionaries, and encyclopedias on the other. It brings together, intellectually, what has traditionally been separate, joining the improbable with the verifiable, coming as close as possible to being a study of the natural world perceived mythologically, and mythology perceived as natural.

The glittering genealogical trees of fantasy contain original beings that constitute the mythological types (or mythotypes) that are the sources of stories and legends still popular today. Creatures familiar to modern readers such as leprechauns, elves, sprites, and gnomes may, then, be considered *mythologically* modified versions of the figures they initially were. This succession of earlier forms makes up the mirrored body of an ideal being, living on through its iconic or discursive reflections. This is a way for us, as readers, to test the intangible volatility of fictional characters, by tracing their incarnations and their process of becoming, not only in terms of surface appearance, but also in the mythogenetic substance of what is written about them, for it is this narrative that determines not only the very existence of an imaginary character, but the ability of a real character to join the ranks of the mythical.

Haunted by "hopeful monsters"—Are wargs not reminiscent of werewolves, and the Nazgúl of Ankou, or any other representation of Death, grinning beneath his hooded cape?—Tolkien's work is the breeding ground where the web of kinships, the mesh of affiliations, the network of shared or acquired traits develops. It is the cosmogonic melting-pot that borrows two singular trees from Genesis, and from the three major monotheistic religions a unique god, Ilúvatar, who creates, firstly, the cohort of 'Holy Ones', the Ainur, among whom are the Valar (higher spirits), who are guardians of the terrestrial world. Celtic legends and Welsh folklore (including *The Four Branches of the Mabinogi*, four medieval stories of Celtic mythology) are intermingling here, along with the language of Norse fairy tales and Scandinavian eddas (13th-century poetic compilations), while the perfume of the Icelandic *Völsunga Saga* (a legendary saga dating from the 13th century and melding love, adventure, and tragedy) blends with that of the Finnish *Kalevala* (an epic poem composed in the 19th century).

A MIXTURE OF LANGUAGES

Tolkien's work is also a linguistic melting pot containing the Old Norse of northern Europe, Gothic writing, Old English poems, and shades of Classical literature. Hesiod's *Theogony* (the Elves' abandonment of Middle-earth to Men also being evocative of *Works and Days*, by the same author, which describes how the race of heroes was supplanted by the race of iron) is juxtaposed with Ovid's *Metamorphosis*; like Orpheus, the Elf Lúthien journeys to the kingdom of Mandos to plead for additional years of life for her lover Beren, who has been mortally wounded by the wolf guarding Morgoth, in exchange for her own immortality. Virgil's *Aeneid* and Homer's *Odyssey* can also be detected between the lines of other stories of descents to hell, in the depictions of the Paths of the Dead in the forest of Dimholt, and of the abyss in the depths of Moria. This collection of traditional themes (including that of the quest, associated with the destruction of the Ring of Power, inspired by the tales of the Golden Fleece, the Holy Grail, and the Niebelungen) provides Tolkien with the elements of an unknown English mythology which, against a background of irrationality, allows its readers to see with new eyes.

Tolkien's creation, as a whole, invites us to bear witness to what is incontestably an act of mythogenesis, bringing together the births, fates, and metamorphoses of supernatural beings, and the task of reconstituting their lineages and the laws governing their literary role is consequently given to us. Hence, Balrogs, including Gothmog, mentioned several times in *The Silmarillion*, recall certain demonic figures connected to fire in numerous mythologies, particularly Scandinavian ones. Likewise, Tolkien's eagles are as reminiscent of the giant raptors described in classical ornithology as they are of the Minka Bird in Australian Aboriginal folklore and the Thunderbird of Native American legend, and even the Roc, the enormous bird of prey that appears in Arabian fairy tales and is closely related to the Simurgh of Persian mythological tradition.[*] During the Great Battle that marks the end of the First Age, pitting Eärendil against the dragon Ancagalon the Black and leading to significant tectonic upheaval and, eventually, the expulsion of Morgoth from Arda (Earth), the text reverberates with Biblical echoes, including the Apocalypse according to St. John.

THE SPECIES OF ORIGIN

Tolkien's stories encompass a vast and ancient heritage, enabling him to address not the origin of species, but the species of origin. His creatures are the common mythological ancestors of families of demons, groups of monsters, and branches of wonders, all stemming from shifts characteristic of imaginary ecosystems. This is why oliphaunts can take on the mythical form of southern mammoths, just as snow-trolls can be related to Yetis, Mongolian almas, and other Sasquatches. The tentacled aquatic creature that rises up at the entrance

---

[*]   See the chapter "Flying Giants? Really?," p. 347.

to Moria harks back to a monstrous cephalopod; the giant squid or octopus, such as the Kraken.

As for the nameless winged creatures ridden by the Nazgûl, they seem related to fossil reptilian species such as Pterosaurs, but in some sort of tera-tological, or malformed, combination. Our suggestion is that they may be subnormal versions of dragons, Sauron not having had either the strength or the time to create true dragons at the end of the Third Age, given the onrushing speed with which events progressed. For, as Tolkien tells us in *The Silmarillion*, it takes time for a dragon such as Glaurung to reach the peak of its power. Is it possible that Sauron, attempting to match the creative hubris of his master Morgoth, managed to produce only a pale and weakened imitation of a real dragon?

Wrongly considered to be synonymous with cryptids (beings that are pre-sumed to exist, but without possessing any scientific proof of their zoological reality), mythomorphs populate folk-tales and fantastical texts, which provide poetic evidence confirming the truth of their existence.

It is Tolkien's aim to be the chronicler of these "mythozoic" times that pre-ceded the great chronological divisions and their traditional geological staging; if the Years of the Trees, which lasted nearly 20,000 years, can be compared to our mythical Golden Age, the Third Age would be, for the Dwarves, the Mithril Age, named for the mysterious white metal behind both their amassing of wealth and their downfall, and which saves Frodo's life. The author is dealing here with a world that has come down to us scientifically and literarily only as distorted echoes—but, whatever the field of knowledge confirming the veracity of this or that fact, it is clear that Tolkien's writings transcend the issue of truth, standing as an encyclopedic inventory of everything we know.

A QUESTION OF TIME

Myth is associated with a becoming, a linear timeline that guides its pro-gression. It has a duration; for example, the time spent by the Teleri Elves on the island of Tol Eressëa is long enough to foster the establishment of a unique language. This duration stems from an absolute origin point, making it identical, at least in this sense, to the natural, foundational historical and biological processes that make up the evolution of beings and species. In fact, the specialist Édouard Kloczko has suggested that those animals endowed with speech in Tolkien's world are actually Maiar (second-tier divinities in the service of the Valar) that have taken the form of beasts. This would be the case, then, with Huan, the Hound of Valinor, the eagle Thorondor, and indeed Sauron himself, who assumes the form of both a werewolf and a bat during the quest of Lúthien and Beren. It would also apply to wargs, which are capable of using a specific language, like the spiders of Mirkwood and the dragon Smaug. The gift of speech is thus linked to the divine essence of the being able to employ it.

The separate "camps" of abstract myth on one hand, and the concrete reality of animals, plants, and minerals on the other, no longer exist. Rather, both organic forms and legendary creatures share a sort of identity matrix; in short, for Tolkien, mythological beings become better identified as members of real species become more mythologized. Thus the reader becomes familiarized with what is supernatural in life and the fantastical elements of reality—in other words, the precise point at which the mythological essence of the living and the biological substance of the mythical coexist.

As protean as they are, mutations and variations remain classic in nature (dwarfism, gigantism) and occur around a sole organizational plan; that of four-limbed beings. It is suggested that the Elves held captive in the fortress of Angband in the First Age and tortured by Morgoth became orcs (in any case, orcs were created by Morgoth in an aping of the creation of the Elves, the Firstborn Children of Ilúvatar). Orcs also count among their numbers a separate and powerful race, the Uruk-hai, which remain largely compliant with the typical mammalian, humanoid form. Similarly, according to the inhabitants of the Forest of Fangorn, trolls are a kind of imitation Ent (troll, moreover, is written in Old English as *etten*, which is close to *Ent*). Stone-trolls are spirits imprisoned in rock by Morgoth, which are deathly afraid of light; other types of trolls include cave-trolls, hill-trolls, snow-trolls, and more. Trolls manifest the impossible condition of being bodies created from noth-ingness, whose 'becoming' reifies them as half-organic, half-mineral beings that dichotomously possess lifeless heartbeats and motionless breathing; they incarnate the insupportable heaviness of the non-being. Despite all of this, though, all trolls are bipedal.

The One Ring contributes actively to evolution and, in addition to increasing the lifespan of its wearer, causes changes that are first moral and then physical: hence, Sméagol becomes Gollum, Saruman becomes Sharkey (based on the Orkish for 'old man'); trolls become Olog-hai; and corrupt Men, the Nazgûl.

HUMAN DEPLETION

In contrast to the burgeoning of monstrous genealogies, human phylogeny becomes depleted with the passage of time, until only Man remains. However, several species of hominids seem to coexist during the First Age; in addition to the Dwarves, mysterious, swarthy men who are short and stocky, with long, powerful arms and hairy chests and faces, dark-haired and dark-eyed, appear in Beleriand. An eastern people that speaks numerous languages, these humans should not be confused with the Drúedain, or Drûgs, who migrated north along with the Haladin. The Drûgs are also called Woses, Wild Men of the Woods, or Rógin by the Rohirrim, who mercilessly drove the last members of their race out of Rohan during the Third Age. Living in small forest tribes, they are a secretive people that live in crude shelters, mistrustful of other races of men. Their lives are short and they rarely reproduce; they are skilled at tracking living creatures, thanks to their sharp senses of sight and smell, and their knowledge of everything that grows in the ground is almost as extensive of that of the Elves. On occasion, they prove to be implacable enemies of the orcs. They have no system of writing, and employ the knapped flint-stones, scrapers, and knives of the distant past.

In *Unfinished Tales of Númenor and Middle-earth*, the Drûgs are described as stocky and broad-shouldered, with deep, guttural voices and hearty laughs. How can their wide faces, sparse hair, flat noses, and deep-set eyes beneath low, protruding brows fail to remind us of Neanderthal Man—or at least the popular image of him with which people were familiar in Tolkien's day? After all, some have seen the discovery of Florès Man as proof of the existence of hobbits;[*] here, perhaps, is the *Homo floresiensis* of Great Britain. The Drûgs were taller, stockier, and less pleasant-faced than hobbits; they had a grimmer and more enigmatic character, and were thought to have supernatural powers very different from those attributed to the Dwarves. After making contact with other peoples, they became engravers of wood and stone, producing images of themselves crouching over dead orcs and placing these images, called "watch-stones," on the edges of pathways. It is said that they are able to place part of their own power into these stones, which temporarily come to life if needed—which is reminiscent, even according to Tolkien himself, of the transfer of Sauron's evil powers into the One Ring.

---

[*]   See the chapter "When a Hobbit Upsets Paleoanthropologists," p. 218.

Creating a tremendous and incredible pool of symbols and images, Tolkien takes us gradually back through the mists of time, turning the resurrection of these lost worlds into a veritable argument against the history of life as we understand it. This novel experience (rendering tangible the chronological incorporeality of time immemorial), gives rise to an unprecedented perception of the stuff dreams are made of and makes continuity an explicative value of (super)natural history. Tolkien's work testifies to this historicity of a continuum feeding into the imaginary; in the sequential nature of his volumes and his cycles he manages to cut through time, to hollow out the thickness of the world so that, in the end, we are given a glimpse at the possibility of a history of history, at the virtually endless track of deep time, at the very first sunrises, the predictive forms.

TOLKIEN, THE ORNITHOLOGIST

ANTOINE LOUCHART, Paleontologist, CNRS
and École normale supérieure de Lyon

olkien, that lover of trees, filled his writings with birds. Let's see if we can spot them; let's listen for their songs and cries along our way. Let's try to glimpse them in the corner of a page, and observe them like Legolas, who, experienced birdwatcher that he is, can identify the raptors wheeling far overhead well before the other members of the Fellowship of the Ring.* As we'll see, in Arda, Tolkien's world, birds are much like the other animals and plants: quite similar to our real-world birds in some ways (and with the same names), but created with an added, imaginary, fantastical dimension. Tolkien's treatment of birds is special even beyond its magicality, however; he is constantly setting them apart from other animals. Let's take a walk through *The Silmarillion*, *The Hobbit*, and *The Lord of the Rings* now, in search of the winged race.

A STROLL IN THE FORM OF A BALLAD

In Tolkien, sounds are incredibly important, giving life to his world—and this, even without mentioning the special attention paid to its various languages. Music, lays, lamentations, melodies, and hobbit-songs, as well as the soundscapes of nature, including the songs and cries of birds, are omnipresent in the

---

* See the chapter "The Eyesight of Elves," p. 233.

narrative. Bird sounds are often evoked in unexpected ways—through their absence, or when a heavy or disturbing silence settles over a scene. This occurs in areas close to the enemy or in the presence of death or imminent danger, as well as in the critical moments preceding a major event, or a showdown with the powers of darkness. Birds, it is implied, are for Tolkien a necessary backdrop to life.

While the mysterious Tom Bombadil whistles "like a tree-full of birds" in *The Fellowship of the Ring*, and Rivendell and Ithilien are filled with birdsong, the absence of birds and their calls makes itself felt in *The Hobbit*, first on the escarpment of Smaug's mountain, and then with Bilbo's nephew Frodo, when he travels through the region of Hollin; at the entrance to the tunnels of Moria, and along the riverbank after he leaves Lothlórien. In some swamps and marshes, what rare bird-calls are present become "wails." On the peak of Amon Hen, it is only after the passage of the Shadow searching for Frodo that the birds begin to sing again, and the sky becomes clear and blue. The situation worsens the closer one gets to battle and to Mordor (where the absence of birds is total), as we see in the foothills of Emyn Muil, in the Dead Marshes, and when leaving Ithilien. In Minas Tirith, just before the fall of the corrupt and disgraced Maia Sauron, time stands still and there is "neither wind, nor voice, nor bird-call." In moments of deepest despair, as when Frodo is held prisoner, condemned to death, in the Towers of the Teeth, it is the merry song of finches that enables Sam to locate his friend, and hope is reborn. Birdsong is synonymous with the joy that persists deep in the hearts of hobbits, even in the most dire of circumstances.

Let's go back to the beginning, now, and tread softly so that we don't frighten the first birds, those connected to Manwë, the greatest of the Valar, master of the winds and the air, in *The Silmarillion*.

THE GENESIS OF BIRDS ON ARDA

Birds were created by the Vala goddess Yavanna in the first days of Arda. For the Elves—especially the Nandor, who know them best—birds are beings endowed with the gift of speech, capable (it is implied) of communicating with the Firstborn of Ilúvatar in particular, but also with each other via music, a language that some Elves are able to learn. The inhabitants of Lothlórien, for example, sometimes communicate with one another via calls like the low whistle of a bird.

Birds in Middle-earth

The first birds we recognize are nightingales, companions of Melian the Maia. This bird evokes the eternal spring of the First Age, similar to the springtime rebirth seen in the courtly poetry of troubadours. It is said that when Melian taught the nightingales how to sing, at the moment when the Two Trees mingled their rays of light, everything stopped; even the fountains "forgot to flow," and the birds of Valinor, the original, first land, fell silent. This peaceful, reverential quiet is utterly opposed to the absence of birdsong in the Third Age mentioned above. Later, the Elf Lúthien, who renounces her immortality for love of Beren, is given the nickname Tinúviel, the Sindarin poetic term for 'nightingale', by her husband. This bird sings before the battle between Huan the Hound and Sauron, its crepuscular (that is, occurring at twilight) and nocturnal song symbolic of the struggle against the forces of darkness and decline. A frequent presence in *The Silmarillion*, the nightingale is mentioned only once in *The Lord of the Rings*, in a song—an illustration, perhaps, of the lost paradise of the First Age.

Birds themselves are combatants on occasion. We quickly become acquainted with eagles and hawks, both also emblematic of the medieval bestiary and symbols of the hunter-warrior aristocracy. These two raptors act as Manwë's eyes and ears, and indeed some real-world hawks have incredible eyesight, able to discern objects up to eight times farther away than human eyes can see. Eagles, associated with the sun and considered the king of birds in the symbolism of many cultures, are also entrusted by Manwë with the task of relaying the words of those who invoke the Valar.

SPACE FOR SWANS

Also in *The Silmarillion* we find white swans, associated with the Elves and with their sailing vessels, which are often made in the image of birds. The swan symbolizes purity and nobility, but also power, with its majestic bearing, its whiteness, and its large size (some modern mute swans are the heaviest birds still capable of flight). Like eagles, hawks, and nightingales, the swan is a bird of the Valar. The Vala Oromë sends "many strong-winged swans" to draw the ships of the Teleri Elves (whose emblem is a swan) across the sea toward the land of Aman. Eärwen, a Teleri Elf who is the daughter of King Olwë, is given the nickname "swan-maiden." The ship of Eärendil has its prow carved in the form of a swan, and this same silver swan will later become the emblem and banner of Dol Amroth, an ally of Gondor in the Third Age. Swans are found elsewhere, too, including the reedy marshes of the Swanfleet river, but most striking, perhaps, are the ones depicted as swimming on the open ocean—a rare practice among real-world swans, which, though they do fly over seas, are freshwater and coastal birds.

The migratory flight of these great birds plays a significant role in a particular episode portrayed in *The Silmarillion*: seven swans flying south prove to be a harbinger of the fall of Nargothrond. This is ornithomancy, or "auspice," which, in antiquity, designated a priest who saw omens of the future by observing birds. The movements of birds are often related to climatic or meteorological conditions, which the author reports in detail. Taken all together, these phenomena often act as harbingers in Tolkien's world.

The gatherings of birds that have come from the south after the death of the dragon Smaug are rightly perceived as anomalous compared to normal migratory movements in winter. These heralding signs, marks of destiny or providence, also contribute, on a larger scale, to the cyclical flow that associates wars and tribulations with the coming of winter. Does the somber role of certain birds help to instill a sense of grimness and melancholy when dark events are set to occur in Tolkien's world?

FROM THE GRIM TO THE GROTESQUE

Gulls and other seabirds are closely linked to the seas, to Elves, and to their nostalgia for Valinor, their place of origin across the ocean—a nostalgia so deep that it eventually overcomes the Elves of Middle-earth, as well as those who have worn the One Ring, and is manifested by the plaintive cries of these gregarious birds. This call becomes a growing obsession for Legolas, until his eventual departure by ship from the Grey Havens. Frodo even manages to hear the wail of seabirds in a far-distant river estuary, and later, with the One Ring on his finger, he sees them in a vision at the summit of Amon Hen. In the First Age, the Vala Ulmo gives Elwing the appearance of a seabird with white and silver-gray wings, like a gull, and she learns the language of the birds on this occasion. Certain avian-inspired embellishments continue to be embraced by the Men of Middle-earth as well, as in the case of the helmets worn by the Guards of the Citadel, which bear the carved white wings of seabirds, and the crown of the kings of the island of Númenor, with its wings of silver and pearls.

The lark, another bird favored by troubadours and symbolic of the renewal of spring, is mentioned in *The Silmarillion* in language that clearly conveys Tolkien's great sensitivity to certain birdsongs: "suddenly [Lúthien] began to sing. Keen, heart-piercing was her song as the song of the lark that rises from the gates of night and pours its voice among the dying stars." The last sound from Lothlórien heard by the Fellowship of the Ring is the distant trill of larks.

Larks, nightingales, and other songbirds are closely associated with femininity. Wherever Lúthien goes, birds sing even in winter, "beneath the snow-clad hills." Another indication of their importance lies in the fact that Elves making an offering in Númenor bring not only flowers, herbs, and a seedling of Celeborn the White Tree, but songbirds as well—and Legolas promises similar offerings for Minas Tirith after the war: birds that sing and trees that do not die.

What a contrast with Gollum's attitude! For him, birds are nothing but prey to be eaten. Orcs joke about birds and sing a cruel ditty to the trapped Bilbo, Gandalf, and the Dwarves: "Fifteen birds in five fir trees," in which the poor characters are birds with "no wings." Hobbits are also likened to impish children that trap birds, shoot at them with slingshots, and knock them from their nests, attracting them by making birdlike chirping sounds: the rural games of yesteryear, common in the countryside so beloved by Tolkien.

BIRDS OF GOOD AND EVIL

Imperceptibly, evil insinuates itself into the makeup of certain birds, where originally there was only good. In *The Silmarillion*, all of the birds mentioned are on the side of good. The nightingale so prominently featured in these tales is all but absent in the accounts of the later ages chronicled in *The Hobbit* and *The Lord of the Rings*. In these books, a number of birds appear for the first time. This is the case with various members of the Corvidae family (ravens, crows, and *crebain*), and with other scavenger birds such as vultures. These birds, which lack nobility and are almost always allied with the enemy and with death (the ravens in *The Hobbit* are an exception to this rule), are indicative of the regression toward a less beautiful, less noble, less eternal world. The tale told by the Ent Bregalad (or Quickbeam) of the corruption of the birds as a momentous historical shift gives us an image of this negative change: "But the birds became unfriendly and greedy and tore at the trees, and threw the fruit down and did not eat it." We are witnessing an estrangement between the trees, so dear to Tolkien, and the birds, which have become more ambiguous. Such birds go over to the side of the enemy in various eras, but none of them were originally bad.

Ravens, used generically to represent all Corvidae, were mainly positive symbols praised for their wisdom in Antiquity and in Celtic and Germanic tradition, as well as Scandinavian legend. It was the influence of Christianity that caused the switch to associating ravens with the forces of darkness and death, undoubtedly in part due to their black color. Bilbo, however, learns that while crows are spies working for the enemy, ravens (in the strict sense of the term) are on the side of good. They are friends and messengers of the people of Thrór, who compensate them for their service with shiny objects, and possess great intelligence and exceptional memories. Bilbo and the Dwarf Balin engage in conversation with a raven called Roäc, who is 153 years old. Like the eagles, ravens are able to communicate with various characters—a clear sign of Tolkien's recognition of their considerable intellectual capacities, and indeed, recent studies of the cognitive abilities of several Corvidae have reached the same conclusions.

Unlike the white swan, the black swan, a powerful symbol of occult power, appears only once, when the members of the Fellowship find themselves in worrisome circumstances. Having departed from Lothlórien, they are traveling south via the river when plaintive sounds, described as that of swan-wings beating the air, are heard. These are black swans. Named *Cygnus atratus* and first discovered in Australia, these birds do exist in real life, and it is perhaps noteworthy that a related species lived in New Zealand, the home country of filmmaker Peter Jackson, until the Polynesian colonization, with the Australian breed having been reintroduced more recently.

FAMILIAR BIRDLIFE

Most of the birds of Arda are at least roughly identifiable with common European species (finches, starlings, etc.). By virtue of the environments they inhabit, they suggest landscapes similar to those of the English coastlines and countryside. Many marsh birds live on the swampy banks of the lake of Lin-aewen in Nevrast, and in the Midgewater Marshes navigated by the Fellowship of the Ring. Humid environment-loving swans, Corvidae, coastal seabirds, and various relatively common songbirds complete these landscapes. In *The Adventures of Tom Bombadil*, Tolkien uses a single bird to symbolize the English countryside and its rivers, ponds, and marshes: this is the kingfisher, from which Bombadil recovers a blue feather for his hat. The most favorable climates for these birds are temperate ones, often with a wintry atmosphere. The birds that live in these areas are non-migratory or partially migratory, meaning that some (Scandinavian) populations winter at lower latitudes, and that, for this reason, the species is present all year in England. So, in the same way that paleontologists use the modern habitats of species discovered in the fossil state to reconstruct past environments, we can apply this practice—which we'll call "terrianism"—to Tolkienology! Terrianism consists, for our purposes, of looking at certain characteristics of elements that exist both in our world and in Tolkien's writings, to deduce characteristics of the Tolkienian world. These elements can be birds that exist in both worlds, but other things as well—living beings, structures, etc.

Alongside these "mutual" birds appear mysterious ones that are more difficult to identify, which have been the subject of particular attention from ornithologists.

MYSTERY BIRDS!

Let's start with the thrush, which plays a major role in *The Hobbit*, keeping a low profile and then, at the last moment, revealing the dragon Smaug's vulnerable spot to Bard the Bowman. Upright in posture and large in size, this bird displays behavior typical of the song thrush, cracking snails' shells with the aid of rocks or against a boulder in order to eat them. Its trills are similar to those of many members of the Turdidae (thrush) family, but its plumage is unlike that of any known thrush, except perhaps for *Zoothera monticola*, the mountain thrush of southern Asia. However, it is unlikely that Tolkien intended to introduce a real-world exotic species into his world, and in his England it would probably be a hybrid between a large male common blackbird (its feathers are "nearly coal black") and a mistle thrush (it has a "pale yellow breast freckled with dark spots"). It is interesting to note that this thrush is part of a species that was tamed by humans. While it does not actually converse with Tolkien's characters, it understands them, and acts quietly to help them. The wizard Radagast is a "tamer of birds," as his sinister colleague Saruman notes with disdain—which may indicate Tolkien's own respect for this practice, which was widespread in the England of his day.

Also belonging to the Corvidae family, the crebain, native to Dunland and to Fangorn Forest, appear in Hollin and are both spies and harbingers of dire events—they are literally "birds of ill omen"! Crebain are "a kind of crow of large size," like certain subspecies of real crows that lived during some of Europe's ice ages and are known by their fossils. Likewise, the swans of Gorbelgod, which are white and extremely large, are mentioned in ancient Númenorean legend. Were these early swans ancestors of the current ones? The line has, perhaps, undergone a reduction in bodily size over the span of many thousands of years, as is the case with eagles[*]—a phenomenon that could be interpreted, once again, as a negative evolution of the world toward less majesty, less nobility.

Another example, found this time in *Unfinished Tales*, is the kirinki. This bird, smaller than a Eurasian wren, or jenny wren, with its scarlet plumage and voice so high as to be inaudible to humans, was native to the island of Númenor. Here again, there are numerous real-world birds with attributes that correspond to the kirinki's, even down to the ultrasonic trill (produced by the guacharo, or oilbird, a large cave-dwelling bird). Some hummingbirds are scarlet-feathered and very tiny and have very shrill cries, but there are others as well. Best not to take our analysis and deconstruction too far, though—for, as Gandalf says to Saruman regarding the breakdown of light, "he that breaks a thing to find out what it is has left the path of wisdom."

Birds often appear as simple evocations, images, or metaphors, without actually having been physically encountered in the course of the story. Gandalf's fireworks take the form of birds, eagles, swans. Tom Bombadil whistles like a starling. Saruman and other traitors, as well as the men under their influence, nickname Gandalf "Stormcrow."

Of the birds that are specifically mentioned, the two sentries that guard the entry to the Towers of the Teeth in Mordor are each formed of three half-living vulture effigies, joined together. They are undoubtedly the most

---

[*]   See the chapter "Flying Giants? Really?," p. 347.

The Raven

negative of all the birds cited in Tolkien. On a different note entirely, this same quality of strangeness and incongruity appears again when the Ents arrive "swiftly" as reinforcements at the end of the Battle of Helm's Deep, "walking like wading herons in their gait, but not in their speed [. . .]." We can imagine Ents taking long, quick strides, but what an odd comparison!

Tolkien borrowed from legend and symbolism both ancient and medieval, Norse and eastern—and often in original and unexpected ways.

MEMORIES OF OLIPHAUNTS

ARNAUD VARENNES-SCHMITT, doctor of paleontology

"Grey as a mouse,/Big as a house,/Nose like a snake . . ." Mûmakil, which hobbits call *oliphaunts*, are huge elephant-like creatures native to the region of Harad, south of Gondor and Mordor. They are used mainly by the Southrons, or Haradrim, as war-steeds. They appear briefly in *The Two Towers* and more extensively in *The Return of the King*, during the Battle of the Pelennor Fields. Just how realistic are these creatures? Comparing their anatomy to that of existing elephants, and also to their extinct relatives, will answer this question. Another point of interest lies in examining their use as war-steeds, by drawing parallels with the very real war-elephants that have been pressed into service throughout history.

A FAMILY PORTRAIT

The three elephant species that still exist on our planet today (the Asian elephant, *Elephas maximus*; the African bush elephant, *Loxodonta africana*; and the African forest elephant, *Loxodonta cyclotis*) belong to the much larger zoological order Proboscidae (from the Greek *proboskis*, meaning "trunk"), which includes elephants and their close relatives. Proboscideans appeared around 60 million years ago and there have been more than 170 species of them, virtually all of which are extinct. As oliphaunts are not elephants, strictly speaking, did Tolkien draw inspiration from fossil Proboscideans? Let's look over the principal representatives of this zoological order.

The oldest known proboscidean fossil was found in Morocco and has been dated at 60 million years old. It is an example of *Eritherium azzouzorum*, a small mammal not much bigger than a cat, and very different from modern elephants. It has neither a trunk nor large ears, and possesses no tusks, and a non-specialist would have a very hard time connecting it to the elephants we know—and yet, on the basis of anatomical comparisons (particularly cranial features), *Eritherium* is, without a doubt, the earliest known representative of the order Proboscidae. There is no shortage of examples of extremely ancient proboscidean fossils that are very different from modern elephants. These include *Phosphatherium*, also found in Morocco, and *Moeritherium*, an animal thought to be largely dependent on aquatic environments. However, these animals cannot have influenced Tolkien in the creation of his oliphaunts, with their anatomies that bear no resemblance to his fictional creatures, and their much smaller size—not to mention the fact that both *Eritherium* and *Phosphatherium* weren't discovered until well after Tolkien's death.

Among the fossil proboscideans, one genus with a very specific set of anatomical characteristics stands apart: these are the Deinotheriidae,[*] which lived between 27 million and 2 million years ago. Their appearance is reminiscent of elephants: a trunk, pillar-like limbs, and large size (up to five meters, or 16½ feet, for *Deinotherium giganteum*). Their most significant characteristic is the possession of tusks in their lower jaw, rather than the upper, meaning that their tusks pointed downward and toward the rear of their bodies. Their size makes them a good candidate for Tolkien's oliphaunt inspiration, but the uniqueness of their tusks doesn't match his description—he says the tusks are "upturned" and "hornlike," which bears no resemblance to the inward- and downward-curving tusks in the lower jaws of deinotheres.

Another striking example of the order Proboscidae is the gomphothere.[†] These animals actually possessed not one, but two pairs of tusks, one in the lower jaw and one in the upper. The most widely accepted hypothesis on this point is that the lower tusks were used to dig in the ground in order to unearth roots to eat. Here again, the double-pair of tusks takes gomphotheres off the list of possible influences on Tolkien, as he would surely have mentioned this particular attribute.

---

[*] A group of fossil proboscideans sporting a pair of downward-curving tusks set into the lower jaw.

[†] A group of fossil proboscideans with two pairs of tusks.

That leaves the best-known family, the Elephantidae, or elephantids,[*] which notably includes the three elephant species currently living, as well as mammoths. The elephantids, with the exception of *Primelephas*, have a single pair of tusks (actually incisors that have transformed over time) located in the upper jaw. This family is the best match for Tolkien's description of the oliphaunt, except for one detail: size. A mûmak is much larger than any elephantid, living or extinct. However, given the evidence, we must conclude that oliphaunts are probably an exaggeratedly large member of the elephant family.

WHERE DOES THE OLIPHAUNT FIT IN?

The description of oliphaunts given in *The Two Towers* is quite vague. Most of the emphasis is placed on their size; they are "big as a house," a "moving hill," a "vast shape." The animal's legs are like trees, which corresponds well to the pillar-shaped limbs of elephants. Their ears are "enormous" and "sail-like," their trunk like a huge serpent, and their "small red eyes raging." Finally, their tusks are "upturned" and "hornlike." Tolkien is rather stingy with details, and it is difficult to put together a more detailed composite sketch of the oliphaunt—but, if we start with the hypothesis that they are indeed elephantids, we can complete Tolkien's anatomical description.

The description that follows is based on modern elephants. Their skeletons possess the same organizational structure common to most mammals, with some particularities. First of all, their mode of locomotion is unique; it consists of putting the sole and phalanges of the foot in contact with the ground at the same time with each step. While the phalanges are in direct contact, the carpal and tarsal bones rest on a fatty cushion located in the rear of each forefoot and hind foot. Additionally, the limbs have a so-called "columnar" orientation, meaning that they extend almost vertically beneath the body, like tree trunks! This characteristic allows them to stand still without expending too much energy. Their skulls are massive, very solid, and yet made lighter by a network of bony, air-filled cells that extends throughout the cranium. In the center of the skull, a large depression marks the area where the trunk is attached. This organ possesses numerous muscles and enables the animal to

---

[*]   The group of proboscideans that includes mammoths and modern elephants.

drink (water is sucked up into the trunk and then emptied into the mouth) and to grasp objects such as food.

Elephants also have large ears that, in addition to giving them an excellent sense of hearing, enable them to regulate their body temperature. Large animals like this have a tendency to accumulate a great deal of heat. The ears, with their networks of blood vessels spread over a large area, allow for better dissipation of heat. African elephants, which live in hotter regions, have larger ears than Asian elephants. Tolkien gives his oliphaunts enormous ears as big as sails, which is relevant to their geographic origin, the region of Harad, with its extremely hot climate. The incredible thickness of oliphaunts' skin is also specifically mentioned in *The Lord of the Rings* ("the triple hide of his flanks"), which again corresponds to the skin of actual elephants, which is so thick that it was the inspiration for the former name (and current nickname) of proboscideans: pachyderms, literally "thick skin."

With regard to dentition, oliphaunts clearly share a particular characteristic with elephantids: the horizontal replacement of their bicuspids and molars. Unlike most mammals (including humans), whose new teeth grow in vertically, elephantids go through several generations of teeth which succeed one another horizontally. This means that the bicuspids and molars of elephants are not all present simultaneously in their jaws. The first teeth begin to wear down and are pushed forward by the appearance of new ones, eventually falling out when they wear out completely. A total of six teeth (three bicuspids and three molars) will emerge during the course of an individual's development. When the sixth tooth is fully worn down and falls out, the elephant ceases to feed itself and dies.

TOO MANY TUSKS FOR OLIPHAUNTS?

Let's step away from Tolkien's books for a moment and look briefly at the choices made in Peter Jackson's film trilogy. The oliphaunts are gigantic in these films, each carrying more than a dozen soldiers in towers mounted on their backs. But they are also represented with not one, not two, but *three* pairs of tusks! There is one pair in the upper jaw, which is very large and looks like a pair of upturned horns, matching Tolkien's description. Then a smaller pair has been added in the lower jaw, recalling the tusks of gomphotheres. And finally, a third pair of tusks, this one very small and positioned laterally in relation to the large upper tusks, completes the set. There is no known proboscidean, living or extinct, equipped with three pairs of tusks, so the question becomes: which pairs are equivalent to known pairs of tusks in real-world proboscideans, and which is an evolutionary innovation developed only by Peter Jackson's oliphaunts?

To answer this question, let's compare the upper dentition of three real species—the Asian elephant (*Elephas maximus*), the wild boar (*Sus scrofa*), and the gomphothere *Gomphotherium augustidens*, a fossil species—and two fictional species, which we'll call *Mumakitherium tolkieni* (J.R.R. Tolkien's oliphaunt) and *Mumakitherium jacksoni* (Peter Jackson's oliphaunt). The dentition of *Elephas maximus* consists of a single incisor modified into a tusk and several molars subject to horizontal dental replacement. Note that modern elephants do not possess canine teeth. The male wild boar possesses three incisors and a highly developed canine called the whetter, which is used to sharpen the lower canine (which is modified into a tusk), four bicuspids, and three molars. The wild boar is not a proboscidean, but its unique dentition is useful in an anatomical comparison of the teeth of the two species of oliphaunts.

Let's look at *Mumakitherium tolkieni* now. As we have seen, the oliphaunt described by the author is undoubtedly a very large elephantid, and so its dentition is probably comparable: one incisor modified into a tusk, and horizontally-replaced molars. As with elephantids, the incisor modified into a tusk is equivalent to the second incisor (I_2) present in other mammals such as the wild boar.

The upper dentition of *Mumakitherium jacksoni* remains the same of *Mumakitherium tolkieni* and the elephantids, except for the presence of a second, smaller set of tusks between the first set and the molars. Two hypotheses may explain these additional tusks: either this is a new modified incisor, in which case, given its position, it can only be the counterpart of the third pair of incisors (I_3) present in other mammals including the wild boar; or it is a highly developed canine similar to the whetters of wild boars, acquired via evolutionary convergence. Both of these possibilities, though, are highly unlikely, as elephantids have lost their canines as well as the I_1 and I_3 incisors. Even though some very ancient proboscideans, including *Phosphatherium*, still possessed three incisors and a canine, their loss in more recent proboscideans renders the cinematic version quite far-fetched—though certainly more effective on the battlefield!

I (incisors) = I
C (canines) = C
P (bicuspids) = B
M (molars) = M

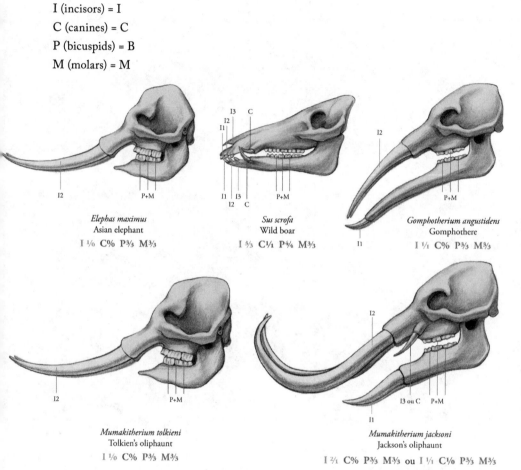

Elephas maximus
Asian elephant
I ¹⁄₀ C⁰⁄₀ P³⁄₃ M³⁄₃

Sus scrofa
Wild boar
I ³⁄₃ C¹⁄₁ P⁴⁄₄ M³⁄₃

Gomphotherium angustidens
Gomphothere
I ¹⁄₁ C⁰⁄₀ P³⁄₃ M³⁄₃

Mumakitherium tolkieni
Tolkien's oliphaunt
I ¹⁄₀ C⁰⁄₀ P³⁄₃ M³⁄₃

Mumakitherium jacksoni
Jackson's oliphaunt
I ²⁄₁ C⁰⁄₀ P³⁄₃ M³⁄₃ ou I ¹⁄₁ C¹⁄₀ P³⁄₃ M³⁄₃

FIGURE I

A comparison of the dentition of three real species
(above) and two fictional species (below).

WEAPONS OF MASS DESTRUCTION

Oliphaunts in *The Lord of the Rings* are basically war machines. They are ridden by the Haradrim (Southrons), who ally themselves with Sauron at the Battle of the Pelennor Fields. Tolkien probably drew inspiration from real facts here, as the few details he supplies are reminiscent of scenes described in historical texts. Indeed, Hannibal's famous attempt to conquer Rome with his elephants after having crossed the Alps tends to overshadow the fact that these animals have been used to make war for more than three millennia, from eastern Asia to North Africa, not to mention the Middle East and Europe.

The goal of the oliphaunts and their mahouts[*] in the Battle of the Pelennor Fields is to storm the gates of Minas Tirith and take the city. Historical data shows that proboscideans could have played several roles. Firstly, as in *The Lord of the Rings*, they created chaos among the opposing ranks and weakened the enemy. Horses are thought to be highly sensitive to the sight and smell of elephants; they are terrified of the beasts, and will stop obeying their riders. This detail is included in *The Return the King* when the horses of the Rohirrim, warriors native to Rohan, refuse to advance toward the oliphaunts and shy away, frightened, at the sight (and smell?) of these strange monsters.

Elephants can inflict colossal damage in the midst of opposing troops, liable to trample enemy soldiers, or seize them and fling them to the ground. Elephants have also historically been used as siege weapons, mainly in India, where they were trained for this function, and defensive fortifications were eventually covered with spikes to keep the powerful animals away. Elephants also aided in preparations for battle by transporting heavy materials, and were used as breakwaters to facilitate river crossings; this is how Perdiccas, a general (or diadochus) of Alexander the Great, employed the animals. Upon the latter's death in 323 B.C., the kingdom of Macedonia and empire over which it ruled found themselves without a leader. War then broke out between the pretenders to the throne, including Perdiccas. Determined to recover Alexander's body, which had been taken to Egypt by Ptolemy, another diadochus seeking to gain the upper hand, Perdiccas found himself

---

[*]   Elephant trainers who also rode the animals in battle, generally astride the back of the neck.

temporarily thwarted by the barrier of the Nile. To cross it, he caused his elephants to be lined up in the river in order to reduce the strength of the current—and yet this endeavor brought its share of disaster anyway, as several thousand soldiers perished in the execution of it.

The mûmakil in *The Lord of the Rings* are caparisoned[*] in scarlet and gold, their tusks adorned with bands of the same colors. Tolkien also equips them with war-towers strapped to their backs, in which the *mahouts* and other Haradrim rode. He is undoubtedly making reference here to the howdah, a sort of platform attached to the back of an elephant that could carry up to four soldiers. In the case of oliphaunts, which are larger than real elephants, these howdahs become "buildings" able to contain several dozen soldiers. Tolkien does not give us any information about the defensive equipment worn by his beasts, but war elephants often wore leather armor, sometimes with metal parts protecting the highly vulnerable forehead and trunk.

The tusks of war elephants could be protected, but also transformed into deadly weapons. For example, those of the elephants of King Bimbisâra of Magadha (558–491 B.C.) were tipped with metal points. Similar objects are recorded as being worn by the elephants employed by the Persian king Darius III in his battle against the armies of Alexander the Great in 331 B.C. And the Turkic ruler Mahmud of Ghazni equipped his elephants' tusks with swords when he repulsed the forces of Ilak Khan, king of Kashgar, in around 1000 B.C.

Do oliphaunts, with their immense size and thick skin, have any weak points? Tolkien does not tell us in detail how the oliphaunts are killed at the Pelennor Fields, but none survive the battle. How is this possible? We have to presume that the eyes are their main Achilles heel; indeed, we learn after the battle that Duilin, son of the lord of Blackroot Vale, and his brother were killed leading an attack on the mûmakil with the intent of shooting them in the eyes with arrows. Real war elephants had several weak points, starting with their trunks. These organs are highly sensitive and, if cut, cause the animals to lose huge quantities of blood. It isn't hard to understand the desire to protect this part of their anatomy.

Elephants are also highly vulnerable on the bottoms of their feet, where the skin is thinner. During the siege of Megalopolis by Polyperchon, Damis

---

[*] A caparison is a piece of ornamental cloth draped over the back of an animal used for riding (usually a horse).

took advantage of this weakness with the ingenious idea of scattering the city's streets with nail-studded boards. When the elephants trod on these, they caused so much pain that the animals could neither advance nor retreat.

Elephants are also easily frightened by loud noises. One strategy consisted of covering pigs with flammable material and setting them alight near the war-elephants. The pigs' shrieks of agony terrified the elephants, which usually fled, causing a chaotic free-for-all. Another use of fire involved throwing naphtha-filled jars which burst into flames when they struck the elephants.

Tolkien's oliphaunts are enormous creatures very similar to the elephants of our world and have been used in the same way in the wars of our past. Yet we might wonder how truly useful they were in battle. Their destructive potential is certainly appealing, but they often cause as much damage to their own camp as to the enemy's. This hasn't stopped other fantasy authors from pressing them into service, however, including George R.R. Martin in his *Game of Thrones* series.

One last point in common between the imaginary creatures and real ones: due to being captured for use as a war machine, the oliphaunt becomes an endangered species. What remains of the proboscideans in Middle-earth is nothing compared to the numbers that once lived—a situation lamentably similar to that of elephants on our planet.

WARGS: WAR-DOGS OF SCANDINAVIAN ORIGIN?

VINCENT DUPRET, doctor of paleontology
ROMARIC HAINEZ, associate professor of
Life Sciences and Earth Sciences

Recent debates over the culling of wolves in regions where they pose a threat to livestock are only the latest episode in the conflictual relationship between humans and wolves. It is a very old story, as the two species have lived side by side since prehistoric times. Tolkien makes use of this fact by featuring hordes of gigantic mutant wolves called wargs, which attack our heroes in both *The Lord of the Rings* and *The Hobbit*. Now here are some creatures that are truly the stuff of nightmares! But what do we know about wargs, exactly? Very little, and a lot, at the same time. As he did for the rest of his universe, Tolkien did a great deal of research . . . but in the case of wargs, not much of it has come down to us. A little detective work is called for here, to unearth the author's source of inspiration. Why and how did Tolkien create his wargs? And why are they so terrifying?

Fear of wolves goes back at least to the Middle Ages, following a campaign of demonization launched by the Christian church, and folk-tales of the era did little to improve their image. Bears suffered the same fate, going from being considered king of the animals to being thought of as stupid and dangerous beasts![*]

---

[*]   See the chapter "Beorn: Man-Bear or Bear-Man?," p. 314.

But the wolf has not always been seen as a malevolent figure. Though it has always been a predator and direct competitor of humans, some civilizations have revered it. Remember the founding myth of Rome, of Romulus and Remus, rescued and nursed by a she-wolf. In other societies, including those of the Native Americans in the northeastern United States, the wolf is a symbol often associated with family values (living in groups and caring for their young), as well as with strength and courage. It can also be symbolic of male or female sexuality, depending on context, as in the expression "wolf whistle."

Wolves and bears are still seen as "harmful" in parts of the rural world. While they are enjoying a resurgence of popularity among city-dwellers, the figure of the Big Bad Wolf, perpetuated by certain children's books, continues to inspire terror in little ones. In Tolkien, it is the wargs who occupy this role. Where do they come from? The authors will have had a multitude of sources to draw on . . .

A WARG'S PEDIGREE

Wargs must be intelligent creatures, because they have allied themselves with the orcs without needing to be trained. Some associate them with the first werewolves in Beleriand, all sired by Draugluin, or with the wolf-dogs of the line of Carcharoth (himself bred from Draugluin) in the First Age. Draugluin was the first werewolf to serve Sauron, living with him in the fortress of Tol-in-Gaurhoth. He was killed by the hound Huan during the Quest for the Silmaril. Later, Beren and Lúthien use his pelt to sneak into Angband.

Tolkien left no notes regarding his choice of name for wargs, but you don't have to look far for some interesting possibilities. For example, 'warg' may suggest a contraction of *war-dog*, after the animals chosen from among a few species selected for their special abilities and used as trackers, messengers, in combat, etc. by the ancient Egyptians, Greeks, Romans, and throughout history even to the present day. One of the earliest battles to feature war-dogs pitted Ephesus, in Asia Minor, against the neighboring city of Magnesia on the Maeander.

The name 'warg' may also be of Scandinavian origin. In Old Norse (and Old Icelandic), *vargr* is the word for 'wolf', and it is still used in modern Sweden, though the term *Ulf* (evocative of the English *wolf*) has become more widespread throughout Scandinavia. The word *Varg* is also associated with the fearsome wolf Fenrir, who brought about the twilight of the gods; Tolkien may have drawn once more on Scandinavian legend to create a wholly evil lupine creature. It is also notable that *varg* would have become *warg* in Old English, and been used to designate particularly bloodthirsty wolves.

A SCANDINAVIAN INSPIRATION?

Who was the Fenrir of Norse mythology, archenemy of the ancient Scandinavian gods? Also called Fenris, or Fenrisúlfr, Fenrir was an enormous wolf, a son of Loki and brother to Sleipnir (the eight-legged horse ridden by Odin), Hel (the goddess of the dead and of the underworld, whose name inspired that of Hell), and Jörmungandr (the gigantic serpent that encircles the globe, causing earthquakes, and is Thor's sworn enemy).

One day, Loki brings the wolf-cub Fenrir to Asgård, the realm of the gods. Týr, the god of war, is assigned the task of training him, but all of Asgård soon realizes the truth: the wolf, which is growing larger and larger, is evil at heart. Moreover, Odin has had a vision of his own end, in which he is devoured by Fenrir during Ragnarök, the twilight of the gods. He must get rid of the wolf at all costs.

To accomplish this, the Æsir (the main gods, of whom Odin is the leader) call on the dwarves, who create Gleipnir, an enchanted binding with a silky and delicate appearance. The magical components of this binding are evocative of its function: the noise of a cat's footsteps, the beard of a woman, the breath of a fish, the roots of a mountain, the nerves of a bear, the spittle of a bird, etc.

The Æsir, incredulous at first, try to tear the binding, but cannot do it; the material resists. Convinced, they summon the wolf to put him to the challenge. But Fenrir, sensing the trap, makes a demand: one of the Æsir must put his hand in the wolf's mouth as a gesture of good faith. The gods hesitate, and eventually Týr volunteers. Fenrir allows himself to be bound, and tries—in vain—to free himself, while the gods jeer and mock him. And because they refuse to liberate him, he bites off Týr's hand.

Fenrir, furious, now tries to bite anyone who approaches him. The Æsir thrust a sword between his jaws and attach it with a rope to the stones Gjöll and Thviti, which they bury deep in the ground. From the saliva (some sources say blood) that flows from Fenrir's propped-open mouth, the river Ván is created (*Ván* means "hope" in Old Norse). Thus Fenrir is condemned

Wargs, terrifying steeds.

to remain bound, howling and slobbering with rage until the binding gives way—because it *will* give way, and the gods will have their twilight. After a series of disastrous events, Fenrir succeeds in devouring Odin, but the latter is avenged in the end by one of his sons, Vidar.

Astonishingly, Tolkien's werewolf Carcharoth undergoes an experience similar to Fenrir's. Raised by Morgoth and guarding the doors of Angbad, he devours the hand of Beren which is holding the Silmaril, and is driven mad by this contact with the stone. Carcharoth is killed by the hound Huan, after he has mortally wounded both the dog and Beren.

The role of steed that has been assigned to wargs is also inspired by Scandinavian mythology, which features numerous examples of wolves being ridden. The Rök runestone in Sweden contains the longest known runic text, dating from the 10th century. On it, the "horse of Gunnr" is mentioned; this phrase actually designates a wolf, with "horse" (*kenning*) being used stylistically here, and should be taken in the sense of a steed and not a member of the Equidae family. In *The Lay of Hyndla*, a poem written in Old Norse, the *völva* (witch-priestess) Hyndla rides a wolf. Likewise, the female giant Hyrrokkin arrives at the funeral of the god Balder astride a wolf, with a venomous serpent used as the reins.

In Tolkien, this idea of riding wolves appeared for the first time in *The Tale of Tinúviel*, an early version of the story of Beren and Lúthien, written in the 1920s and published posthumously in *The History of Middle-earth*. Wargs themselves are mentioned for the first time in *The Hobbit*, Tolkien's first book. He remains evasive, however, in terms of the creatures' physical appearance, describing them merely as "great," "evil wolves" that speak a "dreadful language."

PETER JACKSON'S VISION

On the basis of this scanty information, Peter Jackson and his team created two different "animals" to represent the wargs in *The Hobbit* and those in *The Lord of the Rings*. In the latter—the first trilogy of films, released between 2001 and 2003—wargs are used by the orcs of Isengard and Mordor as steeds, and above all as beasts of war. These creatures resemble massive, muscular hyenas similar in size to a horse, or standing about 1.5 m (5 feet) in height at the shoulder, and 2.5 m (8 feet) long. They are frightful to look at, but do not match the rest of the universe developed by Tolkien, with its roots in the mythology of northern Europe.

In *The Lord of the Rings—The Two Towers Creatures Guide*, the director explains that the hyena was chosen as a visual source over the wolf in order to give wargs a more powerful, more menacing aspect. However, in his commentary on the extended version of the second film of the trilogy, Jackson admits that financial costs and time constraints influenced the final appearance of these animals.

In the *Hobbit* trilogy, wargs are mentioned as being from Gundabad and look much more lupine. Their appearance remains fearsome and terrifying, but is closer to the popular image of the Big Bad Wolf super-predator. One of these wargs is larger than the others and is white while the rest are gray; he is ridden by Azog the Defiler. This particular warg has a surprise in store for us—we'll get back to that later. In the meantime, what can we infer from the formidable morphology of wargs when we apply our knowledge of anatomy, ethology, and evolution?

ANATOMY OF A WARG

The frightening appearance of wargs is due in large part to the exaggeration of their predatory characteristics. They are furry, and so we can classify them as mammals. Compared to large predatory mammals both living and extinct, what tells us that wargs are efficient hunting animals?

Their skulls are short and massive, and their dentition shows the imposing canines associated with powerful muscles, as possessed by the hyena. These elements are proof of a powerful jaw, able to crush bone and prevent prey from escaping. The eyes are small, the ear flaps set far back on the head, and the pelt short and dense, with little for an enemy to grasp on to, except perhaps for the mane, which serves as reins by the warg's rider. The neck is muscular, undoubtedly enabling the animal to tear out large chunks of flesh without difficulty.

The hind legs are shorter than the forelegs, which allows the warg to put on astonishing bursts of speed over short distances. Surprisingly, in the films, the wargs seem capable of running long distances with a rider on their backs. Maybe they've been given steroids . . .

Finally, the wargs in The Hobbit, like wolves and cats, have eyes that shine in the dark. From this we can infer that they are equipped with a *tapetum lucidum* (or "bright tapestry"), a reflective layer of cells behind the retina, which humans lack. The *tapetum lucidum* improves night vision by sending ambient light to the retina, to the detriment of clarity of vision. So, wargs must hunt as well by starlight as they do during the day.

These characteristics make wargs specialists in hunting big game and prey that is slow or weak. They are very effective at catching prey that moves more slowly than they do, but which can prove to be dangerous sometimes—such as Dwarves, Elves, or even hobbits!

What real-world mammalian predator is most similar to the warg? For size, the only candidate among living fauna is the bear, but other super-predators roamed our lands in the past. For example, we could compare the warg to *Canis dirus*, the dire wolf, a canid that lived in North America 100,000 years ago. It was slightly larger than the modern gray wolf, and its skull displays all the traits of a typical carnivore's. *Canis dirus* was first unearthed in Indiana in the mid-19th century, and given its name in 1854; Tolkien may have heard it spoken of when he was writing *The Hobbit*.

The warg's intimidating musculature is more than a visual fantasy meant to impress. Cattle breeders are well acquainted with a natural mutation that results in muscular hypertrophy and is caused by the mh (for "muscular hypertrophy" gene). The non-mutated gene codes for a protein, myostatin (hence the other name for the mh gene, the "mstn gene"), which limits the growth of muscular tissue. Because muscles are the most valuable part of a bovine carcass, breeders have selected animals displaying a mutation of the mh gene, enabling greater muscle development. The result is an animal poetically described as "double-muscled," of which the Belgian Blue is one example—and also representative of the kind of prey an adult warg would love to feast on!

So, Saruman and the orcs, like cattle breeders, will have been careful observers, able to select and crossbreed individuals displaying the traits best suited for the use they have in mind—in this case, crunching heroes' bones.

ABNORMAL AGGRESSION

In addition to musculature, warg-breeding orcs will have looked for a specific type of behavior: aggression. The aggression they display toward their prey requires no further discussion, but this behavior also erupts, and on more than one occasion, among themselves.

Wargs are imaginary, but an experiment was conducted with silver foxes in the USSR beginning in 1959, and which is still ongoing today. Dmitry Belyayev and his team, studying the domestication of dogs from wolves around 14,000 years ago, bred wild silver foxes by selecting either aggression or docility in the face of human contact. In just ten generations (around a dozen years), they obtained two types of foxes: very fearful animals that bit the researchers, and tame animals that sought contact with humans. The docility obtained is linked to significantly lower adrenaline levels.

Surprisingly, the trait selected went along with phenotypical changes such as pelt color (more uniform in wild individuals and spotted in docile animals, a duality also seen in wolves and dogs). The adrenaline-color link is still poorly understood; recent research suggests that a slower development of neural crests (early, temporary, embryonic structures that give rise to various tissues, including pigment cells and the adrenal glands, which produce adrenaline) is associated with this process of domestication.

Finally, both types of foxes were studied genetically. They were differentiated by the expression of only 40 genes, while the foxes in the experiment, as a whole, differed in the expression of 2,469 genes compared to a wild population. So, naturally aggressive or docile behaviors can be fairly easily selected in just a few generations and involving only a small number of genes—and without the need to understand the intricacies of genetic science.

Earlier, we mentioned the fearsome steed of Azog the Defiler, the famous white warg. As we told you, there is something surprising about this warg, revealed in *The Lord of the Rings—The Two Towers Creatures Guide*, which is full of secrets about Peter Jackson's film. The book refers to this creature as the "matriarch of the wargs." She's a female!

The social structure of wargs, then, is matriarchal, like that of spotted hyenas, whose hierarchy is topped by a dominant female and then by all the other females in a complex order based on relationships of dominance and submission. Males, including the dominant one, appear only below the weakest female. Packs can contain up to 70 individuals organized according to a very strict social hierarchy, where each one has a specific place. Young hyenas inherit their social rank from their parents. Reflecting her dominant position, the warg-matriarch in *The Hobbit* does not attack, letting her subordinates do the work instead. The reasons for this are hormonal: female hyenas have levels of testosterone (the male hormone) that are at least as high as those of males.

At the end of the day, more than with any other member of the Middle-earth bestiary, Tolkien the storyteller created a rare pearl of animalhood in the wargs, whether deliberately or otherwise. In terms of the choice of their name, their physical appearance, and their behavior, all in the context of northern legend, the author gave life to a fearsome set of adversaries, wild and rampaging, at the sight of which our heroes shudder . . . and so do we.

BEORN: MAN-BEAR OR BEAR-MAN?

VINCENT DUPRET, doctor of paleontology
ROMARIC HAINEZ, associate professor of
Earth Sciences and Life Sciences

Many stories and myths feature therianthropes (from the Greek *therion*, "wild beast," and *anthropos*, "man")—that is, humans capable of transforming into animals, to a greater or lesser degree. These include many of the Egyptian gods, which were endowed with animal heads (falcon, hippopotamus, jackal, ibis, etc.) atop human bodies. Lycanthropes, or werewolves (*Harry Potter*, *An American Werewolf in London*, etc.), also fall into this category. Tolkien's books are no exception, but they have the particularity of placing therianthropes in both camps. For example, in addition to the werewolves* of *The Silmarillion*, Tolkien gives us the character of Beorn, the man-bear who can be categorized as an arctanthrope (arctos being the Greek for "bear"). Rarer than lycanthropy, arctanthropy is nevertheless at the heart of two Disney-Pixar films: *Brother Bear* (2003) and *Brave* (2012).

Beorn is depicted as a very tall, muscular man with a thick beard and black hair, dressed in a long black tunic that falls to his knees, all of which lend his appearance a bearlike quality. He is not of a very sociable nature, but in fairness, he is the last of his species, the rest of whom have been massacred. Subsisting on only cream and honey, Beorn is a sympathetic

---

* See the chapter "Wargs: War-Dogs of Scandinavian Origin?," p. 305.

character for readers—but despite all that he is no teddy bear, as attested to by his first encounter with the heroes of *The Hobbit*. In Beorn, Tolkien has created a nuanced character capable of mood swings and unexpected violence, drawing—as we will see—on the deep medieval fears created by Viking raids. Beorn's complex psychology reflects the author's desire to add shading to his characters' personalities, as further proven by the notes he left behind.

THE UPS AND DOWNS OF THERIANTHROPY

Therianthropes are an uncommonly diverse bunch! Depending on place, era, society, and even romantic fashion, therianthropic transformations happen consciously or unconsciously, deliberately or uncontrollably, and are the result of either a blessing or a curse (or a simple state of being that is neither good nor bad; the character is just "born this way," to quote Lady Gaga!). Transformations also involve varying degrees of completeness, varying from a final appearance that is still humanoid to a total metamorphosis. Likewise, these changes are accompanied by a wide range of physical sensations, ranging from slight awareness to excruciating pain. The therianthrope's consciousness and capacity for judgment may remain human, or they may give way to the most bestial impulses. The same is true for articulated language and communication. The variations are infinite, and have fed countless tales and legends throughout the course of history. According to the medievalist Thomas Alan Shippey, writing in 2002, Beorn was largely inspired by the character of Bödvar Bjarki in *Hrólfr Kraki's Saga*, a sixteenth-century Icelandic legend, who is cursed and forced to transform into a bear.

Tolkien's notes and working copies show that in the earliest drafts of *The Hobbit*, Beorn was called Medwed, which means "bear" in the non-Germanic Slavic languages, as pointed out by the scholar John D. Rateliff in 2007. Yet the name Beorn, which is of Scandinavian origin, was chosen in the end; it is worth noting that the first names Bjorn, Bertrand, and Bernard share these ursine roots. Finally, according to the medieval literature specialist Thomas Shippey, Tolkien may also have drawn inspiration from the famous warrior-hero Beowulf, a "bee-wolf" . . . in other words, a bear!

Beorn, a real bear.

A WARRIOR-PEASANT

Beorn, the man-bear, harbors more than one duality. Echoing his physical transformation, he has a double personality (though he is far from suffering from the same schizophrenia as Gollum[*]): wild and friendly, home-loving and violent. Which is due to his Viking heritage?

The popular image cobbled together of the Viking raids of the 9th century (reaching as far as Paris!) is a traumatizing one: villagers slaughtered, women raped, churches pillaged. Very few of the population could read or write in those days, and our main sources are chronicles written by monastic communities, which were regularly targeted by these raids. This may explain the emphasis placed on the violence associated with Vikings in the accounts that have survived. It wasn't until just a few years ago, thanks to comic books like *Thorgal* and, more recently, certain TV series inspired by real figures like Ragnar Lothbrok, that Vikings began to reclaim a bit of their former glory, being shown as peaceful farmers and merchants with needle-sharp business instincts, possessing an elaborate and complex civilization and customs. Yet the raids, in which berserker warriors constituted the front line, were very real. Who were these berserkers? Historians have shared our curiosity.

Berserker means "bear shirt": *ber särk* in Old Norse, meaning "coat made of bear skin," as first explained by the linguist Elof Hellquist in 1922. Alternatively, it may mean "without protection," with *berr särk* corresponding to "bare shirt," as claimed by the academic Benjamin Blaney in 1972 (the confusion over the word *ber* has persisted into modern English, with "bear" referring to the animal and "bare" meaning naked). The question is still open for discussion, so here's our contribution: what if it were a play on words? This would make both origins plausible. After all, the Vikings were great poets who made enthusiastic use of rhymes and other stylistic formulas, as their sagas attest.[†] And people thought at the time that the simple act of wearing an animal's skin could confer that animal's strength on the wearer, and even its appearance. It is even a viable possibility that the myth of the werewolf has some part of its roots in the berserkers, especially given their ceremonial rites!

Berserkers are depicted as fearsome Viking warriors who destroyed everything in the path. The appearance, or announcement of the imminent arrival

---

[*]   See the chapter "Gollum: The Metamorphosis of a Hobbit," p. 226.

[†]   See the chapter "Wargs: War-Dogs of Scandinavian Origin?," p. 305.

of a troop of berserkers struck terror into the hearts of the opposing camp. Imagine a horde of tall, muscled men running toward you, draped in bear, wolf, or boar skins. And before they attacked, they let out terrifying yells, slobbering and foaming at the mouth and biting their shields. Truly ferocious animals! The linguist Régis Boyer explained in 1992 that berserkers were warrior-beasts that entered into an altered-consciousness state of extreme rage (berserksgangr in Old Norse, meaning "gait" and "speed of the warrior-beast"). This state of consciousness gave them superhuman strength, making them capable of legendary feats.

The animal-skins that covered them also acted as semaphores for allies, since, blinded by their destructive rage, berserkers stopped paying attention to their immediate environment, striking out at enemies and allies indiscriminately. Even worse, if you did manage to wound one of these godlike warriors, his fury would only increase! This state, called *furor*, was notably depicted by Masami Kurumada's manga *Knights of the Zodiac*, first published in January 1986. This anger-fueled insanity recalls that of Beorn hurling himself into combat during the Battle of the Five Armies, in which his decisive intervention routs the goblin army. Berserkers do not actually transform biologically into animals, of course, but it is said that one of their powers was to blur the vision of their enemies, who saw only wild beasts and not humans.

Historians believe that berserkers made up kings' personal security detail, as suggested in the Icelandic saga of Grettir, in which King Harald's warriors are called "wolf-skins" (*úlfhédnar*, from *ulf*, "wolf," and *hedinn*, "pelt" in Icelandic). They may also have acted as priests of Odin. Or they may, in everyday life, merely have been brave, "benign" peasants, as in the 2011 novel *Wolfsangel* by M.D. Lachlan, with ordinary goals and concerns (harvests, the education of children, marriages, increasing their social standing). We are a long way from sociopaths here!

How did the berserkers attain that second state, which rendered them so fearsome?

A CERTAIN STATE OF MIND

From a mythological perspective, it's all in the skin: as soon as he put on his animal-skins, the berserker felt himself to be endowed with its power: the beast's spirit "molded itself" to the inside of the human body and took control of it. Some historians, such as Claude Lecouteux, believe that the word *hamr*, meaning "skin," may be taken in this particular context to mean "soul."

The real reasons, however, are quite different. But on this point too, historians disagree. Some believe the use of mind-altering substances was involved, that berserkers were drugged to induce a sort of trance. But with what? The fly agaric mushroom, or *Amanita muscaria*, known for its psychotropic properties (which are linked to the presence of a molecule called muscimol), may have been consumed by berserkers, according to a theory first suggested by the physician Howard Douglas Fabing in 1956. The mushrooms would have been cut, dried, and ground up before being mixed with alcohol (drugs and alcohol—a combination to be avoided). The resulting potion would have taken effect in about thirty minutes, with initial symptoms consisting of muscular spasms, sweating, and hypersalivation (side effects include memory loss, fainting, vomiting, nausea, flushing, and loss of balance). This hypothesis has been widely embraced, despite a discrepancy between physiological and behavioral effects. However, a recent study by Lily Florence Lowell Geraty (following in the wake of Blaney) totally refutes the drug theory, as the effects of ingesting bufotenine, scientifically tested on humans in the 1950s, do not correspond in any way to those described by the contemporary historians chronicling Viking raids. On the contrary, the test subjects were apathetic, and far from being enraged. It appears that Fabing's observations were good, but the conclusions he drew from them to make this behavior fit with that of berserkers were not valid.

Multiple specialists, including the medievalist and philologist Vincent Sampson and the doctor of medieval British literature Eve Siebert in a 2001 study, have also denied the use by Vikings of mushrooms, the effects of which would have made them unfit for combat. Moreover, historians point out that no contemporary account exists mentioning the use of the fly agaric mushroom or any other substance: the secret appears to have been very well kept. But it is notable that the word *berserk* forms part of the Icelandic term *berserkjasveppur*, meaning "mushroom." The physiological origin of the beserkers' behavior remains a mystery, then; however, we might guess at a combination of several factors: power of persuasion, chanting, or designation of a goal could have acted as catalysts for violent behavior in combination with some sort of intoxication. Illusionists like Derren Brown have even shown that it is possible to influence the actions of Mr. and Mrs. John Doe without their knowledge—and without drugs.

In 1990, the historian Richard Gabriel suggested that berserkers suffered from some form of post-traumatic stress disorder. A similar theory was proposed and added to by the historian Thomas Heebøll-Holm in 2014, based on a study of 12th-century chronicles dealing with the First Crusade and describing the violence carried out by lords in various situations in the field and back at home. According to Heebøll-Holm, there are two sources for these extreme behaviors: either they are caused, from our modern perspective, by post-traumatic stress disorder, or they are merely typical behavior for the era in times of war, driven by escalating violence and power-struggles between different lords. The bias and one-sidedness displayed in the written sources available to us make it hard to choose between these two possibilities.

Change in form and loss of awareness of oneself are reminiscent of shamanic trances, in which the shaman's spirit travels to the other world and communicates with the spirits there. However, this idea, suggested by Régis Boyer, is disputed by Vincent Samson and the Scandinavian specialist François-Xavier Dillmann.

Another theory related to trance is that of autohypnosis (or self-hypnosis), in which the mind takes complete control of the body. Daniel McCoy, a specialist in Norse and Germanic mythologies, confirmed in 2016 that fasting, intense heat, and ceremonial dances were traditional practices in German shamanic rituals. Autosuggestion and controlled metabolism would have opened the floodgates of testosterone and adrenaline, rendering berserkers both ultra-powerful and indiscriminate.

The "impervious to pain" quality displayed by berserkers during combat has to do with the brain's management of fear. In life-or-death situations, the brain assigns priority to self-preservation and survival of the individual (this is the energy of desperation). Once safety is regained, awareness of one's injuries triggers awareness of pain (though this point is highly subjective, as some people will be safer than others), paradoxically leaving the berserker in a state of great exhaustion and extreme vulnerability, as described in *Egil's Saga*, translated in 1977 by Bernard Scudder.

To these plausible explanations we would add religious fanaticism: as mentioned earlier, berserkers were also priests/representatives of Odin—and what better means of being raised to sit alongside him in Valhalla than to achieve dazzling feats and die on the battlefield? Rejoicing, feasts, and good-natured brawling are promised to those happy ones chosen for ascension to Valhalla—until Ragnarök, the twilight of the gods, that is. And if one must die, all the better to do so for the glory of the gods, and therefore all the more reason to fling oneself headlong into the heart of battle! This physical, chemical, and psychological cocktail would have transformed a berserker into the ultimate soldier.

THE GRANDEUR AND DECADENCE OF THE BERSERKERS

Berserkers were initially the personal guards of the Scandinavian kings, as was the case for the hero Bödvar Bjarki, who, despite his ursine transformation, remained at the side of King Hrólfr during his battle against King Hjörvard. Then, in the Icelandic sagas of the late Middle Ages, they became combatants full of *furor*, repulsive to look at, howling like ferocious beasts, biting their shields, impervious to blows—in brief, invisible super-warriors straight out of a fantasy film, terrorizing the enemy camp.

Vincent Samson recounts how, from being regarded as heroes and living legends, they went (after the Christian conversion of Scandinavia) to being seen as coarse and brutish—and, even worse, easily defeated by a young and clever hero. It is tempting to imagine, though difficult to prove, that this withdrawal of esteem had some connection to the end of the bear's reign as king of the animals.[*] The term *berserker* has survived in modern English, with *to go berserk* meaning to become insanely furious, to lose control, a change in mental state previously equated with the assumption of power by a spirit animal (usually a bear, but sometimes also a wolf or boar) over the host body, investing it with supernatural strength.

In 1015, the jarl[†] Eirikr Hákonarson of Norway decreed that berserkers were outlaws, and by the twelfth century these organized warrior factions had vanished.

Tolkien gives us no real answer as to Beorn's origins, and only adds to the mystery by having Gandalf muse: "Sometimes he is a huge black bear, sometimes he is a great [. . .] strong man [. . .]. I cannot tell you much more. . . ."

---

[*]   See the chapter "Wargs: War-Dogs of Scandinavian Origin?," p. 305.

[†]   The equivalent of a king or earl; a chieftain.

THE BESTIARY OF ARTHROPODS

ROMAIN GARROUSTE, (paleo)entomologist
CAMILLE GARROUSTE, illustrator

t is a battle against a Titan, and a key event. Guided by Gollum, the hobbits Frodo and Sam take the pass of Cirith Ungol ("Spider's Cleft" in Sindarin) through Ephel Dúath, the Mountains of Shadow. There, they are surprised by the host of the premises, the giant spider Shelob. She is wounded by Sam, who is well-equipped for the occasion with the Phial of Galadriel and the sword Sting—but not, unfortunately, before she stings Frodo with her venom and wraps him in a web.

Shelob is one of the most spectacular representatives of Tolkien's bestiary in terms of arthropods (insects, arachnids, myriapods, and crustaceans). Reading the author's work, we can see that some specimens are described in great detail while others are sketched more briefly, which denotes a deliberately chosen interest. Shelob and the other giant spiders are favored in literary terms and are described with extreme precision, and are also given much attention in the graphic and film interpretations of Tolkien's books. But whether on the page or the silver screen, liberties have been taken in comparison to the zoological realities. By examining the zoology of arthropods in Tolkien's work and in its various adaptations, we will try to pin down what influenced him, from the zoology of the 19th and early 20th centuries to cryptozoology. Despite his sometimes monstrous bestiary, Tolkien's work is not devoid of a certain poetic ecology—a sort of ecofantasy, in which all living beings have their place, even the most humble ones, in accordance with what we see in real-world ecosystems. First (and spookiest) things first: let's begin with the spiders.

TOLKIEN VS. JACKSON

In Tolkien's writings, spiders are zoomorphic incarnations of characters in their own right, and some of them are even key characters in the author's cosmogony as set out in *The Silmarillion*. Even the most malevolent and powerful entities (Melkor and Sauron) fear these creatures and what they represent. They cannot get rid of them, and enlist them as allies.

The name *Shelob* is a contraction of *she* (the feminine pronoun) and *lobbe* ("spider" in Old English). She is a descendant of the spider-spirit Ungoliant, whom we encounter in *The Silmarillion*. Galadriel, the powerful ruler of the Wood-elves, and Shelob have a literary relationship within Tolkien's work; are we meant to see in it a reference to the Greek myth of Circe? Circe, the daughter of Helios, is a powerful enchantress who crosses paths with Ulysses. An expert in poisons, she aids Ulysses after having killed part of his crew. The characters of Galadriel and Shelob seem connected by their ancient origins (spoken of in *The Silmarillion*), as if they were two facets of a single, dual figure.

The importance assigned to these spider-characters is probably indicative of Tolkien's aversion to the real ones. Indeed, he indicates in a letter that he hasn't liked spiders since he was bitten by one as a little boy in South Africa, but insists that he doesn't hate them. More broadly, arachnophobia, which is extremely common, explains the success of giant spiders in visual representations of the books, first in various illustrations and then in the film adaptations. Yet, to our knowledge, Tolkien never drew one of his own spiders—and, indeed, almost none of the animals in his world were given the honor of being portrayed by his pencil or paintbrush.

In his descriptions of spiders, Tolkien deviates from the zoological model, probably for reasons of literary style. Here is what he says about Shelob, whose depiction, unlike her ancestor's, is richly detailed:

> *Most like a spider she was, but huger than the great hunting beasts, and more terrible than they because of the evil purpose in her remorseless eyes. Those same eyes that he had thought daunted and defeated, there they were lit with a fell light again, clustering in her out-thrust head. Great horns she had, and behind her short stalk-like neck was her huge swollen body, a vast bloated bag, swaying and sagging between her legs; its great bulk was black, blotched with livid marks, but the belly underneath was pale and luminous and gave forth a stench. Her legs were bent, with*

A snack for Shelob.

great knobbed joints high above her back, and hairs that stuck out like
steel spines, and at each leg's end there was a claw. [. . .] But Shelob was
not as dragons are, no softer spot had she save only her eyes. Knobbed
and pitted with corruption was her age-old hide, but ever thickened
from within with layer on layer of evil growth [. . .] bloated and grown
fat [. . .]. Slowly he raised his head and saw her, only a few paces away,
eyeing him, her beak drabbling a spittle of venom [. . .]

How does Shelob fare in Peter Jackson's film adaptation? She is probably one of the best-ever 3D animations of a spider, even though, here again, she does not strictly conform to the zoological reality. She combines observations of position, precise behaviors in terms of movement and appearance, not of any one species, but rather of an amalgamation.

Shelob's battle with Sam and Frodo in the film does not correspond to the zoology of spiders, as they do not really have venomous stingers on their abdomens. A spider injects its venom with its chelicerae, fangs located in front of the mouth. But the animation of this spider, with its precise anatomical and behavior detail, is still a great success. The director has explained that the model on which he chose to base Shelob is a New Zealand native, *Porrhothele antipodiana*, the black tunnelweb spider, a large and aggressive species.

A BIT OF ZOOLOGY
AND A BIT OF CRYPTOZOOLOGY

What do real spiders look like? Arachnids, including scorpions and mites, are arthropods possessing four pairs of legs and a segmented body in two parts, the abdomen and the opisthosoma (which is composed of the head and the thorax), as well as multiple pairs of eyes set symmetrically in the head and, as we have seen, a set of two chelicerae, which are often quite powerful and always venomous. Wholly distinct from insects, the vast majority of spiders are carnivorous. Their sometimes-aggressive behavior and occasionally toxic venom have earned spiders a bad reputation that they don't deserve; most of them are harmless, "calm," and useful as predators.

Cryptozoological folklore virtually everywhere in the world mentions giant spiders, corresponding to the interpretation of rumors, based on explorers' accounts or gathered from local populations. For the latter, these are often local legends exploited by some scientific explorers (though it would generally be more accurate to say "pseudoscientific"!) seeking sponsorship for voyages to more or less distant regions, thus perpetuating these rumors and further spreading tales of "secret" animals. These exploratory voyages are not wrong in themselves, but using science to justify them is not acceptable, except for the ethnological dimension it may represent.

The most striking examples of these mythical creatures are the Himalayan yeti and the famous Loch Ness monster—and spiders have their yeti too! This is the J'ba Fofi (in the language of the Baka, spoken in central Africa). First "seen" in the 1930s in the Congo, this spider of more than a meter in diameter has been reported in several parts of equatorial Africa and even the Amazon rainforest since then. Giant, outsized webs have also been mentioned as further proof of the J'ba Fofi's existence.

What is the truth? Let's look at size first. The largest spider in the world is *Theraphosa leblondi*, the Goliath birdeater tarantula, native to South America, which measures 30 cm (12 inches) in diameter (including the legs, but its body is still very large) and can weigh as much as 350 grams (just over 12 ounces). This spider is well known in Guyana and the northern parts of the South American continent, where it is hunted to be tamed—or, in the case of some local populations, eaten. It has recently been noted that the size of domesticated specimens is becoming progressively smaller, and that it is necessary to farther and farther outside easily-accessible hunting areas (into the deep forest, where it digs its characteristic dens) to find the larger specimens, which indicates that the species is becoming rarer.

Occasionally, science backs up cryptozoological arguments. Known large fossil specimens are sometimes used to justify expeditions in search of survivors from bygone eras, prehistory, the Jurassic period, or even vanished continents such as Gondwana. These lost worlds were celebrated in the early 20th century by Arthur Conan Doyle in his Professor Challenger stories. And, indeed, in 1980, in Argentina, paleontologists unearthed a moderately well-preserved specimen of *Megarachne*, literally "giant spider," a Carboniferous period creature that could reach one meter in diameter. This isn't too surprising, for the Carboniferous was the period of "giant" arthropods, with its atmosphere being more oxygen-rich than today (an explanation that is often cited but not necessarily scientifically valid): discoveries from this time have included dragonflies with 80-cm wingspans, cockroaches and other insects from 40 to 50 cm in length, centipedes one meter in length or even longer, and more.

Unfortunately, a recent careful examination of the *Megarachne* fossil has shown that it is not actually a spider, but rather a representative of Euryptida, an extinct order of arthropods similar to scorpions, but aquatic. As of now, there are no giant spiders known to date from the Carboniferous period, which is bad news for J'ba Fofi hunters!

As for giant webs, it's worth pointing out the enormous ones woven by social spiders; these webs can reach up to several dozen meters in diameter. Social spiders live in groups and raise their young together, and they add their webs together generation after generation. This is the case with the spider *Anelosimus eximius*, which is widespread in Guyana and all of South America, whose webs can occupy 100 cubic meters and accommodate thousands of individual spiders. But these are no giant spiders either, measuring only 5 millimeters in diameter.

THE SECRET INHABITANTS OF MIDDLE-EARTH

Let's look now at the other arthropods featured by Tolkien. We'll begin with the Neekerbreekers, which make it impossible to sleep, according to Tolkien:

> They spent a miserable day in this lonely and unpleasant country. Their camping-place was damp, cold, and uncomfortable; and the biting insects would not let them sleep. There were also abominable creatures haunting the reeds and tussocks that from the sound of them were evil relatives of the cricket. There were thousands of them, and they squeaked all round, neek-breek, breek-neek, unceasingly all the night, until the hobbits were nearly frantic.
>
> —The Fellowship of the Ring

These are orthopterans, without a doubt. These insects, which include crickets and grasshoppers, emit sounds which are sometimes very audible in order to communicate, and sometimes sing in chorus. These sounds generally result from the rubbing together of the front wings, which are modified for this use.

The half-man, half-bear character of Beorn[*] raises bumblebees to produce honey. These social hymenopterans, belonging to the genus *Bombus*, are

---

[*]   See the chapter "Beorn: man-bear or bear-man?," p. 314.

larger than their cousins the honeybees and live in colonies that can be very large. Tolkien tells us in *The Hobbit* that "the bands of yellow on their deep black bodies shone like fiery gold." This description corresponds to a wide variety of European species that are vital for pollination, to the point that some bumblebees are effectively raised like livestock. In greenhouses, *Bombus terrestris* visits two times more flowers per minute than the honeybee. And, happily for Beorn, bumblebees do make honey, but only in small quantities which they use to feed themselves, and they are not exploited for it.

Giant butterflies appear in *The Hobbit* (but not in *The Lord of the Rings*), as shown in the sequence in Peter Jackson's film when the wizard Gandalf speaks to one of them, depicted as a large *Saturniidae*, and asks it to summon help, which arrives in the form of a great eagle.

This scene was probably added in order to provide a sort of cinematic abbreviation compared to the written text. The book explores in far greater depth the relationships between characters, their mystique, and even their ontogeny (a real obsession on Tolkien's part—a professional distortion of linguistics, perhaps). In *The Silmarillion*, giant butterflies, eagles, and wizards seem to be linked, which explains their apparent bond. In another incident in *The Hobbit*, a cloud of butterflies, pursued by bats, prevents the heroes from making a fire.

"Giant" butterflies do exist, such as the giant moth *Thysania aggripina*, also called the white witch; the huge Ornithoptera of Papua, and the spectacular *Saturniidae* with their large eyespots. The latter are found in tropical Asia, but also in Europe. In England, the small emperor moth *Saturnia pavonia*, bears a slight resemblance to Gandalf's messenger in the film.

TOLKIEN AND TAXONOMY

Tolkien's world has long been a source of inspiration for taxonomists. There are species named in reference to Tolkien's books among living and fossil vertebrates as well as invertebrates. Before discussing a few of the species that fall into the latter category, let's remember what taxonomy is.

Taxonomy is the science of describing living and fossil species and their belonging to the categories of systematics (the categories of living things and their study via their relationships of kinship or phylogeny). Systematics, which includes taxonomy, also looks at things from an evolutionary perspective, taking into account species' histories and ecosystems. This is a vital task in biology, for the position of a species in classification alone immediately pro-vides a great deal of information about its biology. This great and permanently ongoing cataloging of living things is not just an academic exercise, but an indispensable branch of biology. However, neither its importance nor the rigor it demands have kept certain liberties from being taken, particularly when it comes time to name a species and the different taxa (genus, family, order) to which it belongs. To sum up, a taxonomist is a biologist who describes the biodiversity of species.

AN INEXHAUSTIBLE SOURCE

Most authors favor names that characterize morphology, biology, or geo-graphic distribution, as these provide a sort of descriptive support, an extra bit of characterization. The first descriptions were Latin, and Latin, Greek, and the ancient mythologies have continued to be a source of inspiration, as have toponyms and names stemming from local folklores having to do with the geographic distribution of species.

The principal rule is a binomial name (genus and species), accompanied by the name of the person(s) describing the creature, and finally the year. The next part is where space is left for the imagination, tributes, and subtle winks. A few examples: the Cretacean pterosaur *Arthurdactylus conandoylensis*; the dipteran insect *Campsicnemius charliechaplini*, the bivalve *Abra cadabra*, the "helmeted" New Caledonian cricket *Lepidogryllus darthvaderi*, the insect genera *Mozartella* and *Beethovena*, the Madagascan ant *Proceratium google* . . .

Tolkien's universe is a well-established source in taxonomy, with dozens of taxa bearing names drawn from this modern mythology. These are often somewhat significant or extraordinary insects or fossil species, and scientific authors are frequently multiple offenders—as well they might be, so immense is Tolkien's repertoire of characters and toponyms. Another rich source can be found in several of the languages invented by Tolkien, such as Elvish.

References have been made to dwarves, hobbits, elves, and dragons, as well as Gandalf himself; few place names have actually been used. But the real star of the show is undoubtedly Gollum, alias Sméagol, whose name is often used to describe cave-dwelling animals with unusual forms.

It will come as no surprise that it is among the insects that we find the most Tolkenian species or taxa. It is in this group that descriptive efforts have been the most intense since the second half of the 20th century, and continue to be so. The task is immense, and there will be more Tolkienian species; that is certain. To date we might mention *Beorn*, a genus of tardigrade (small arthropod) first described in 1964 after its fossil was discovered in Cretaceous amber in Canada, and which has also given its name to the family *Beornidae* in homage to a character who transforms into a bear in *The Hobbit*.

But the grand prize goes to the tiny proturan (an order of hexapods very similar to insects) *Gollumjapyx smeagol*, a cave-dwelling, semitranslucent, spherical-headed creature described by Sendra and Ortuño in 2015. It bears a genus name dedicated to Gollum and the species name of Sméagol, which is the original name of the same character before his transformation—caused by the One Ring—and his subsequent life as . . . a cave-dweller. The animal lives in the caves of western Spain.

Khamul gothmogi, a tiny wasp described by Gates in 2008, is another very Tolkienian species. The genus *Khamul* refers to the only one of the Nazgûl whose name we are told, and the species name *gothmogi* corresponds to Gothmog, lord of those ancient and powerful demonic creatures, the Balrogs.

Paleontologists have come in for their share of inspiration too, often searching for references to the past. One example is the fossil insect *Gallogramma galadrieli*, named for Galadriel.

Since the release of Peter Jackson's films, the association of Tolkien with New Zealand has ceased to be a surprise, as it is the director's country of birth and the place where the six movies were filmed beginning in 1999. Yet it was back in 1980 that F.M. Climo gave his newly discovered order of marine mollusks the name *Smeagolida*, and this was predated by the 1973 description of the small shark *Gollum attenuatus*. Another fish, this time a freshwater-dweller, *Gallaxias gollumoides*, was described in 1999. What if it was because of taxonomy that a New Zealand filmmaker became interested in Tolkien?

And, of course, many species have been dedicated to Tolkien himself, with the Latin species name *tolkieni*. Hence we have the small wasp *Shireplitis tolkieni*, the rove beetle *Gabrius tolkieni*, the aforementioned *Khamul tolkieni*, the tiny crustacean *Leucothoe tolkieni*, and *Martesia tolkieni*, a bivalve mollusk.

For Henry Gee, English paleontologist and great Tolkien specialist, Tolkien's passion for nomenclature (geographic and linguistic) and euphonic names (that sound beautiful when spoken) could not have failed to attract taxonomists—and indeed, these types of references say more about the scientists themselves, perhaps, than about the biology of the species they're naming.

SMAUG, GLAURUNG, AND THE REST: MONSTERS FOR BIOLOGISTS TOO

STÉPHANE JOUVE, paleoherpetologist,
Sorbonne University

olkien didn't invent anything! Dragons are fantastical animals that have sprung from the imaginations of men since ancient times. They are fictional, yes—but there is nothing to keep us from studying and classifying them among other living beings, just like biologists do with real animals. Let's go to it!

There are many dragons in Tolkien's writings, but only five are given names: Glaurung, Ancalagon the Black, Smaug (also called Smaug the Golden), Scatha, and Chrysophylax Dives. Following the example of his dragons, symbols of avarice and stinginess, Tolkien gives us hardly any descriptions, and their appearance and morphologies remain extremely sketchy. But they all share a reptilian appearance, with bodies covered with scales (some of which are thick enough to be made part of a cuirass, or breastplate), and they can all breathe fire. We can separate these dragons into two main groups: flightless dragons, which lack wings, called "Great Worms" or "Long-worms," present in the First Age only, and winged dragons, the first representative of which is Ancalagon the Black; these dragons appear in the First Age and last through the Fourth.

Reference to the real-world classification of living species tells us that Tolkien's dragons, due to their breathing of air (via lungs) and the presence of two pairs of legs in flightless dragons, are tetrapod vertebrates. Their slow and continuous growth and the absence of hair on their bodies in favor of thick, scaly skin suggests that they are not mammals. Can we be more precise about their taxonomy? Do they belong to a single family of reptiles—and do they all belong to the same family? Can we connect them with other, real, living or extinct animals?

We will concentrate our efforts on the creatures described and drawn by Tolkien himself, and leave out the versions depicted by so many other authors and illustrators. This will help us to avoid the inconsistencies and contradictions that might appear, among the different versions of Glaurung and Smaug in particular, with representations of the latter varying especially widely. Now, let's start with the dragons of the First Age, and begin with Glaurung.

GREAT WORMS

The dragons of the First Age are different from dragons in the traditional European sense. First of all, they lack wings, and Tolkien describes them as Great Worms. Are we to see, in this appellation of "worm," the drawing of inspiration from, and literal use of, the terms *worm* from German mythology, *wyrm* in Middle English, and *wurm* in Old High German, all meaning "serpent"? It would hardly be surprising if so, for, remember, from 1919 to 1920, Tolkien's first job after the war ended was studying the history and etymology of words of Germanic origin, and starting with *W*, for the *Oxford English Dictionary*. The fact also remains, though, that the description of his "worm-dragons" is a good match for the traditional reptilian dragon.

Glaurung, the only worm-dragon to benefit from a brief description scattered throughout the text and a drawing by Tolkien himself, seems to be immense in size. His body is snakelike and his skin is corrugated, a trait found in many reptiles. His back, unlike his belly, is armored. All of the dragons described by Tolkien have scales, and the dexterity and flexibility of their bodies tells us that these scales are articulated, or connected by moveable joints. The gleaming, metallic appearance of these scales suggests that they are covered with a thick layer of keratin, the protein that makes up scales, horns, fingernails, fur, and hair. Armor made of thick scales is fairly common in reptiles, including some dinosaurs, most crocodyliformes (crocodiles and their close relatives), numerous lizards, and many other fossil and living species. In these animals, this armor is composed of individual bony plates (called osteoderms) that are more or less articulated with one another and covered with a layer of keratin—all traits that coincide with those of Glaurung. On the other hand, osteoderms are unknown among snakes, and there is no mention in the text of any possible limbs possessed by this creature, which seems to move by crawling. Nevertheless, Tolkien's illustrations, published in 1977 in *The Silmarillion Calendar 1978*, show Glaurung with four legs, which are relatively short and positioned laterally on the body. This so-called transverse anatomy is typical of lizards but does not correspond at all to that of many archosauriformes (a fossil order of reptiles), particularly land-dwelling crocodyliformes (fossil cousins of modern crocodiles) and dinosaurs, in which the limbs are positioned parasigitally; that is, vertically on the body.

One of Glaurung's most striking traits is the absence of eyelids, known as ablephary, a characteristic found in many modern reptiles. It is found in snakes and in most geckos and skinks (a family of lizards). It is, more precisely, the lack of a movable eyelid. In most cases, the lower lid, endowed with a transparent covering, has fused with the dorsal part of the eye socket. In other words, the eye is protected by a scale, especially from dessication, or drying out. This fused eyelid is present only in squamates, the order of reptiles that regularly shed their skin. The clear scale protecting the eye is renewed at each molt, thus retaining its transparency and resistance. In other reptiles that do not molt, such a trait would be impossible.

Based on this characteristic, we can limit our search for a group in which to classify worm-dragons to the squamates—an order which still contains 95% of living reptiles. Other characteristics further shrink the field of possibilities. The nostrils at the very tip of the snout eliminate a large family of lizards, the varanoids, whose nostrils are set far back on the head. The non-forked tongue, as seen in Tolkien's drawing, eliminates the snakes.

Some characteristics are unique to this species: its enormous adult size, and buccal pyric production (it spits fire) make it an uncontestably new species, which we will call *Wirmidraco glaurungi* and present as naturalists do on the next page.

Next, let's look at the winged dragons.

WHO IS SMAUG?

The second type of dragon is, despite appearances, very different from the first. These dragons are also sometimes called worms, but this appellation seems to have been inherited from the worm-dragons of the First Age, which we have just discussed. Though Ancalagon the Black seems to have been the most powerful of them all, we are given no description of him except for the fact that he possesses wings. On the other hand, we are given some details about Smaug, the last great dragon, who has taken over the Lonely Mountain and creates a lot of trouble for Bilbo and his Dwarf companions: "There he lay, a vast red-golden dragon, fast asleep; thrumming came from his jaws and nostrils, and wisps of smoke [. . .]. Smaug lay, with wings folded like an immeasurable bat [. . .]" (*The Hobbit*). His wings, similar to those of a bat, are therefore composed of a dermal membrane (or patagium). His eyes have lids; his tail is immense, and he has legs, though the exact number is not specified. It would have been more realistic biologically if Smaug's wings were modified forelegs, but drawings by Tolkien himself clearly show him with four legs plus two wings.

Like *Wirmidraco glaurungi*, Smaug possesses dorsal osteoderms, but where he differs is in the fact that his whole body is covered with them, not just his back. This trait is present in several real reptile species, notably in some small fossil crocodyliformes and in helodermatids like the Gila monster (*Heloderma suspectum*). Here again, Tolkien's illustrations provide some additional information. Smaug's neck is longer than that of *Wirmidraco glaurungi*; his legs are very short and also positioned transversally on his body. His snout is much longer, and his tongue is forked. This trait occurs only in squamates and—astonishingly—hummingbirds. In squamates, the tongue serves to collect scent particles and bring them to the chemosensory organs and the Jacobson's organ, which is located in the roof of the mouth. Differences in the concentration of scent particles perceived by the two parts of the tongue help the animal to orient itself toward its prey.

The presence of osteoderms and a forked tongue mean that Smaug is undoubtedly a squamate reptile. However, his winged back and ability to produce flames also make him a unique animal, and thus a new species, which we will call *Dracoaliger smaugi*.

A BIT OF GENEALOGY

When they discover new species, biologists classify these species on the basis of their morphological characteristics. Each group, on all rungs of the living scale, is defined by a number of traits theoretically possessed by all the members of this group and inherited from a common ancestor. These traits are called apomorphies. The more traits the members of a group have in common, the closer their common ancestor is, and the closer their biological relationship or kinship. To determine this kinship, biologists use several characteristics, which are first defined with different states indicating variations during the course of evolution. For example, eyelids are either mobile (state 0 of character 314), or fused (state 1 of character 314). All of the character states for each species are coded in this way, often with a 0 or a 1, according to their precise definition. This data is then entered into a software program which calculates the possible relationship trees, grouping species hierarchically according to the maximum number of shared derived characteristics.

Since we have been able to classify both dragons as squamates, we can now use the phylogenetic analysis published by Conrad in 2008 to test their biological relationship to the other members of this order. This analysis includes 223 species and 364 morphological characteristics, to which we have added our two taxa, and a 365th characteristic: the capacity for buccal pyric production, which is either absent (0) or present (1). The software used tested 101,157,073,954 combinations of trees in two hours, forty-seven minutes, and fourteen seconds, and generated four trees. *Dracoaliger smaugi* is, of the two dragons, the one whose relationships are clearer. It is included in the clade *Monstersauria*, similar to modern heloderms, and in *Gobiderma*, an extinct genus of fossil lizards from the Cretaceous period, examples of which have been found in the Gobi Desert in Mongolia. These animals are fearsome predators, with sharp cutting teeth and a body completely covered in osteoderms, even on the head, where they are fused to the skull! In other words, they are, like Smaug, true walking armored tanks! However, the particularly long snout of *Dracoaliger smaugi* distances it from the clade *Monstersauria*—though *Eosaniwa koehni*, a small squamate from the Eocene epoch in Germany (around 48 million years ago), also has a relatively long snout and is still classed among the monstersaurids. Could *Dracoaliger smaugi* be a cousin, somewhere between this creature, with its long snout, and the heloderms with their short ones? Perhaps, for heloderms have short tails, though Smaug's tail is very long, as in the first monstersaurids similar to *Eosaniwa koehni*.

The stratigraphic distribution of *Monstersauria* shows that the oldest members of the clade date from the Upper Cretaceous period (around 100 million years ago). This would mean that the species *Dracoaliger smaugi* would have

Wirmidraco glaurungi

Class: REPTILIA Laurenti, 1768, *sensu* Modesto & Anderson, 2004

Order: SQUAMATA Oppel, 1811

Infraorder: SCINCOMORPHA Camp, 1923

Genus: *Wirmidraco*; new genus.

Etymology: from *wirmis*, serpent in Latin, and *draco*, dragon in Greek.

Diagnosis: as for the type species and the only known species of the genus.

Type species: *Wirmidraco glaurungi*; new species.

Species: *Wirmidraco glaurungi*; new species.

Etymology: from Glaurung, the largest representative of the species.

Iconotype: "Glaurung sets forth to seek Turin", J.R.R. Tolkien, 1977, in *The Silmarillion Calendar*, 1978.

Diagnosis: squamate possessing the following unique characteristics: extremely large size in adulthood (probably more than 25 meters); buccal pyric production; presence of very thick and dense dorsal osteoderms covered with a thick layer of keratin.

Dracoaliger smaugi

Class: REPTILIA Laurenti. 1768, *sensu* Modesto & Anderson, 2004

Order: SQUAMATA Oppel, 1811

Infraorder: MONSTERSAURIA Norell & Gao, 1997

Genus: *Dracoaliger*; new genus.

Etymology: from the Latin *drago*, dragon, and *aliger*, winged

Diagnosis: as for the type species and the only known species of the genus.

Type species: *Dracoaliger smaugi*; new species.

Species: *Dracoaliger smaugi*; new species.

Etymology: from Smaug, the last representative of the species.

Iconotype: an illustration of Smaug, J.R.R. Tolkien, 1937, in *The Hobbit*, first edition.

Diagnosis: squamate possessing the following unique characteristics: extremely large size in adulthood; buccal pyric production; one pair of wings, bringing the total number of pairs of limbs to 3.

broken off from the other members of the clade earlier, at least by the end of the Lower Cretaceous (145 to 100 million years ago), and that its direct ancestors would have lived at the same time as the dinosaurs and even outlived them. How? Hard to say, but the dinosaurs themselves haven't completely disappeared—certain, specific ones remain: birds.

Wirmidraco glaurungi belongs to the infraorder *Scincomorpha*, a group of reptiles that includes skinks, lizards with reduced or absent legs (amphisbaenians, or worm lizards), and snakes. The dragon shares with other derived scincomorphs both a lack of ventral osteoderms and fused eyelids. The first scincomorphs, to whom *Wirmidraco glaurungi* is very similar, have a morphology that corresponds closely to that of Glaurung, with a long tail, short legs, and a crawling mode of movement. Phylogenetic analysis shows that the evolutionary line of *Wirmidraco glaurungi* would have diverged from the other scincomorphs at least by the end of the Lower Cretaceous period; that is, during the same epoch as the separation of *Dracoaliger smaugi*'s line.

So, dragons can trace their origins back to well before the appearance of Man. This reinforces Glaurung's closeness to snakes, justifying his frequent appellation of "worm-dragon." And as for Smaug, his classification as a monstersaurid only confirms his bloodthirsty tendencies . . .

FLAMES *AND* WINGS: COULD IT REALLY HAPPEN?

STÉPHANE JOUVE, paleoherpetologist,
Sorbonne University

What animal with a reptilian body flies and breathes (or spits) fire? All together now: the dragon! These traits have been unchanging in their depiction in Europe since the Middle Ages. But these animals have peopled the legends and myths of many other civilizations the world over, since the dawn of time. We find them in Asia, the Middle East, and elsewhere. Even in the New World, the feathered serpent Quetzalcóatl can be interpreted as a kind of dragon. These creatures also occupy a prominent place in Tolkien's work, and particularly in those writings concerning the First Ages of Middle-earth, the period that precedes the first appearance of Sauron. Where do they come from?

A TOUR OF THE WORLD OF DRAGONS

Though the origin of the myth remains unknown, representations of fantastical animals resembling dragons are present in almost every ancient culture. The oldest dates from around 6,000 years ago and was found in a Neolithic burial in Xishuipo in Henan, China. Mušḫuššu, a mythological Mesopotamian creature that can also be equated with a dragon, appears on the Ishtar Gate, dating from 2500 B.C., guarding the entrance to the ancient city of Babylon. Often these depictions are not clearly identified animals, but rather a mixture of various real-world predators contemporary to the civilization concerned. For example, Mušḫuššu has the body, horned head, and forked tongue of a serpent, the forelegs of a lion, and the hindquarters of an eagle.

Classical Greek and Roman dragons are quite different from the medieval and Nordic dragons that inspired Tolkien. They are mythical serpents that sometimes possess extraordinary attributes or abilities, such as poisonous breath (the Colchian dragon) or multiple heads (the Lernaean Hydra). The word *dragon* also comes from the Latin *draco*, which is itself derived from the ancient Greek δράκων (*drákōn*), which stems from the verb δέρκομαι (*dérkomai*), meaning "to see," "to watch." These dragons are usually guardians of treasures or sacred places, and the word *drákōn* in Greek texts designates both the immense serpent and the guardian, attributes that may have been inherited by the later dragons that inspired Tolkien. Western tradition turned the dragon into a winged and fire-breathing reptilian creature that heroes must face in one-to-one combat. Christianity made it the incarnation of evil, of Satan, and the Gospel according to St. John made it one of the actors of the apocalypse.

TOLKIENIAN DRAGONS

Tolkien was inspired by dragons very early, and various aspects of his narrative, as well as certain characters, seem to be drawn from well-known tales or myths. These include the Scandinavian legend of Sigurd and Fáfnir, which influenced his 1930 book *The Legend of Sigurd and Gudrún*. In the original tale, Fáfnir is a dwarf who, after taking possession of a cursed treasure hoard, transforms into a dragon in order to protect it, and is eventually killed by the hero Sigurd. His death—he is stabbed by Sigurd, who is hiding in a pit—evokes that of Glaurung, who is eviscerated by the dwarf Túrin in *The Silmarillion*.

A second probable source of inspiration for Tolkien's dragons is the Anglo-Saxon epic poem *Beowulf*. This fable, composed between the first half of the 7th century and the 10th century, recreates a Germanic saga in verse, narrating the exploits of the hero Beowulf. It is one of the oldest written accounts in Anglo-Saxon literature, and occupied a special place in Tolkien's life, as he worked on translating it during his youth. He studied the text all his life, and produced a number of very important analyses of it. The dragon in *Beowulf*, the first fire-breather, is the "father" of European medieval dragons. Like

Smaug, this dragon burns down a village after an object is taken from the treasure-hoard he is guarding.

Largely inspired by popular medieval legends and literature, Tolkien's dragons are, understandably, endowed with traits typical of these fantastical animals, and European dragons in particular, including wings and the ability to breathe fire. But are these attributes plausible from a biological perspective?

THINGS ARE ABOUT TO HEAT UP!

The ability to breathe fire, and thus to emit pyric substances, is a unique characteristic. It is not found in any reptile—or, indeed, any animal species. How could an animal produce flames? Two things would be necessary: fuel, and an igniter.

For fuel, the simplest solution would be the utilization of methane and hydrogen. These highly flammable gases are naturally produced by bacteria in the intestines and are also present in the stomach. In humans and other animals, these gases are expelled by belching, flatulence, and other gassy emanations. For dragons, we might imagine a specific pouch or sac for the collection and storage of the methane and hydrogen produced by bacteria in the digestive system, and controlled expulsion in the event of need.

The problem of the spark to initiate combustion of the methane and hydrogen, which do not catch fire spontaneously, is much thornier. Some have suggested that platinum acts as the catalyst to ignite the gases, and platinum does cause hydrogen to combust. The only hitch is that platinum is an extremely rare ore, thirty times rarer than gold! The same reaction can be obtained with other ores, including osmium, iridium, ruthenium, rhodium, and raladium—but these are just as rare as platinum, which is currently extracted from nickel ore, which contains the former at the average rate of around 2 grams per ton. In South Africa, which produces more than 75% of the world's platinum, the Bushveld Igneous Complex, where platinum is "highly concentrated," contains up to 5 grams per ton. It's difficult to imagine hordes of dragons spending their days chewing rocks to extract the platinum required to obtain the smallest spark . . .

Other substances could do the job, too. Phosphorous, in its so-called "white" form, combusts spontaneously in the open air. This substance is used in some bombs, notably to ignite napalm. Phosphorous was discovered by the German

Mušḫuššu, a Mesopotamian dragon.

alchemist Hennig Brand in 1669, while he was attempting to transmute metal into gold by distilling urine. After distilling considerable quantities of urine, he obtained not a single ounce of gold, but rather an extremely flammable, white, luminous substance. This white phosphorous can also be produced by heating bone or phosphate in the presence of carbon. But, here again, very large amounts of material must be heated or distilled in order to obtain a small quantity of phosphorous.

Could dragons have developed other natural mechanisms to ignite hydrogen, methane, or other gases? This isn't a totally outlandish suggestion. A tiny beetle, the bombardier beetle, utilize a chemical reaction to expel a jet of high-temperature toxic substances. It's possible that venom glands, which are present in many reptiles, could have been gradually modified to eject a chemical substance that would ignite hydrogen and methane. But what about flight?

THE DUBIOUS FLIGHT OF DRAGONS

Flight is not rare in the history of reptiles, and this ability has developed independently more than once. The best-known group of flying reptiles is undoubtedly . . . birds. These animals are, in fact, descendants of the dino-saurs, which were themselves reptiles: birds, then, are reptiles. The body part dedicated to flight is formed by the forelimbs, some of the fingers of which have fused. The lifting surface is composed of feathers, which are the result of the transformation of dermal scales. None of this has anything to do with the wings described for Smaug. Other reptiles, such as the pterosaurs of the Mesozoic era, flew thanks to forelimbs whose fifth, smallest finger (the auricular) had disappeared, while the fourth was significant elongated and supported a membrane of skin stretched between this finger and the side of the body.

The legs are the main wing-supports in vertebrates. For example, they are elongated and support a membrane in the flying gecko (*Ptychozoon kuhli*) and flying frogs (countless species of the families *Hylidae* and *Rhacophoridae*), but this membrane also connects their oversized fingers to the front and rear limbs in mammals such as chiropterans (bats). Flight in mammals has also appeared in a similar way in two different groups of rodents, often wrongly lumped together under the appellation "flying squirrels," colugos (a species similar to primates) and the Petauridae family of marsupials. Though they all have a membrane (the patagium) between the front and rear limbs that enables them to glide, this capacity appeared independently in all of these species.

Smaug's wings, as described by Tolkien, represent appendages that are supplementary to his two pairs of legs, and therefore are not the result of a transformation of his forelegs, or of any membrane stretched between his front and hind legs. We can also clearly distinguish several bones supporting the wing membrane, separate from the limbs. This characteristic can be seen in two groups of reptiles, including *Coelurosauravus*, the oldest flying vertebrate ever to have existed. To fly, this animal extended a sail the length of its body, supported by bony dermal rods. These unarticulated rods did not permit flapping flight, but they helped stiffen the sail and improved buoyancy. Common flying dragons (*Draco volans*), small lizards about twenty centimeters long, have developed similar appendages, but the bony rods are in fact the animal's ribs, which project laterally.

While the sails of these animals may resemble what winged dragons possess, they are not articulated, and the animals can only glide. How can dragons rise into the air without flapping flight? The hydrogen and methane gases we discussed earlier are much lighter than air. The BBC documentary *Did Dragons Really Exist?* suggests that these gases could have helped dragons to fly. However, this is an argument difficult to support. Take hydrogen, the lightest of gases: 14 times less dense than air, its density is 0.08988 kg per cubic meter, compared to 1.2 for air. To determine the volume of hydrogen necessary to lift Smaug, we must first know his weight. The animal whose form is closest to Smaug's is the Komodo dragon, a large lizard that can measure up to three meters long and weigh 70 kg. If we estimate Smaug's size at 20 meters, that brings his weight to 470 kg, not counting the fact that he is caparisoned with osteoderms, which are not possessed by monitor lizards (of which the Komodo dragon is one), significantly increasing his weight—so we have probably greatly underestimated this value, because a six-meter-long crocodile already weighs almost a ton.

But let's keep the 470 kg value anyway, supposing that, like birds and bats, dragons have lightweight skeletons with hollow bones. To obtain the volume of hydrogen necessary to lift Smaug, we simply have to divide his weight by the difference between the density of air and that of hydrogen. The result? It would take nearly 60 cubic meters of gas to lift a Smaug weighing under 500 kg, or the equivalent of a tank 2.5 meters in diameter by 13 meters in length, filled with highly explosive gas. A disproportionately large volume, even for an animal the size of Smaug!

This problem could be ameliorated by the utilization of flapping flight combined with a lower volume of gas. But can Smaug bat his wings? Of all the vertebrates, only birds and bats can engage in flapping flight. This action is obtained by a transformation of the front limbs into wings, while Smaug clearly has forelegs separate from his wings. In modern real-world vertebrates, only birth defects produce specimens with two pairs of front limbs, with the additional pair stemming from a fused (Siamese) twin. These limbs are positioned haphazardly on the body, cannot really be controlled, and are not hereditary. So, this possibility can be abandoned. Is there any other scenario that would explain a surplus pair of limbs? Mutations of homeotic genes, which regulate bodily structure and organization during development, are most frequently studied in the common fruit fly, in which the mutation of a homeotic gene (called a bithoraxoid mutation) enables the development of an additional pair of wings. Could a mutation of this type occur in vertebrates?

The answer isn't clear, because the structure and organization of vertebrates' bodies during development are more complex and, especially, less sectorized. An additional pair of wings would require a complete reorganization of the bones and muscles of the trunk. Also, these surplus limbs, which would start identical to the forelegs, would have to evolve into wings over the course of time. Is this plausible? It has never happened to any vertebrate in the hundreds of millions of years of their evolution, except dragons. But maybe Sauron was a dab hand at genetic manipulation, and added extra limbs to dragons, like modern geneticists are actually doing to fruit flies . . .

Another size-related problem: the size of the wings. Could even the most enormous wings really bear the weight of an animal weighing more than 400 kg? The largest bird in our world, the wandering albatross, has a wingspan of only 3.7 meters (11 feet) for a weight of 12 kg (26 pounds) at most. Among fossil birds, *Pelagornis sandersi* had a wingspan of 7 meters, but for a probable weight of less than 30 kg. Other reptiles fare better, such as *Quetzlacoatlus*, the late Cretaceous pterosaur, with its 10-meter wingspan for a maximum weight of 250 kg! Specialists in biomechanics estimate that this pterosaur reached the size limit for a flying animal, and some even think it was too large to fly, or that its weight did not exceed 100 kg. Lifting into the air in flight, even gliding flight, seems impossible for an animal weighing 400 kg.

The dragons of European myth and legend, and Tolkien's dragons too, are always characterized by the ability to breathe fire, and often by the presence of dorsal wings. Since the Middle Ages, these two attributes have been inseparable from the image of the dragon in the Western imagination. Yet, both qualities are incompatible with the biological reality, and must therefore remain confined to myth and literature. And, in the end, we should probably be glad about this. Can you imagine a 500-kilogram pigeon or seagull flying overhead?

FLYING GIANTS! REALLY?

ANTOINE LOUCHART, paleontologist,
CNRS and École normale supérieure de Lyon

f the birds that haunt Tolkien's writings, eagles, which appear frequently, are among the most impressive due to their gigantic size. Other than their dimensions, these enormous eagles are set apart by their manner of carrying characters on their backs or in their talons. How plausible is this? Could such giant birds really fly? Could they carry people? The study of fossils, and also modern and extinct eagles, mythological ones, and even cryptozoological birds—that is, those that straddle the line between myth and reality—all contribute elements to an answer that cannot fail to surprise. Inspirations and explanations can be suggested for all of these fantastical creatures. In every case, Tolkien revisited the knowledge and traditions of various periods and lands to create a wholly original work.

Giant eagles, present in all three Ages, are the birds that appear most often in Tolkien's writings. Though he may have been inspired by the golden eagle (*Aquila chrysaetos*), which nests in the British isles, Tolkien's giant eagles are not really comparable with this species—for, not content with being merely the king of birds, his creations are sentient beings, able to speak, endowed with extensive powers, and immortal!

Specializing in the transport of characters, usually to extract them from a bad situation (as in the cases of the dead Fingolfin and of Fingon and Maedhros, Húrin and Huor, Beren and Lúthien in *The Silmarillion* and later the wizard Gandalf, the Dwarves and Bilbo, Frodo and Sam), the eagles are also characters in their own right, who have names, converse, take part in

Guardians of the skies.

decision-making, and, most importantly, spy on the enemy. They provide for new capitals, and their arrival alone seems to change the course of history. For example, their appearance is greeted by cries—"The Eagles are coming!"—that herald the denouement of *The Lord of the Rings* as an apocalypse. Who announces the fall of the corrupt Maia Sauron at Minas Tirith? An eagle. In the First Age, who informs Turgon of the fall of Nargothrond, the murder of Thingol and Dior, and the ruin of Doriath? Again, an eagle!

Eagles sometimes participate actively in combat, as when they send the goblins tumbling from the mountain-slopes during the Battle of the Five Armies. Like semi-legendary eagles such as the Thunderbird—to whom we'll return later—the wings of giant eagles make "the noise of the winds of Manwë. Some eagles arrive in Númenor from the west and announce its fall, "[bearing] lightning beneath their wings."

Not all of the eagles are models of virtue; some are cowardly and cruel. But those who belong to the ancient race of the northern mountains, the largest of them all, are noble-hearted. These eagles are often providential, but they arrive like destiny, outside of any possible planning by other actors. The half-serious query "Why doesn't Frodo just fly to Mount Doom on an eagle's back?" has been greeted with a large number of explanations, stretching to the most outlandish. The fact is that they *did* save Frodo and Sam in the nick of time, but let them complete their mission, perhaps because it wasn't completely hopeless. Or would they have intervened sooner in the case of failure, or problems more serious than the ones that actually arose? The eagles keep this aspect of the motivations of their actions a mystery—so now let's linger for a moment on their most astonishing quality: their size!

INCREDIBLE DIMENSIONS—BUT ARE THEY REALISTIC?

The largest of Tolkien's eagles are truly enormous, with the greatest of them, the Lord of the Eagles, Thorondor (*The Silmarillion*), boasting a wingspan of around 30 fathoms (180 feet), about the same as a B-52 bomber! So giant eagles are far, far bigger than real-world golden eagles, and bigger than any extinct eagle as well, even the largest specimens known to have existed. These extinct eagles lived on islands, where they hunted other giant birds and large mammals. *Harpagornis moorei*, the largest of these species, native to New Zealand, had a 2.5 to 3-meter wingspan and weighed up to 15 kg (33 pounds); that is, similar to golden eagles in size, but much heavier. Undoubtedly very powerful, these eagles were known by the Maori as *pouākai*, "man-eating bird"! Similarly-sized extinct eagles populated the Caribbean islands as well as Madagascar, illustrating an increase in size through evolution in an insular context, a phenomenon that repeated itself with many birds.

We can see traces of another evolutionary phenomenon in Tolkien's eagles: as the various Ages progress, they become smaller, though remaining immense. This phenomenon evokes the reduction in size that occurred during the millennia after the last ice age in many species on Earth, due to the warmer global climate.

Conversely, adaptation to the cold often results in increased size, as an animal's volume increases proportionally to the cube of its linear dimensions, meaning that body heat (which is related to volume) is much more effectively produced and retained in a larger organism. This is Bergmann's rule, which also takes into account the larger size of the individual members of a species that dwell closest to the poles. And don't Tolkien's giant eagles make their nests in the northern mountains?

Fossils reveal that two extinct birds surpassed real eagles in size. One was *Argentavis magnificens*, a kind of giant South American vulture that had a wingspan of 7 meters, double that of the largest condors, and reached the record weight of 70 kg. The other, *Pelagornis sandersi*, had the appearance of a giant albatross; it, too, had a wingspan of 7 meters (again double that of the largest albatrosses) and weighed 30 kg. This second species was equipped with a very recognizable feature: long, sharp denticles in the beak, making *Pelagornis sandersi* a pseudotooth bird (one possessing bony "teeth").

But these birds look tiny in comparison to *Quetzalcoatlus northropi*. This giant North American pterosaur, which lived during the late Cretaceous period (just before the extinction of the pterosaurs and the dinosaurs) had a wingspan of 10.5 meters, or 35 feet (the largest known wingspan of any animal ever to have existed) and was as big as a giraffe! The various models proposed for this creature have predicted that such a wingspan, in conjunction with the latest estimations of its weight (around 250 kg, or 550 pounds), would not have enabled it to fly—and yet, a close examination of the wing-bones of *Quetzalcoatlus* incontestably proves the opposite! The theoretical models, usually based on living birds, are therefore wrong, and incapable of "predicting" the flight of creatures that are extinct but were very real. It is even possible that even larger pterosaurs populated the skies of the Cretaceous.

According to the current model, and all things being equal otherwise, weight increases proportionally to the cube of wingspan. The wingspan of Tolkien's giant eagles would probably therefore correlate to a prohibitive body weight. But we have to imagine birds with far more developed muscular power, different proportions, minimized weight, and an optimized usage of winds and upward drafts. There are potentially many extremely diversified ways of flying, and other possible flying creatures can be imaged with a much wider wingspan, different forms of wings, lighter weight, and a specific way of utilizing winds or managing energy. So, why not imagine enormous, powerful eagles twenty meters long and with a 30-meter wingspan, or even greater? It's not too big a leap from there to Thorondor and his 180-foot wingspan.

The propulsion of airplanes, moreover, enables far greater wingspans, and this is, of course, without flapping flight. The current record-holder is the H4-Hercules hydroplane, with a wingspan of 97.54 meters (320 feet)! We should mention, however, that this plane flew only once, reaching a height of 21 meters above the water and traveling a distance of 1600 meters despite a powerful motor. Planes are built to transport people, and even heavy merchandise, so why not eagles?

WELCOME ABOARD AIR THORONDOR!

One characteristic of Tolkien's giant eagles is their ability (and willingness) to carry people on their backs. There is no known historical instance of real eagles—or any other flying bird—transporting humans on their backs, but there are numerous stories told all over the world of babies and small children being carried off in the claws of eagles or other raptors. Some of these accounts are very real, like the 1932 taking by a white-tailed eagle (*Haliaeetus albicilla*, a sea-eagle similar in size to the golden eagle) of a three-and-a-half-year-old girl, Svanhild Harvigsen, on the Norwegian island of Leka. The eagle carried her over a mile from the place where she was playing, to a mountainside 180 meters higher in altitude, where she was found alive by three men. She must have weighed nearly 10 kg (22 pounds), and it is believable that the bird was able to carry her, helped by the wind, despite some skepticism regarding the story. The three men who found Svanhild stood categorically by their account, and the girl retained her memory of the incident, and the psychological effects of it, all her life.

Thorondor, 25 times larger than a golden eagle, and his descendants would have had no difficulty transporting a human (or two hobbits), and their dimensions enable these passengers to be carried on their broad backs. Giant eagles can also carry humans in their talons, and when Gwaihir (a descendant of Thorondor) rescues Gandalf, he says to him, "Light as a swan's feather in my claw you are." Gwaihir is able to carry Gandalf "many leagues"; carrying loads seems to be as nothing to Tolkien's eagles, in view of their strength and power, and when he departs for Mount Doom with his fellows to save Frodo and Sam, Gwaihir tells Gandalf: "I will bear you whither you will, even were you made of stone." "The North Wind blows," he adds, "but we shall outfly it." The eagles are incredibly fast, faster even than the Nazgûl and their steeds, and indeed, faster than any other creature of Arda, it seems, whether carrying a load or not.

Nature, ever inventive, has favored means other than backs and claws to transport loads by air: for example, *Heliornis fulica*, the sungrebe, carries its young in flight beneath its wings, in pouches developed for this purpose! Other birds do carry their young on their backs, including grebes, for example, but only when they are floating in the water. And large birds such as swans can carry a human baby on their backs (provided both parties are in an agreeable mood, of course), but here again, it is only when the bird is floating on water, as certain 19th-century photos can attest! An outsized wingspan, a willingness to carry people, and a desire to act like demi-gods: where did Tolkien find these ingredients?

DISTANT ROOTS, AND WINGS

Among the giant eagles of legend, and those who transport characters in particular, we must absolutely mention the Roc, or Rukh, which appears in various eastern mythologies, notably the legend of *Sinbad the Sailor*, the eponymous hero of which is carried in the bird's claws. In this story-cycle from the *One Thousand and One Nights*, the Roc not only transports an elephant, but lays an enormous egg. Some have tried to link this mythical creature to the extinct elephant birds of Madagascar, which vanished around three thousand years ago, and which laid extremely large eggs. These were not eagles, however, but rather a sort of giant ostrich.

The Malagasy crowned eagle, *Stephanoaetus mahery,* also lived on the island of Madagascar. It had a wingspan of between 2 and 2.5 meters (6 to 8 feet), while the Roc is described as having a wingspan of 15 to 30 meters (50–100 feet) or more! A recent, fairly simplistic estimate of the Roc's dimensions started from the principle that an eagle can carry up to half of its own weight, and that a Roc able to carry a four-ton elephant would have to weigh eight tons, and the total weight to be carried by the wings would suggest a lifting/bearing surface of around 250 square meters, resulting in a wingspan of 50 meters (164 feet)—almost exactly that of Thorondor, so he would have been able to carry a young elephant in his talons! Once again, these calculations run up against the improbability of all the rules of proportion remaining the same for all these parameters with both a normal eagle and an immense Roc. In reality, these proportions would have to be very different for the Roc, but we don't know quite how. These allometrics are due to the limits of resistance to mechanical stress, particularly that due to a giant animal's own weight, which increases in proportion to the cube of its linear dimensions. This explains why elephants need legs so thick and strong to support their weight.

The Middle East does not lack for giant eagles: along with the Roc we have the creatures Garuda, of Buddhist and Hindu legend (in his zoomorphic form, that of an immense eagle), and Simurgh, of Persian myth. Some have speculated that the origins of these myths lay in fossils of dinosaurs or other giant creatures, seen by the local populations in regions where they were abundant. A claw here, a beak there, vertebrae taken for those of an immense serpent, etc. In any case, given the legendary status of these illustrious examples, Tolkien may have drawn on them for inspiration. Simurgh is of particular interest, for, in most of the incidences where he transports men or beasts, it is with his claws—but sometimes it is indeed on his back, like when he helps Prince Gauhar and his companion to escape from prison by bearing them away on his wings.

The gryphon (half-eagle, half-lion) of Egyptian antiquity and the Mediterranean region also carries people on its back. Closer to home, the hero of *Gulliver's Travels* is transported by a giant eagle—but in a special traveling-box clutched in the bird's claws, rather than on its back. Nils Holgersson, the hero of Swedish author Selma Lagerlöf's *The Wonderful Adventures of Nils*, shrunken by a magic spell, travels all over Sweden on the back of a bird.

Even more "fantastically," we might also regard travel on an eagle's back from an allegorical perspective. According to shamanic and totemic tradition, individuals can be carried away by an eagle or a similar raptor. When they return to the world of men, they are considered to have attained magical stature. In Hebraic tradition, divine protection is compared to that of the eagle carrying its young on its back, where other birds clutch their young in their talons, for the eagle fears only the arrows of men and prefers to offer itself as a target rather than risk the lives of its babies. Mythical as this allegory is, it is an important biblical reference in the overall context that may have inspired Tolkien in his frequent depictions of transportation on the back of a giant eagle. Between legend and well-documented reality lies a gray area, one that Tolkien might well have plumbed.

ON THE BORDERS OF REALITY

A number of "giant" eagles inhabit the murky area of cryptozoology, somewhere between legend and fact. These include the Kungstorn of northern Europe, which may have been simply a golden eagle or white-tailed eagle, and the Ngoima of central Africa, possibly a martial eagle or crowned eagle. Sometimes birds look larger than they are in reality, and fictionalized or romanticized visions tend to be exaggerated in transmission. An example of this would be the case of the Thunderbird, an enormous North American bird, and more particularly the Pennsylvanian Thunderbird, whose wingspan has been estimated at 7 meters (23 feet). The successor to Native American legend, this bird, similar in appearance to the eagle or the condor, is a shadowy figure, sightings of which have been reported relatively recently, and its wingspan is compatible with the largest known (real) fossil birds.

Giant eagles were unquestionably the leaders of Arda's birds. But in Tolkien, other winged creatures rival them in certain categories.

BIGGER AND MORE BIZARRE

The winged steeds of the Nazgûl in *The Lord of the Rings*, called Fell beasts, hell-hawks, or Nazgûl-birds, bear the Black Riders after their horses are killed as they try to ford the river Bruinen. These beaked creatures, which resemble pterosaurs (they have the same membranous wings), have the look of enormous vultures and are even more enigmatic than the eagles. They are never given individual names, and only scantly described. They are the descendants of unidentified beings that must be extremely ancient on Arda, and have been corrupted by the evil Sauron, because no animal starts out bad. This resemblance to the pterosaur was not intentional on Tolkien's part, but he didn't deny that these steeds might be their descendants. Said to be larger than "any other living creature," and this must include even the largest of the giant eagles, the Fellbeasts' wingspan can reach 60 to 70 meters, or even more—and here again, nothing prevents them from flying.

At the other end of the spectrum we find the balrogs, those extremely ancient and malevolent creatures already in existence when *The Silmarillion* begins. The question of balrogs' wings is never fully resolved, the description of them specifying only shadows "like the shape of great wings" surrounding them. Neither true wings nor possible flight are ever mentioned, which does not prevent them from moving extremely fast. Supposing that they do indeed possess wings, balrogs would illustrate the regression of wings in dragonoid beings, much like birds that have lost the ability to fly and have shorter wings. Here, as elsewhere, Tolkien leaves a share of mystery, toward which science gladly allows itself to be carried . . . on the back of an eagle.

THE WATCHER BETWEEN TWO WATERS

JÉRÉMIE BARDIN and ISABELLE KRUTA,
paleomalacologists, Sorbonne University

he Lord of the Rings, part I, book II, chapter IV, "A Journey in the Dark": on the way to Mount Doom, the members of the Fellowship of the Ring find themselves stymied by the entrance to the mines of Moria, beneath the mountain of Caradhras, next to a lake. The door is sealed by a riddle that Gandalf undertakes to solve. Once he manages this, the Fellowship starts into the depths of the mountain, and it is at this moment that a creature partially emerges from the water and seizes Frodo's leg. After a desperate struggle, the company manages to free him and enters the mines. The creature in question is one of the most mysterious in the universe created by Tolkien: the Watcher in the Water. The confrontation just mentioned is actually the only direct encounter with this guardian-being, who is only summarily described: "Out from the water a long sinuous tentacle had crawled; it was pale-green and luminous and wet. Its fingered end had hold of Frodo's foot and was dragging him into the water. [. . .] Twenty other arms came rippling out. The dark water boiled, and there was a hideous stench."

Devotees of Tolkien's world have long ruminated on the identity of the Watcher, usually associating it mainly with mythological creatures such as the Kraken, or a highly modified dragon, both partly inspired by real animals. The world depicted by Tolkien seems to be governed by the same laws as those of classic biology, and most of Middle-earth's creatures can therefore be associated with organisms we already know, but the morphological characteristics of the Watcher, taken as a whole, do not readily enable us to connect it with any known species.

However, we can utilize a comparative approach in order to place this creature within both the real biosphere and Tolkien's imagined one. Another interpretation of the Watcher's description may be that it is not one individual, but several. Though this theory has not previously been explored in depth, Frodo himself considers it: "I felt that something horrible was near from the moment that my foot first touched the water [. . .]. What was the thing, or were there many of them?"

A COMPOSITE SKETCH OF THE WATCHER

There is very little morphological data available for the Watcher; only the aboveground parts of the creature are seen, with the rest remaining submerged. This partial view limits how far we can theorize in terms of taxonomic attribution.

We can count twenty-one appendages, of which only one, called a "tentacle" by Tolkien, is described in any detail. The impression given by these appendages to the Fellowship is one of swimming snakes, which means that they are long and made up of a series of curves. The end of at least one of these appendages is subdivided into parts like prehensile "fingers," with which it seizes Frodo. The whole is pale green in color and emits light, probably through the phenomenon of bioluminescence.

What do we know about the behavior of the Watcher in the Water? The creature's behavior leaves no doubt that it is aggressive, but its motivations are unknown. Is it a predator, or merely defending its territory or the Doors of Durin? It reacts to the opening of the door, or, more probably, to the throwing of a stone into the lake by Boromir. After the initial attack, movement is seen in the water coming from the far southern end of the lake, seemingly caused by the movement of other appendages, and twenty of them break the surface a moment later.

Nothing in the original text gives us a definite idea of the Watcher's size, but an interpretation of the events in which it is involved may give us an overall idea. Based on the data provided in the book about the dimensions of the lake of Moria, we can estimate that the distance from the Doors of Durin to the southern tip of the lake is over one hundred meters (328 feet). Frodo is seized

The Watcher in the Water.

by the first "arm" before the companions even glimpse the twenty others on the other side of the lake, suggesting that the creature, if it is a single individual, is at least one hundred meters long. There is no animal on Earth that has attained such an enormous size, so the Watcher is a giant creature belonging to Tolkien's world rather than ours. Another explanation of the distance of at least one hundred meters between the first appendage and the others could be that there are multiple, smaller individuals in the lake.

The body of water in question, located at the foot of the Misty Mountains, is formed by the river Sirannon. Because this is a mountain river, the lake is composed of fresh water, probably stagnant and dark in color.

AN UNCERTAIN IDENTIFICATION

The anatomical characteristics of the Watcher have allowed several theories to be formulated about its identity. For example, it is often thought to be a Kraken, a mythological creature related to cephalopods and, more precisely, to a large squid or octopus that can be dangerous to humans. But can the Watcher's traits really be equated with these cephalopods? There is substantial evidence to support this hypothesis.

First of all, Tolkien's use of the word "tentacle" seems to be a reference to the arms of cephalopods, which are often sinuous and rather viscous, giving them a mobility compatible with the Watcher's attack on the Fellowship of the Ring. Next, bioluminescence is characteristic of many cephalopods, particularly those that live in deep water. Certain organs, such as photophores, contain specialized cells called photocytes, which emit or modify light emissions. Luminescence can also be produced by bacteria living in a symbiotic relationship with the cephalopod. However, in this case, the bacteria are associated with the ink sac rather than the tentacles. Ink is also secreted and released into the environment by some cephalopods, and may explain the very dark water of the lake of Sirannon. If the Watcher is a Kraken, its length of more than one hundred meters would explain its ability to produce enough ink to blacken a lake the size of the one in Moria. And finally, in the popular imagination, Krakens behave in a dangerous manner, using their tentacles to attack or to defend their territory, which is precisely what the Watcher does.

However, there are also some attributes possessed by the Watcher that are incompatible with its being a cephalopod. These include the digits at the end of each tentacle. Some cephalopods, such as cirrate octopods, have a pair of filaments or strands on either side of each of the suckers on their arms. These organs, called cirri, are sensory organs, and probably used in feeding, but their exact function remains a subject of debate. We might imagine that the digits on the Watcher's tentacles are overdeveloped cirri concentrated at the tips of the appendages.

Another hindrance to the Watcher's being equated with a cephalopod is its large number of appendages, specifically twenty-one. The only real-world creature with this many arms is the nautilus, which can have more than thirty digital tentacles. However, the overall morphology of the nautilus (its external shell, the shape of its arms, its manner of movement) is not compatible with a Kraken, and indeed, all known representations of this mythical animal serve to link it with octopi (which have eight arms) or squid (which have ten).

Finally, a major argument against the "cephalopod" interpretation is based on the environment in which the Watcher lives. In reality, all cephalopods, both living and extinct, are found exclusively in salt water (oceans and seas).

Could the Watcher be a hydra, or a giant polyp? Hydras belong to the phylum Cnidaria, which also includes jellyfish, anemones, and corals. Because the hydra is the only cnidarian that lives in fresh water, it is a good candidate for the Watcher.

The morphology of polyps consists of a trunk or base resting on a foot, surmounted by a crown of tentacles. The sinuous form and texture of the Watcher are compatible with those of freshwater hydras. The snakelike appendages of the Watcher could be bases, and the terminal digits may correspond to a polyp's tentacles.

There are two possible interpretations for the large number of appendages. The first would be to regard each one of the Watcher's tentacles as an individual hydra, making the "creature" actually a colony of hydras. The other explanation has to do with the asexual mode of reproduction of these cnidarians, which takes place via budding. A bud forms on the base and produces, through cellular division, a new individual that can either separate from the parent or remain fixed. In this case, the Watcher would be a single giant hydra with multiple buds still attached to a single base.

The color green occurs in the green hydra (*Chlorohydra viridissima*), while bioluminescence, though absent in the hydra, exists in many cnidarians. So, these two properties are also compatible with a hydra-Watcher.

Taking its name from the Lernaean Hydra, the seven-snake-headed creature from Greek mythology defeated by Hercules, the hydra is known for its regenerative capacity, which confers on it a sort of quasi-immortality. Like the heads of the Lernaean Hydra, the cut tentacles of a hydra regrow. Even though none of the Watcher's tentacles are severed, it is indirectly characterized by Gandalf as a very ancient creature.

As with the Kraken, there are some factors that do not support the hydra argument. For example, the hydra is a virtually stationary animal, living fixed to a surface or allowing itself to drift on the water's surface. Hydras' tentacles can capture prey, but seem to have reduced mobility. This is all very unlike the Watcher, which is capable of rapid movement in both swimming and attacking.

Another counter-argument is that, unlike its mythological namesake, the hydra is a very small animal, reaching a maximum length of 2 cm (eight-tenths of an inch), and it would take a case of gigantism by a factor of at least 5,000 to produce a hydra as monstrous in size as the Watcher.

A third possibility put forth by devotees of Tolkien's world is that of a dragon-Watcher. These arguments stem from a particular interpretation of the succession of events in the Tolkienian timeline, and from the existence of aquatic dragons. Supposing that dragons can be linked to the reptiles that we know, here are a few points for and against the theory of the Watcher as a reptile.

If the Watcher is an aquatic dragon resulting from the transformation of some other form of dragon, then it is a derived tetrapod (a category that includes most known four-legged animals: frogs, lizards, horses . . .). For a tetrapod to produce the Watcher's sinuous movements, its limbs would have to have lost their bones. One consequence of an organism's loss of its skeleton is the problem of muscle insertion, as the muscles move during bodily motion in such a way as to make the bones to which they are attached move. For tetrapods, then, without a skeleton or other rigid internal structure, there can be no muscular movement. Moreover, classic dragons have six limbs (four legs and two wings), while the Watcher has at least twenty-one, implying a significant multiplication of limbs (a serial homology, for biologists). Biologically speaking, this theory is difficult to defend, because of the large number of biophysical modifications and inconsistencies.

The idea that the Watcher is a snake would easily explain several of its characteristics. The undulating movements of the arms are compatible with those of a snake thanks to the numerous vertebrae that make up its body.

The number of "fingers" on the Watcher's arms is unknown, but a wide-open snake's mouth could be compared to a prehensile pincer with two digits. The color and texture of the Watcher's arms are also common in snakes, and finally, this hypothesis supports the inferral of a population of individuals, which resolves the paradox of size.

Still, not all of our questions have been answered. First of all, if each arm is a snake, why don't the members of the Fellowship identify the Watcher easily? For such immediate recognition not to happen, the "snakes" would have to be highly modified, with no eyes, teeth, or mouth.

At the end of the day, it seems extremely difficult to give a firm and definitive answer to the question of where the Watcher in the Water of the mountain lake in Moria belongs on the tree of life. Biologically speaking, it does not wholly fit with any known animal. Other direct eyewitness accounts of it would be useful in supporting or refuting some of our theories. Many of Tolkien's as-yet unpublished handwritten notes may help to solve the mystery, and situate the creature in the author's systematically laid-out universe. According to the available data, however, though it has never really been favored, a logical interpretation of the text favors the populational hypothesis as the most probable one. As for the morphological traits of the Watcher, it is very possible that the mosaic of characteristics shown by the creature were inspired by both living creatures and mythological ones. And indeed, though Tolkien was principally inspired by Norse mythology, the influence of the monstrous Lernaean hydra, born of Greek mythology, may have combined with that of the Kraken and the fear of tentacles so deeply ingrained in the collective imagination to result in the creation of the Watcher of Moria.

THE CRYPTOZOOLOGICAL BESTIARY OF J.R.R. TOLKIEN

BENOÎT GRISON, biologist and scientific sociologist,
Université d'Orléans

ryptozoology, founded in the 1950s by the Franco-Belgian zoologist Bernard Heuvelmans, is an outlier in the discipline of zoology, somewhere between an inventory of animal biodiversity and a kind of zoological mythology. Ideally, the goal of this discipline is to discover new animals, relying on the knowledge of local populations (known as "ethnoknowledge"), eyewitness accounts, and, more rarely, material clues such as body parts. On the basis of this definition, we might ask ourselves: does Tolkien's work have links to the world of cryptozoology? It seems incongruous at first, and yet . . .

First of all, let's restore the reputation of cryptozoology. Often misrepresented and subverted by cranks and weirdos, the cryptozoological approach has actually made possible some remarkable discoveries in the last few decades, among them *Pseudoryx nghetinhensis*, the northern Vietnamese saola, an odd-looking animal halfway between a goat and an antelope, and the Indonesian coelacanth (*Latimeria menadoensis*), one of only two survivors, with its cousin the West Indian Ocean coelacanth, of the subclass Actinistia, the rest of which is thought to have become extinct 70 million years ago.

In reality, the cryptozoology championed by Heuvelmans constitutes a form of ethnozoology, touching on both the question of the very specific naturalist knowledge possessed by certain communities and the cultural production of mythical creatures. Its interdisciplinary methodology has led, for example, to the "deconstruction" of the famous Loch Ness Monster, an illusion created on the surface of the Scottish lake through optical mirages, "windless waves," and wandering seals. Now, let's turn our attention to the cryptozoological bestiary of J.R.R. Tolkien.

THE GREAT SERPENTS OF THE NORTH

We'll begin with that "great unknown of the oceans," the Sea Serpent. Tolkien mentions this creature as early as 1927, in *Roverandom*, which, though ostensibly a children's story, is littered with sophisticated references to Nordic literature and culture. In it, the wizard Artaxerxes (named for the great Persian king!) describes a snakelike monster, immense compared to Roverandom, the canine hero of the tale: "primordial, prehistoric, [. . .] fabulous, mythical, and silly." Here we can detect allusions to Jörmungandr, the Midgard ("Middle-earth"!) or World Serpent from Viking mythology which, when the End Time arrives, emerges from the abyss to take part in the downfall of the gods, or Ragnarök; and also to the Soe Orm, or "Sea Worm," of the ancient Scandinavians. The latter was often depicted in Norwegian churches but, contrary to common misconception, rarely on the prows of their longships. Similar sea monsters are mentioned in the Anglo-Saxon epic poem *Beowulf* (7th–8th century), a major source of inspiration for Tolkien's writing, and in his *Etymologies* he gives them the name *Lingwiléke*, an Elvish term he translates, logically, as "Sea Serpent."

In the 17th and 18th centuries, the Italian voyager Francesco Negri and the bishop Pontoppidan of Bergan, two avid scholars of natural history, collected a large number of accounts, from sources considered to be reputable, of belief in the Soe Orm, the "Great Serpent" of the north. For them, it was a real flesh-and-blood creature—but certain improbable characteristics given to the creature (immense size, and aggressiveness worthy of a dragon) clearly indicate that these tales stem from oral tradition, rather than actual, substantiated evidence as we conceive of it today!

The real, naturalist history of the matter of the Great Sea Serpent does not truly begin until the 19th century. More precisely, the year 1817 marks the birth in New England of scientific debate around a sea monster: that year, reports of sightings increased in number in that part of the northern hemisphere, particularly in the harbor of Gloucester. Various accounts, including many by persons considered well-educated (clergymen, captains of sailing vessels) all echoed one another: they had seen a very large sea creature, unquestionably more than ten meters (33 feet long), dark in color and with multiple humps. Its speed, comparable to that of some large whales, was reported at several dozen kilometers per hour.

The Linnaean Society of New England, founded in Boston and made up of eminent scholars, was intrigued enough by the reports to launch an investigation based on a "standard questionnaire" well ahead of its time. Soon, the English intelligentsia began in its turn to take stories of sea serpent sightings seriously, frequently coming as they did from sea captains who were seen in British society as pillars of an unimpeachable institution: the Imperial Navy. During the reign of Queen Victoria, researchers as distinguished as Charles Lyell, a pioneer of modern geology, and Thomas Henry Huxley, passionate defender of Darwin's theories, took an enthusiastic interest in these eyewitness accounts of the famous sea monster.

Rather than tailing off during the Edwardian period (1901–1913), which coincided with the years of Tolkien's youth, popular interest in the "great unknown of the ocean" remained high. The pet theory of enthusiasts of the day, galvanized by recent paleontological discoveries, was that the sea monster was a sort of marine reptile, a living relic of the Mesozoic era—a hypothesis, however, which was wholly at variance with the descriptions given by eyewitnesses!

Tolkien could not ignore these speculations, aimed at naturalizing the ancient and legendary Scandinavian animal. One of his major influences as a researcher, the Scottish folklorist and anthropologist Andrew Lang, put forth the idea in a number of his writings that the "cryptic survival" of species thought to be extinct could explain some of the sightings of mysterious animals reported by explorers. In this context, it will come as no surprise that the adolescent Tolkien, finding a jawbone fragment from a fossil vertebrate on the coast at Lyme Regis, declared, tongue firmly in cheek, that it must be from a dragon!

What persists today of these attempts to rationalize the myth of the sea serpent? More than you might think. Of the some 400 seemingly credible sightings reported over the last two centuries, many are clearly suggestive of rare marine animals such as the giant oarfish (or ribbonfish, the world's longest bony fish), the whale shark, and even the leatherback sea turtle.

However, a great many reports continue to emanate from the Pacific coast of North America, once again perpetuating the consistent image of a huge, unknown marine mammal with a long neck, several meters in length, which the press has dubbed "Cadborosaurus" in reference to Cadboro Bay in British Columbia. The oceanographer Paul LeBlond has analyzed this corpus of sightings, and does not exclude the possibility that they might refer to a new species. In terms of biodiversity, the sea might just have a few surprises in store for us yet . . .

FROM THE ASPIDOCHELONE TO THE KRAKEN

Displaying a marked predilection for the theme of aquatic monsters, Tolkien paid special attention to the specific cases of two of them, the aspidochelone and the Kraken.

The aspidochelone appears in his poem "Fastitocalon," first written in the late 1920s and later reprinted in *The Adventures of Tom Bombadil*. This tale, said to be well-known to hobbits, evokes the treacherous nature of an island on which gulls like to bask in the sun, a providential shelter for sailors, which is revealed to be the back of a monstrous creature liable to swallow them at any moment! In the author's mind, the monster in question is clearly a giant turtle, and indeed the term *Fastitocalon* is a medieval Anglo-Saxon corruption of the word *aspidochelone*, a mythical creature first mentioned in an ancient text titled the *Physiologus*, the ancestor of every Christian bestiary, probably written in the 2nd century in Egypt. *Aspidochelone* translates literally to "snake-turtle," and it is clearly implied in the text that it is the shell of this hybrid being that is often taken by sailors for an island, leading them to make landfall there and build a fire . . .

From the perspective of the *Physiologus*—which is in no way a zoological text in the modern sense—the aspidochelone is a symbol of evil temptation. Because this Greek bestiary mostly discusses real animals, though these are certainly still used as pretexts for moral lessons, we might wonder if the aspidochelone is the distorted echo of some immense, very real marine animal. Of course, if we take the composite portrait of it word for word, it does suggest some sort of sea serpent resembling a plesiosaur, as mischievously pointed out by Bernard Heuvelmans.

After all, during a sixth-century voyage, the Irish monk and skilled navigator Saint Brendan was said to have celebrated Easter on an "undulating island" which was none other than the great serpent Jasconius, undoubtedly a trace memory of the Midgard Serpent so dear to the peoples of the North. More seriously, it is possible that the leatherback sea turtle inspired the aspidochelone, at least in part; this turtle, the largest living species, can reach three meters in length and weigh up to 850 kg (1,874 pounds). It is a cosmopolitan animal, found in both tropical seas and the Mediterranean, and in the Atlantic Ocean as well as the seas of the north.

This leads us to the tricky question of the Watcher in the Water, the tentacled, lake-dwelling creature which strikes terror into the hearts of the unfortunate hobbits at the entrance to the dwarven city of Moria in *The Fellowship of the Ring*.* This image is likely an echo of the Scandinavian Kraken, itself also immense and many-tentacled.

The mythos of this sea monster grew slowly beginning in the Middle Ages. Its origins lay in the legend of the enormous fish Hafgufa, the largest in Creation, somewhat shapeless in appearance, endowed with an insatiable appetite and surrounded by masses of smaller fish—as well as in stories of another monster, Lyngbakr, or "heather-back," the biggest whale in the world, which, like the aspidochelone, is an early example of the "island-beast" stories common in the Middle Ages. Gradually, these two mythical creatures, initially separate, merged together and gave rise in the Renaissance to the legend of the Kraken. As described by various scholars, the Kraken is an immense animal with indistinct contours, often mistaken by decent fishermen for a shallow sea-bottom and liable to destroy vessels (through accident or malevolence), rising slowly from the depths. This creature, both strange and terrifying, has a roughly cylindrical body and large "horns." During the process of creation it was assigned a motley variety of traits borrowed from whales and various invertebrates (cnidarians, echinoderms, crustaceans); it's worth remembering that *Kraken* is sometimes also written as *Kraxen* or *Krabben*—that is, crab.

Examples of the giant squid *Architeuthis*, which floats to the ocean's surface as it is dying, indisputably contributed to the formation of the Kraken's legend. These extraordinary creatures, the largest known of which reached around 17 meters (56 feet) and which can weigh over 300 kg (660 pounds), are far and away the largest known cephalopods. They have eyeballs that can be as big as volleyballs; these spectacular orbs were developed to help them escape sperm whales, which dive to a depth of 1,000 meters below the ocean's surface to hunt them in their habitat.

The giant squid was not discovered until 1853, and so for many years, sailors' accounts of their immense size and their battles with whale sharks were taken for superstitious grotesqueries. That said, Tolkien does commit a major biological gaffe with his Watcher: there can be no such thing as a freshwater Kraken!

---

*   See the chapter "The Watcher Between Two Waters," p. 356.

MASTERS OF WILD ANIMALS AND WILD MEN

Beorn himself, the "man-bear" from *The Hobbit*,* has his own links to the field of cryptozoology, recalling as he does the universal myth of the wild man. The oldest incarnation of this figure is perhaps the giant Enkidu, who, draped in animal skins, commands the animals of the forest in the Babylonian *Epic of Gilgamesh*. Likewise, the immense Etruscan deity Sylvanus, bearded and hirsute, continued to haunt Roman forests along with his companion Maia—much like the Basajaun, the "wild man" of Basque legend, found in the woods of the Pyrenees. In every case, the wild man is master of the animals, defending Nature against the excesses of hunters.

But it was medieval culture, perhaps, that developed most extensively the potential symbolism inherent in the myth of the wild man, a radical figure of otherness. As the historian Richard Bernheimer has shown, the woodland-dwelling man, brutish and without language, was fundamentally conceived as the antithesis of both the knight, with his adherence to an elaborate code of honor, and the practitioner of courtly love, serving as a sort of comic foil to the latter. Shying away from cultivated areas and "domestic life," the wild man makes his home in deep forests, mountains, and marshes, subsisting on raw meat. He is armed with a massive club or tree trunk and does not hesitate to capture young maidens.

Mythical traits such as these have sometimes been associated with some of the supposedly unknown large primates with which cryptozoology often concerns itself, including the Yeti, and the Almasty of the Caucasus Mountains, regarding which no less esteemed a scholar than Marie-Jeanne Kauffmann has gathered hundreds of consistent eyewitness accounts, exciting the interest of the paleontologist and paleoanthropologist Yves Coppens—an avenue of research that is not unpromising in itself, as real animals are sometimes associated by various populations with the myth of the wild, hairy man; this has been true of the bear, and was also true of the gorilla and the orangutan before their official discovery.

As the author Jorge Luis Borges loved to insist, the zoology of dreams cannot compare with what is to be found in real life. Fascinated by the extraordinary diversity of the natural world, Tolkien would not have disagreed.

---

* See the chapter "Beorn: Man-Bear or Bear-Man?," p. 314.

"The stew is ready."

BIBLIOGRAPHY

The passages originally quoted in the chapters "Language and Evolution in Tolkien," "Landscapes in Tolkien: A Geomorphological Approach" (for passages from *The Lord of the Rings*), "Invisible to Sauron's Eyes?," "Tales of a Young Doctor . . . in Middle-earth," "Why do Hobbits Have Big Hairy Feet?," "Gollum: The Metamorphosis of a Hobbit," "The Eyesight of Elves," "Are Dwarves Hyenas?," "Could an Ent Really Exist?," "Saruman's GMOs (Genetically Modified Orcs)," "The Evolution of the Peoples of Middle-earth: A Phylogenetic Approach to Humanoids in Tolkien" (for passages from *The Lord of the Rings*), "Mythotyping Origins," "Tolkien the Ornithologist," "The Bestiary of Arthropods" and "Flames *and* Wings: Could it Really Happen?" are taken from the following French-language editions of *The Lord of the Rings* translated by F. Ledoux:

Tolkien, J. R. R. *Le Hobbit*. Translated by F. Ledoux. Paris: éditions Stock, 1969.
Tolkien, J. R. R. *Le Seigneur des Anneaux*. Translated by F. Ledoux. Paris: Christian Bourgois éditeur, 1972–1973.

The passages quoted in the chapters "Archaeological Remains and Hidden Cities," "Landscapes in Tolkien: A Geomorphological Approach" (for passages from *The Hobbit*), "A Chemical History of the One Ring," "The Evolution of the Peoples of Middle-earth: A Phylogenetic Approach to Humanoids in Tolkien" (for passages from *The Hobbit*) and "Memories of Oliphaunts" are taken from the following editions translated by Daniel Lauzon:

Tolkien, J. R. R. *Le Hobbit*. Translated by D. Lauzon. Paris: Christian Bourgois éditeur, 2012.
Tolkien, J. R. R. *Le Seigneur des Anneaux*. Translated by D. Lauzon. Paris: Christian Bourgois éditeur, 2014–2015.

TOLKIEN AND THE SCIENCES: A RELATIONSHIP WITH MANY FACES, by Isabelle Pantin

Aristotle. *Metaphysics*. Translated by J. Tricot. Paris: Vrin, 1991.
Bachelard, Gaston. *La Formation de l'esprit scientifique*. Paris: Vrin, 1938.

Carter, Lin. *Tolkien: A Look Behind The Lord of the Rings.* New York: Ballantine Books, 1969.

Gee, Henry. *The Science of Middle-Earth: Explaining The Science Behind The Greatest Fantasy Epic Ever Told!* Cold Spring Harbor, NY: Cold Spring Harbor Laboratory Press, 2004; undated Kindle e-book edition.

Larsen, Kristine. "Sauron, Mount Doom, and Elvish Moths: The Influence of Tolkien on Modern Science." *Tolkien Studies* 4, (2007): pp. 223–234.

Pantin, Isabelle. "Le conteur en *Janus bifrons. Le double courant du temps* in la création de Tolkien." *Revue Europe*, 2015, special issue edited by Roger Bozzetto and dedicated to Lovecraft and Tolkien.

Tolkien, J. R. R. "The Lost Road." In *The History of Middle-earth.* Translated by D. Lauzon. Paris: Christian Bourgois, 2008.

Tolkien, J. R. R. "On Fairy-Stories." In *The Monsters and the Critics, and Other Essays.* Edited by Christopher Tolkien. Translated by C. Laferrière. Paris: Christian Bourgois éditeur, 2006.

Tolkien, J. R. R. "The Notion Club Papers." In *The History of Middle-earth.* Edited by Christopher Tolkien. London: Allen & Unwin, 1983.

Tolkien, J. R. R. "'The story of Kullervo' and Essays on *Kalevala.*" Edited by Verlyn Flieger. *Tolkien Studies*, no. 7 (2010).

Tolkien, J. R. R. *On Fairy-Stories. Expanded edition, with commentary and notes.* Edited by Verlyn Flieger and Douglas A. Anderson. London: HarperCollins, 2008.

Tolkien, J. R. R. *Lettres.* Edited by Humphrey Carpenter with Christopher Tolkien. Translated by V. Ferré and D. Martin. Paris: Christian Bourgois éditeur, 2005.

TOLKIEN: SCHOLAR, ILLUSTRATOR ... AND DREAMER, by Cécile Breton

Collectif. *Vénus et Caïn, figures de la préhistoire, 1830-1930.* Exhibition catalogue. Musée d'Aquitaine et Réunion des musées nationaux, 2003.

Coulombe, D. "*Bilbo The Hobbit*, un livre d'enfant pour adultes" [online]. Web supplement to the journal *Solaris*, no. 135 (Autumn 2000). Available at: https://www.revue-solaris .com/articles/solaris-135-article-bilbo-le-hobbit/.

Delevoy, Robet L. *Le Symbolisme.* Paris: Flammarion, 1982.

Hammond, W. G. and C. Scull. *J. R. R. Tolkien, artiste et illustrateur.* Translated by J. Georgel. Paris: Christian Bourgeois éditeur, 1995.

Harding, J. *Les Peintres préraphaélites.* Paris: Flammarion, 1977.

Haeckel, E. *Kunstformen der Natur.* Edited by Olaf Breidbach. 1899 and 1904.

Le Pichon, Y. *L'Érotisme des chers Maitres.* Paris: Denoël, 1986.

Lilford, T. L. P. *Coloured Figures of the Birds of the British Isles.* London: Porter, 1885–1897.

Marchal, H. (ed.). *Muses et ptérodactyles, La poésie de la science de Chénier à Rimbaud.* Paris: Seuil, 2013.

Sauvayre, R. "Comment la science alimente les croyances. La surprenante dialectique entre convocation et disqualification du discours scientifique" [online]. Éditions Matériologiques. Sciences et pseudoscience. Regard des sciences humaines, pp.81–93, 2014. Available at: http://materiologiques.com/sciences-philosophie-2275-9948/186 -sciences-et-pseudo-sciences-regards-des-sciences-humaines-9782919694709.html.

Tolkien, C. *Pictures by J. R. R. Tolkien*. London: George Allen & Unwin, 1979 (reissue, 1992).

Tolkien, J. R. R. *Mr. Bliss*. London: George Allen & Unwin, 1982.

Tolkien, J. R. R. *The Father Christmas Letters*. London: George Allen & Unwin, 1976.

Tolkien, J. R. R. *Lettres*. Edited by Humphrey Carpenter with Christopher Tolkien. Translated by V. Ferré and D. Martin. Paris: Christian Bourgois éditeur, 2005.

TOLKIEN AND SOCIOLOGY: FACING THE LOSS OF A WORLD, by Thierry Rogel

Durkheim, Émile. *De la division du travail social*. Paris: PUF Quadrige, 2013 (1st ed. 1895).

Elias, Norbert. *Qu'est-ce que la sociologie?* Translated by Y. Hoffmann. La Tour-d'Aigues: Éditions de l'Aube, 1991.

Klein, Gérard. "Bilbo ou le prophète du *Seigneur des Anneaux*." In *Fiction* no. 188, Opta, August 1969.

Nisbet, Robert. *La tradition sociologique*. Paris: PUF Quadrige, 1984.

Renan, Ernest. "Qu'est-ce qu'une nation?" In *Discours et conférences*. 1882.

Simmel, Georg. "Excursus sur l'étranger." In *Sociologie—Études sur les formes de socialisation*. Paris: PUF Quadrige, 2010.

Tönnies, Ferdinand. *Communauté et Société—Catégories fondamentales de la sociologie pure*. Paris: PUF, 2010, 1re éd. 1887.

MYTHOLOGY VS. MYTHOLOGY: TOLKIEN AND ECONOMICS, by Thierry Rogel

Graeber, David. *Dette: 5000 ans d'Histoire*. Paris: Les liens qui Libèrent, 2013.

Keynes, John Maynard. *Perspectives économiques pour nos petits-enfants*, Paris, NRF, 1930.

Orléan, André. "Crise de la souveraineté et crise de la monnaie: l'hyperinflation allemande des années 1920" [online]. 2008. Available at: http://www.parisschoolofeconomics .com/orlean-andre/depot/publi/crise.pdf.

Orléan, André. *L'Empire de la valeur*. Paris: Seuil, 2011.

Polanyi, Karl. *La Grande Transformation*. Paris: Gallimard, 1983, 1st ed. 1944.

Samuelson, Paul. *L'Économique*. Paris: Armand Colin, 1972.

Simmel, Georg. *Philosophie de l'argent*. Paris: PUF Quadrige, 1999.

Veblen, Thorstein. *Théorie de la classe de loisir*. Paris: Gallimard, 1978.

Von Hayek, Friedrich. *La Route de la servitude*. Paris: PUF Quadrige, 2013, 1st ed. 1944.

FAMILIES, POWER, AND POLITICS IN *THE LORD OF THE RINGS*, by Thierry Rogel

Ghassarian, Christian. *Introduction à l'analyse de la parenté*. Paris: Points-Seuil, 1996.

Godelier, Maurice. *Métamorphoses de la parenté*. Paris: Flammarion, 2010.

Lamaison, P. "La notion de maison: entretien avec C. Lévi-Strauss." *Terrain*, no. 9, 1987.

Lévi-Strauss, Claude. "L'organisation sociale des Kwakiutl." In *La Voie des masques*. Paris: Pocket, 2004.

Lévi-Strauss, Claude. "Maison." In Bonte and Michel Izard. *Dictionnaire de l'ethnologie et de l'anthropologie*. Paris: PUF, Quadrige, 2000.

Lévi-Strauss, Claude. *Les structures élémentaires de la parenté*. Paris: PUF, 1949.

Lévy-Leblond, Jean-Marc. "L'avenir de la science: progrès ou régrès?" [video recording]. 25th Fleurance Astronomy Festival, 2015. Available at: http://www.dailymotion .com/video/x3cxgx8.

Martuccelli, Danilo. *Les sociétés et l'impossible —Les limites imaginaires de la réalité*. Paris: Armand Colin, 2014.

Radcliffe-Brown, Alfred Reginald. "Étude des systèmes de parenté" (1941). In *Structure et fonction dans la société primitive*. Paris: Points-Seuil, 1972.

THE DEFENSE AND DEPICTION OF PHILOSOPHY IN TOLKIEN,
by Michaël Devaux

Some parts of this article are taken from the class "Tolkien and Philosophy" at the Collège des Bernardins, specifically from sessions occurring between October and December 2016.

Anson Fausset, Hugh (l'). *Studies in Idealism*. London: Dent, 1923. Reissued 1965 and 1982.

Arduini, Roberto and Testi, Claudio A. (eds.). *Tolkien and Philosophy*. Zurich: Walking Tree, 2014.

Aristotle. *Éthique à Nicomaque*. Translated by R. Bodéus. Paris: GF, 2004.

Biggar Gordon, Mary C. *The Life of George S. Gordon, 1881–1942*. Oxford: Oxford University Press, 1945.

Boethius. *Consolation de la Philosophie*. Translated by J. Y. Guillaumin. Paris: Les Belles Lettres, 2002.

Burton, Robert. "Amour de l'érudition ou abus d'étude. Avec une digression sur la misère des hommes de lettres et la raison de la mélancolie des muses." In *L'Anatomie de la mélancolie*. Translated by B. Hœpffner and C. Goffaux. Paris: José Corti, 2000, texts 2, 3, and 15.

Disxaut, M. *Le Naturel philosophe. Essai sur les dialogues de Platon*. Paris: Vrin, 1985.

Fitzgerald, J. "A 'Clerkes Compleinte.' Tolkien and the Division of Lit. and Lang." *Tolkien Studies*, 6, 2009.

Kreeft, Peter J. *The Philosophy of Tolkien. The Worldview Behind The Lord of the Rings*. San Francisco: Ignatius Press, 2005.

Marquet, Jean-François. *Miroirs de l'identité. La littérature hantée par la philosophie*. Paris: Hermann, 1996.

Plato. *The Republic*. Translated by P. Pachet. Paris: Folio, 1993.

Ryan, J. S. "The Oxford Undergraduate Studies in Early English and Related Languages of J. R. R. Tolkien (1913–1915)." In *Tolkien's View*. Zurich: Walking Tree, 2009.

Tolkien, J. R. R. "The Clerkes Compleinte." *The Gryphon*, December 1922, NS 4/3, p. 95; reissued "*The Clerkes Compleinte*: Text, Commentary and Translation." Edited by T. Shippey and A. Stenström. *Arda 1984*, 4, 1986; *facsimile* of the Leeds manuscript in "*The Clerkes Compleinte* Revisited." *Arda 1986*, 1990.

—, *On Fairy-stories*. Edited by V. Flieger and D. A. Anderson. London: HarperCollins, 2008.

Tolkien, J. R. R. "Philosophical Thoughts." *Parma Eldalamberon*, no. 20, 2012, pp. 113–115. (Our translation. Arden R. Smith wrongly standardizes the spelling of Fausset as Fawcett, without noting this error.)

Tolkien, J. R. R. *Sauron Defeated*. London: HarperCollins, 1992.

Tolkien, J. R. R. *The Peoples of Middle-earth*. London: HarperCollins, 1996.

—, "The Shibboleth of Feänor," *The Peoples of Middle-earth*. Edited by C. Tolkien. London: HarperCollins, 1996.

Tolkien, J. R. R. *Lettres*. Edited by Humphrey Carpenter with Christopher Tolkien. Translated by V. Ferré and D. Martin. Paris: Christian Bourgois éditeur, 2005.

Tolkien, John and Priscilla. *The Tolkien Family Album*. London: HarperCollins, 1992.

ARCHAEOLOGICAL REMAINS AND HIDDEN CITIES,
by Vivien Stocker

Beare, Rhona. "Goblin Graffiti." *Tyalië Tyelelliéva*, no. 14, March 2000.

Beare, Rhona. "Painted Caves and Yggdrasil." *Tyalië Tyelelliéva*, no. 9, October 1996.

Beare, Rhona. "Painted Caves." *Tyalië Tyelelliéva*, no. 5, April 1995.

Carpenter, Humphrey. *J. R. R. Tolkien, une biographie*. Translated by P. Alien. Paris: Pocket, 2004.

Forest-Hill, Lynn. "The Inspiration for Tolkien's Ring." *History Today*, vol. 64, no. 1, January 2014.

Omotunde, Jean-Philippe. *Manuel d'études des humanités classiques africaines*, vol. 1. Paris: Menaibuc, 2007.

Plutarch. *Vie d'Alexandre*. Translated by V. André and R. Betolaud. Paris: Hachette, 1863.

Sabo, Deborah. "Archaeology and the sense of history in J. R. R. Tolkien's Middle-earth." *Mythlore*, no. 99/100, Autumn/Winter 2007, pp. 99–112.

Scull, Christina and Hammond, Wayne G., *J. R. R. Tolkien Artiste et illustrateur*. Translated by J. Georgel. Paris: Christian Bourgois éditeur, 1996.

Scull, Christina and Hammond, Wayne G. *Reader's Guide, The J. R. R. Tolkien Companion and Guide*. New York: HarperCollins Publishers, 2006.

Scull, Christina. "The Influence of Archaeology & History on Tolkien's World." *Scholarship & Fantasy, Proceedings of The Tolkien Phenomenon*. Turku, Finland: Anglicana Turkuensia, 1992, pp. 33–51.

Stocker, Vivien. "Les Númenóréens, des Égyptiens en Terre du Milieu." *L'Arc et le Heaume* (magazine of the Tolkiendil Association), vol. 3.

Tolkien, J. R. R. "Appendix I: The Name 'Nodens'," *Report on the Excavation of the Prehistoric, Roman, and Post-Roman Site in Lydney Park, Gloucestershire*, R. E. M. Wheeler and T. V. Wheeler, Oxford University Press for the Society of the Antiquaries, 1932, pp. 132–137. Reprinted in *Tolkien Studies*, vol. 4, 2007, pp. 177–183.

Tolkien, J. R. R., *Contes et légendes inachevés*. Translated by T. Jolas. Paris: Christian Bourgois éditeur, 2005.

Tolkien, J. R. R., *Le Seigneur des Anneaux*. Translated by F. Ledoux. Paris: Christian Bourgois éditeur, 1992.

Tolkien, J. R. R. *Le Seigneur des Anneaux, La Fraternité de l'Anneau*. Translated by D. Lauzon. Paris: Christian Bourgois éditeur, 2014.

Tolkien, J. R. R. *Le Silmarillion*. Translated by P. Alien. Paris: Christian Bourgois éditeur, 2004.

Tolkien, J. R. R. *Lettres du Père Noël*. Translated by G.-G. Lemaire and C. Leroy. Paris: Christian Bourgois éditeur, 2004.

Tolkien, J. R. R. *Sauron Defeated*. New York: HarperCollins Publishers, 2002.

Tolkien, J. R. R. *Lettres*. Edited by Humphrey Carpenter with Christopher Tolkien. Translated by V. Ferré and D. Martin. Paris: Christian Bourgois éditeur, 2005.

HISTORY AND HISTORIOGRAPHY IN MIDDLE-EARTH,
by Damien Bador

Tolkien, J. R. R.
    Sauron Defeated. Edited by Christopher Tolkien. New York: HarperCollins, 1992.
    Contes et légendes inachevés. Edited by Christopher Tolkien. Translated by T. Jolas. Paris: Christian Bourgois éditeur, 1993.
    Le Silmarillion. Translated by P. Alien. Paris: Christian Bourgois éditeur, 1993.
    Morgoth's Ring. Edited by Christopher Tolkien. New York: HarperCollins, 1993.
    The War of the Jewels. Edited by Christopher Tolkien. New York: HarperCollins, 1994.
    The Peoples of Middle-earth. Edited by Christopher Tolkien. New York: Harper-Collins, 1996.
    Finn and Hengest: the Fragment and the Episode. New York: HarperCollins, 1998.
    Le Livre des contes perdus. Edited by Christopher Tolkien. Translated by A. Tolkien. Paris: Christian Bourgois éditeur, 2001.
    Lettres. Edited by Humphrey Carpenter with Christopher Tolkien. Translated by V. Ferré and D. Martin. Paris: Christian Bourgois éditeur, 2005.
    Les Lais de Beleriand. Translated by D. Lauzon. Paris: Christian Bourgois éditeur, 2006.
    La Formation de la Terre du Milieu. Translated by D. Lauzon. Paris: Christian Bourgois éditeur, 2007.
    Les Enfants de Húrin. Translated by D. Martin. Paris: Christian Bourgois éditeur, 2008.
    The Lost Road and Other Writings. Translated by D. Lauzon. Paris: Christian Bourgois éditeur, 2008.
    Le Seigneur des Anneaux, La Fraternité de l'Anneau. Translated by D. Lauzon. Paris: Christian Bourgois éditeur, 2014.
    Le Seigneur des Anneaux, Les Deux Tours. Translated by D. Lauzon. Paris: Christian Bourgois éditeur, 2015.
    Le Seigneur des Anneaux, Le Retour du Roi. Translated by D. Lauzon. Paris: Christian Bourgois éditeur, 2016.

LINGUISTICS AND FANTASY
by Damien Bador

Bador, Damien. "Elfiques, langues." In Ferré, Vincent (ed.). *Dictionnaire Tolkien*. Paris: CNRS éditions, 2012, p. 171–175.
Gilliver, Peter, Jeremy Marshall, and Edmund Weiner. *The Ring of Words: Tolkien and the Oxford English Dictionary*. Oxford: Oxford University Press, 2009.
Tolkien, J. R. R. "Du conte de fées." Edited by Christopher Tolkien. In *The Monsters and the Critics, and Other Essays*. Translated by C. Laferrière. Paris: Christian Bourgois éditeur, 2006.
Tolkien, J. R. R. "'The story of Kullervo' and Essays on *Kalevala*." Edited by Verlyn Flieger. *Tolkien Studies*, no. 7, 2010.

Tolkien, J. R. R. "Words, phrases and passages in various tongues in *The Lord of the Rings*." Edited by Christopher Gilson. *Parma Eldalamberon*, no. 17, 2007.

Tolkien, J. R. R. "Qenyaqetsa, The Qenya Phonology and Lexicon together with The Poetic and Mythologic Words of Eldarissa." Edited by Christopher Gilson, et al. *Parma Eldalamberon*, no. 12, 1998.

Tolkien, J. R. R. *The Lost Road and Other Writings.* Translated by D. Lauzon. Paris: Christian Bourgois éditeur, 2008.

Tolkien, J. R. R. *Le Seigneur des Anneaux, La Fraternité de l'Anneau.* Translated by D. Lauzon. Paris: Christian Bourgois éditeur, 2014.

Tolkien, J. R. R. *Le Seigneur des Anneaux, Le Retour du roi.* Translated by D. Lauzon. Paris: Christian Bourgois éditeur, 2016.

Tolkien, J. R. R. "From Quendi and Eldar, Appendix D." Edited by Carl Hostetter. *Vinyar Tengwar* no. 39, 1998, pp. 4–20.

Tolkien, J. R. R. "The Rivers and Beacon-hills of Gondor." Edited by Carl Hostetter. *Vinyar Tengwar* no. 42, 2001, pp. 5–31.

Tolkien, J. R. R. "The Alphabet of Rúmil," Edited by Arden Smith. *Parma Eldalamberon* no. 13, 2001, pp. 5–89.

Tolkien, J. R. R. *The Peoples of Middle-Earth.* Edited by Christopher Tolkien. New York: HarperCollins, 1996.

Tolkien, J. R. R. *Lettres.* Edited by Humphrey Carpenter with Christopher Tolkien. Translated by V. Ferré and D. Martin. Paris: Christian Bourgois éditeur, 2005.

Tolkien, J. R. R. "Eldarin Hands, Fingers & Numerals and Related Writings—Part Two." *Vinyar Tengwar*, no. 48, 2005, pp. 4–34.

THE LORD OF THE RINGS: A MYTHOLOGY OF CORRUPTION AND DEPENDENCE, by Thierry Jandrock

Foster, Robert. *The Complete guide to Middle-Earth.* New York: Del Rey, 1978.

Goethe, W. E. *Faust.* Paris: Garnier Flammarion, 1964.

Gross, Benjamin. *L'aventure du langage: l'alliance de la parole dans la pensée juive.* Paris: Albin Michel, Présence du Judaïsme, 2003.

Jandrok, Thierry. "Des Anneaux du Désir au Désir de l'Anneau: Les avatars de la relation d'objet." In *Actes du Colloque du CRELID, Fantasy: Le Merveilleux Médiéval Aujourd'hui.* Edited by Anne Besson and Myriam White-Le Goff. Paris: Bragelonne, Collection Essais, 2007, pp. 59–71. Available at: http://www.tolkiendil.com/ssays /colloques/colloque_crelid/t_jandrok.

Tolkien, J. R. R. *The Book of Lost Tales 1.* New York: Del Rey, 1992.

Tolkien, J. R. R. *The Book of Lost Tales 2.* New York: Del Rey, 1992.

Tolkien, J. R. R. *The Children of Húrin.* London: HarperCollins Publishers, Special edition, 2007.

Tolkien, J. R. R. *The Lays of Beleriand.* New York: Del Rey, 1994.

Tolkien, J. R. R. *The Lord of the Rings, 50th Anniversary Edition.* New York: Houghton Mifflin, 2004.

Tolkien, J. R. R. *The Lost Road and Other Writings.* New York: Del Rey, 1996.

Tolkien, J. R. R. *The Shaping of Middle-Earth.* New York: Del Rey, 1995.

Tolkien, J. R. R. *The Silmarillion.* New York: Del Rey, 2004.

LANDSCAPES IN TOLKIEN: A GEOMORPHOLOGICAL APPROACH,
by Stephen Giner

Bru, Alice. "Des développeurs danois modélisent la Terre du Milieu du Seigneur des Anneaux." *Slate*, January 30, 2014. http://www.slate.fr/life/82849/developpeurs -danois-modelisent-terre-du-milieu-seigneur-des-anneaux-tolkien.

Curless, A. *A Tolkien bestiary*. Worthing, England: Littlehampton Book Services Ltd, 1979.

For a virtual walk through Middle-earth [online], *Slate.fr*, 30 January 2014. Available at: www.slate.fr/life/82849/développeurs-danois-modelisent-terre-du-milieu-seigneur -des-anneaux-tolkien.

Turlin, J. *Promenades au pays des Hobbits*. Dinan, France: Éditions Terre de Brume, 2012.

"Númenor et la Cottoniana—enquête sur une île en forme d'étoile à cinq branches." Edited by D. Willis. In Collectif, *Tolkien, le façonnement d'un monde. Vol. 2—Astronomie & Géographie*, 2014, pp. 141–154.

A GEOLOGICAL STROLL THROUGH MIDDLE-EARTH,
by Loïc Mangin

Howes, M. "The elder ages and the later glaciation of the Pleistocene epoch." *Tolkien Journal* vol. 4, no. 2, 1967, pp. 3–15.

Reynolds, R. "The geomorphology of Middle-Earth." *The Swansea Geographer*, 11, 1974, pp. 67–71.

Sarjeant, W. "The geology of Middle-Earth." Proceedings of the J. R. R. Tolkien Centenary Conference, 1992.

SUMMER IS COMING: THE CLIMATE OF MIDDLE-EARTH,
by Dan Lunt

Fonstad, K. W.: *The Atlas of Middle-earth*, Revised Edition, Houghton Mifflin, 1991.

IPCC: Climate Change 2013: The Physical Science Basis, Cambridge University Press, http://www.climatechange2013.org/images/uploads/WGIAR5 WGI-12Doc2b FinalDraftAll.pdf, 2013a.

IPCC: Climate Change 2013: The Physical Science Basis. Summary for Policymakers, Cambridge University Press. http://www.climatechange2013.org/images/uploads /WGIAR5SPMbrochure.pdf, 2013b.

Tolkien, J. R. R. *The Lord of the Rings*. London: George Allen and Unwin, 1954.

Tolkien, J. R. R. *The Shaping of Middle-earth*. London: Unwin Hyman Limited, 1986.

SUBTERRANEAN WORLDS IN TOLKIEN: AN UNDERGROUND HISTORY,
by Sylvie-Anne Delaire

Henon, Priscille. "Dédales en Terre du Milieu." In *Tolkien, le façonnement d'un monde Astronomie et Géographie*, vol. 2, 2014, pp. 155–172.

Kuntz, Thomas H. *Ecology of bats*. New York: Springer-Verlag, 2011.

PRECIOUS STONES: JEWELS OF MIDDLE-EARTH by Erik Gonthier and Cécile Michaux

Encyclopédie Tolkiendil [online]. Available at: https://www.tolkiendil.com/encyclo.

Flieger, Verlyn and Carl F. Hostetter. *Tolkien's Legendarium*. Westport, CT: Praeger Publishers Inc, 2000.

Les Archives de Gondor [online]. Available at: https://www.archivesdegondor.net/.

Steyer, Jean-Sébastien and Roland Lehoucq. "L'imaginaire cristallisé." *Pour la Science*, no. 463 (May 2016), pp. 84–85.

MEDIEVAL-FANTASTICAL METALLURGY, by Jean-Marc Joubert and Jean-Claude Crivello

Byko, Maureen. "Fabricating the Weapons and Armor of *The Lord of the Rings*" [online]. *Journal of the Minerals*, Metals & Materials Society. (November 2002). Available at: www.tms.org/pubs/journals/jom/0211/byko-0211.html.

Encyclopédie Tolkiendil [online]. Available at: https://www.tolkiendil.com/encyclo.

Ferré, Vincent (ed.). *Dictionnaire Tolkien*. Paris: CNRS éditions, 2012.

Gee, Henry. *The Science of Middle-Earth: Explaining The Science Behind The Greatest Fantasy Epic Ever Told!* Cold Spring Harbor, NY: Cold Spring Harbor Laboratory Press, 2004.

Gschneider, K.A. Jr. "The magnetocaloric effect, magnetic refrigeration and ductile intermetallic compounds." *Acta Materialia*, no. 57(1).(2009), pp. 18–28.

Nasmith, Ted [online]: https://www.tednasmith.com/.

Scull, Christina and Wayne G. Hammond. *Reader's Guide, The J. R. R. Tolkien Companion and Guide*. New York: HarperCollins Publishers, 2006.

The One Wiki to Rule Them All [online]. Available at: https://lotr.fandom.com/wiki/Main_Page.

Tolkien Gateway [online]. Available at: http://www.tolkiengateway.net/wiki/Main_Page.

A CHEMICAL HISTORY OF THE ONE RING, by Stéphane Sarrade

"Réacteur nucléaire naturel d'Oklo" [online]. Available at: https://fr.wikipedia.org/wiki/R%C3%A9acteur_nucl%C3%A9aire_naturel_d%27Oklo.

Blazy, Pierre and El-Aid Jdid. "Métallogénie: les différents types de gisements d'or" [online]. Techniques de l'ingénieur, 10 décembre 2006. Available at: https://www.techniques-ingenieur.fr/base-documentaire/materiaux-th11/elaboration-et-recyclage-des-metaux-de-transition-42649210/metallurgie-de-l-or-m2403/metallogenie-les-differents-types-de-gisements-d-or-m2403niv10004.html.

Centre national de ressources textuelles et lexicales. "Transmutation" [online]. Available at: https://www.cnrtl.fr/lexicographie/transmutation.

Ferré, Vincent (ed.), *Dictionnaire Tolkien*. CNRS éditions, 2012.

MineralInfo, "Exploration minière," collection "La mine en France." Available at: www.mineralinfo.fr/sites/default/files/upload/tome_04_exploration_miniere_final24032017.pdf.

WHY DO HOBBITS HAVE BIG HAIRY FEET? by Jean-Sébastien Steyer

Dollé et al. "Hox-4 genes and the morphogenesis of the mammalian genitalia." *Genes & Development* 5 (1991), pp. 1767–1776.

Meijer et al. "The Fellowship of the Hobbit: the fauna surrounding *Homo floresiensis.*" *Journal of Biogeography*, no. 37 (2010), pp. 995–1006.

Myhrvold et al. "What is the use of elephant hair?" *PLoS ONE* 7(10), e47018, 2012.

Tolkien, J. R. R. *The Annotated Hobbit.* London: Ed. Anderson D., 1937.

Wilson, Edmond. "Oo, those awful orcs! A review of *The Fellowship of the Ring*" [online]. *The Nation*, 14 April 1956. Available at: https://www.jrrvf.com/sda/critiques /The_Nation.html.

Collectif. *Fantasy, l'origine des mondes.* Special issue, December 2012.

GOLLUM: THE METAMORPHOSIS OF A HOBBIT, by Jean-Sébastien Steyer

Panafieu, J. B. (de) and C. Renversade. *Métamorphoses Deyrolle—Histoires surnaturelles.* Toulouse, France: éditions Plume de carotte, 2016.

Tolkien, J. R. R. "Teeth! teeth! my preciousss; but we has only six!" Edited by Douglas A. Anderson. In *The Annotated Hobbit.* Boston: Houghton Mifflin Company, 1988.

ARE DWARVES HYENAS? by Sidney Delgado and Virginie Delgado-Bréüs

Ameisen, J. C. *Dans la lumière et les ombres*, Chapter 24. Paris: Éditions Points, 2011.

Austad, S. N. and M. E. Sunquist. "Sex-ratio manipulation in the common opossum." *Nature* 324, 58–60 (1986).

Cunha et al. "Development of the external genitalia: Perspective from the spotted hyena (*Crocuta crocuta*)." *Differenciation*, 87, 4–22 (2014).

Dloniak, S. M. et al. "Rank-related maternal effects of androgens on behaviour in wild spotted hyena." *Nature* 440, 1190–1193 (2006).

Drea, C. M., N. J. Place, M. L. Weldele, E. M. Coscia, P. Licht, and S. E. Glickman. "Exposure to naturally circulating androgens during foetal life incurs direct reproductive costs in female spotted hyenas, but is prerequisite for male mating." *Proc. R. Soc. Lond.* B 269, 1981–1987, 2002.

Gould, S. J. and R. C. Lewontin, "The Spandrels of San Marco and the Panglossian Paradigm: A Critique of the Adaptationist Programme." *Proc. Roy. Soc.* London B 205 (1979), pp. 581–598.

Gould, S. J. *Ontogeny and Phylogeny.* Cambridge, MA: Harvard University Press, 1977.

Goymann, W., M. L. East, and H. Hofer, "Androgen and the role of female 'hyperaggressiveness' in spotted hyenas." *Hormones and Behaviour*, 39 (2000), pp. 83–92.

Hamilton, W. D. "Extraordinary sex ratios. A sex-ratio theory for sex linkage and inbreeding has new implications in cytogenetics and entomology." *Science* 156 (3774): 477–88, 1967.

Holekamp, K. E. and L. Smale. "Rapid change in offspring sex-ratios after clan fission in spotted hyena." *The American Naturalist*, 1995.

Muller, M. and R. Wrangham. "Sexual mimicry in hyenas." *The Quarterly Review of Biology*, vol 77, no. 1 (2002).

Raff, R. "Evo-Devo: the evolution of a new discipline" [online]. *Nature Reviews Genetics*, 1 (2000), 74–79.

Tolkien, J. R. R. *The History of Middle-earth*, Book XI, Chapter XIII. London: HarperCollins, 2002.

Tolkien, J. R. R. *The Lord of the Rings*, "Appendix A—Durin's Folk." Translated by F. Ledoux. Paris: Christian Bourgois éditeur, 1992.

Tolkien, J. R. R. *Le Silmarillion*. Translated by P. Alien. Paris: Christian Bourgois éditeur, 2004.

MEMORIES OF OLIPHAUNTS, by Arnaud Varennes-Schmitt

Gheerbrant, E. "Paleocene emergence of elephant relatives and the rapid radiation of African ungulates." *Proceedings of the National Academy of Sciences*, 2009, 106 (26).

Gheerbrant, E., J. Sudre, and H. Cappetta, "A Paleocene proboscidean from Morocco." *Nature*, 1996, 383, 68–70.

Gheerbrant, E., and P. Tassy, "L'origine et l'évolution des lephants." *Comptes Rendus Palevol*, 2009, 8, 281–294.

Kistler, John M. *War Elephants*. Lincoln, NE: Bison Books, 2007.

Nossov, Konstantin. *War Elephants*. Oxford, UK: Osprey Publishing, 2008.

Tolkien, J. R. R. *Le Seigneur des Anneaux, Le Retour du roi*. Translated by D. Lauzon. Paris: Christian Bourgois éditeur, 2016.

Tolkien, J. R. R. *Le Seigneur des Anneaux, Les Deux tours*. Translated by D. Lauzon. Paris: Christian Bourgois éditeur, 2015.

WARGS: WAR-DOGS OF SCANDINAVIAN ORIGIN? By Vincent Dupret and Romaric Hainez

Belyaev. *The genetics and phenogenetics of domestic behaviour. Problems in General Genetics.* Proceedings of the 14th Annual Congress on Genetics. Moscow: Mir Publishers, 1980, pp. 123–137.

Brawn, D. *The Lord of the Rings—The Two Towers—Creature Guide.* Translated by A. Delarbre. Paris: Gallimard jeunesse, 2002.

Clinquart, A., J. L. Hornick, C. Van Eenaeme, and L. Istasse. "Influence du caractère culard sur la production et la qualité de la viande des bovins Blanc Bleu Belge." *INRA Prod. Anim.*, 11(4), 285–297. Available at: https://www6.inra.fr/productions -animales/1998-Volume-11/Numero-4-1998/Influence-du-caractere-culard-sur-la -production-et-la-qualite-de-la-viande-des-bovins.

Falconer, D. *The Hobbit Chronicles: Creatures & Character. An Unexpected Journey.* London: HarperCollins, 2013.

Gélinet, P. and A. Kobylak, "Le Loup." In "2000 ans d'histoire" [audio recording]. April 30, 2016, France Inter.

Guerber, H. A. *Myths of the Norsemen from the Eddas and Sagas.* London: Gearge G. Harrap & Company, 1909.

IFL Sciences! "Why so many domesticated mammals have floppy ears." Available at: http://www.iflscience.com/plants-and-animals/why-so-many-domesticated -mammals-have-floppy-ears.

Jussiau, R., J. Rigal, and A. Papet. *Amélioration génétique des animaux d'élevage*. Paris: Educagri, 2013.

Linberg, J., et al. "Selection for tameness has changed brain gene expression in silver foxes." Correspondence, vol. 15 (Nov. 2015), issue 22, p. 915-916. Available at: http://www .cell.com/current-biology/retrieve/pii/S0960982205013278.

Polin, S. *Le chien de guerre: utilisations à travers les conflits*, thèse de doctorat. 2003. Available at: http://theses.vet-alfort.fr/telecharger.php?id=467.

Trut, L. and L. A. Gugatkin. "Comment transformer un renard en chien." *Pour la science* no. 479 (25 August 2017). Available at: https://www.pourlascience.fr/sd/zoologie /comment-transformer-un-renard-en-chien-9807.php.

Wang, X. et al. *Genomic responses to selection for tame/aggressive behaviors in the silver fox* (Vulpes vulpes). PNAS, October 2018, 115 (41). Available at: https://www.pnas.org /content/115/41/10398.

"Belyaev Experiment: Agressives Foxes" [video recording]. Available at: https://www .youtube.com/watch?v=9fC7l6gW05k.

"Belyaev Experiment: Docile Foxes" [video recording]. Available at: http://www.youtube .com/watch?v=mzTcmE-pMLU.

BEORN: MAN-BEAR OR BEAR MAN? by Vincent Dupret and Romaric Hainez

Anonymous. *Egil's Saga*. Translated by H. Palsson and P. Edwards. New York: Penguin Classics, 1977.

Blaney, B. *The Berserker: His Origin and Development in Old Norse Literature*. University of Colorado, 1972.

Boyer, R. *Sagas islandaises*. Paris: Bibliothèque de la Pléiade. Gallimard, 1992.

Dillmann, F. *Les magiciens dans l'Islande ancienne*. Doctoral thesis submitted to the University of Caen in 1986, supervised by Frédéric Durand.

Fabing, H. D. "On going berserk: a neurochemical inquiry." *American Journal of Psychology*, (1956), 113.5:409–415.

Gabriel, R. A. *No More Heroes: Madness and Psychiatry In War*. New York: Hill & Wang, 1990.

Heebøll-Holm, T. K. "Apocalypse then? The first crusade, traumas of war and Thomas de Marle." In Hundahl, K., L. Kjær, and N. Lund (eds.). *Denmark and Europe in Middle Age. c. 1000-1525*. Farnham, UK: Ashgate Publishing, 2014.

Hellquist, E. *Svensk etymologisk ordbok*. Gleerups, Lund, 1922.

Hirst, M. Vikings [enregistrement vidéo]. 2013, Shaw Media, Octagon Films, Take 5 Productions, MGM Television.

Jones, G. *Eirik the Red and Other Icelandic Sagas*. Oxford World's Classics, Oxford University Press, 2009.

Kurumada, M. *Saint Seiya, Les Chevaliers du Zodiaque*, vol. 9. Kana, 1998.

Lachlan, M. D. *Wolfangel*. Amherst: Pyr, 2011.

Lecouteux, C. *Fantômes et revenants au Moyen Âge*. Paris: L'Arbre à mémoire, 1986.

McCoy, D. *The Viking Spirit: An Introduction to Norse Mythology and Religion*. CreateSpace Independent Publishing Platform, 2016.

Miura K. *Berserk*. Paris: Glénat, 1989.

Ödman, S. *Försök at utur Naturens Historia förklara de nordiska gamla Kämpars Berserka-gang*. Sweden: Kongliga Vetenskaps Academiens nya Handlingar, 5:240–247, 1784.

Price, N. S. *The Viking Way: Religion and War in Late Iron Age Scandinavia*. Thesis, Department of Archaeology & Ancient History, Uppsala University, 2002.

Rateliff, J. D. *The History of The Hobbit. Part One: Mr. Baggins*. New York: Houghton Mifflin Harcourt, 2007.

Samson, V. *Les Berserkir. Les guerriers-fauves dans la Scandinavie ancienne, de l'âge de Vendel aux Vikings (vie-xie siècle)*. Villeneuve-d'Ascq: Presses Universitaires du Septentrion, 2011.

Shippey, T. A. *J. R. R. Tolkien: author of the century*. Boston: Houghton Mifflin Company, 2002.

Siebert, E. "Science Gone Berserk." 2016. Available at: https://skepticalhumanities .com/2011/05/03/science-gone-berserk/.

THE BESTIARY OF ARTHROPODS, by Romain Garrouste and Camille Garrouste

Climo, F. M. "Smeagolida, a new order of gymnomorph mollusc from New Zealand based on a new genus and species." *New Zealand Journal of Zoology*, 7(4): 513–522, 1980.

Jóźwiak, P., T. Rewicz, and K. Pabis, "Taxonomic etymology—in search of inspiration." *ZooKeys* 513: 143–160, 2015.

Kaila, Lauri. "A revision of the Nearctic species of the genus Elachista s.l. III. The *Bifasciella, Praelineata, Saccharella* and *Freyerella* groups (*Lepidoptera, Elachistidae*)." *Acta Zoologica Fennica*, 211: 1–235, 1999.

Morrone, J. J. "Systematics, cladistics, and biogeography of the Andean weevil genera Macrostyphlus, Adioristidius, Puranius, and Amathynetoides, new genus (*Coleoptera*: Curculionidae)." *American Museum Novitates*, no. 3104 (1994).

Sendra, A., V. M. Ortuno, A. Moreno, S. Montagu, and S. Teruel, "Gollumjapyx smeagol gen. n., sp. n., an enigmatic hypogean japygid (Diplura: Japygidae) from the eastern Iberian Peninsula." *Zootaxa*, 1372, 35–52, 2006.

SMAUG, GLAURUNG, AND THE REST: MONSTERS FOR BIOLOGISTS TOO, by Stéphane Jouve

Conrad, J. L. "Phylogeny and systematic of Squamata (Reptilia) based on morphology." *Bulletin of the American Museum of Natural History*, 310 (2008).

Goloboff, P. A., J. S. Farris, and K. C. Nixon. "TNT, a free program for phylogenetic analysis." *Cladistics: The International Journal of the Willi Hennig Society*, 24 (2008): 774–786.

Goloboff, P. A., J. S. Farris, and K. C. Nixon. *TNT: Tree Analysis Using New Technology*. London: Willi Hennig Society, 2003.

Hudspeth, E. B. *Le Cabinet du docteur Black*. Paris: Le Pré aux Clercs, 2014.

Tolkien, J. R. R. *The Hobbit*. Boston: Houghton Mifflin Co., 1938.

Tolkien, J. R. R. *The Silmarillion Calendar 1978*. London: George Allen and Unwin Ltd, 1977.

FLAMES *AND* WINGS: COULD IT REALLY HAPPEN? by Stéphane Jouve

Hardy, J. *Dragons' World: A Fantasy Made Real* [video recording]. Sony Pictures Home Entertainment, 2004.

Hill, Kyle. "Smaug Breathes Fire Like A Bloated Bombardier Beetle With Flinted Teeth" [online]. *Scientific American*, 2 January 2014. Available at: https://blogs.scientificamerican.com/but-not-simpler/smaug-breathes-fire-like-a -bloated-bombardier-beetle-with-flinted-teeth/?redirect=1.

Hudspeth, E. B. *Le Cabinet du docteur Black*. Paris: Le Pré aux Clercs, 2014.

Tolkien, J. R. R. *The Hobbit*. Boston: Houghton Mifflin Co., 1938.

THE WATCHER BETWEEN TWO WATERS, by Jérémie Bardin and Isabelle Kruta

Mangold, K., A.M. Bidder, and A. Portmann. "Organisation générale des céphalopodes." In Grassé, P. (eds.). *Traité de zoologie*, Tome V, Fascicule 4 p. 7, 1989.

Schneidewind, F. "The biology of Middle-earth." In Drout, M.D.C. *J. R. R. Tolkien Encyclopedia: Scholarship and Critical Assessment*. Abington-on-Thames: Routledge, 2007.

THE CRYPTOZOOLOGICAL BESTIARY OF J.R.R. TOLKIEN, by Benoît Grison

Carpenter, Humphrey. *J. R. R. Tolkien, une biographie*. Translated by P. Alien. Paris: Christian Bourgois éditeur, 2002.

Day, D. *Créatures de Tolkien*. Paris: Octopus-Hachette, 2002.

Grison, B. *Du Yéti au calmar géant: le bestiaire énigmatique de la Cryptozoologie*. Paris: Delachaux & Niestlé, 2016.

Heuvelmans, B. *The Kraken and the Colossal Octopus*. London: Kegan Paul, 2003.

Lang, A. *The Complete Works*. Delphi Classics, 2015 [e-book].

Page, R. I. *Mythes Nordiques*. Paris: Seuil, 1993.

Zucker, A. (ed.). *Physiologos, le bestiaire des bestiaires*. Grenoble, France: Éditions Jérôme Millon, 2004.

BIOGRAPHIES

Edited by Roland Lehoucq, Loïc Mangin,
and Jean-Sébastien Steyer

Born after the Great War between the Elves and the orcs, **Roland Lehoucq** is an astrophysicist at the French Atomic Energy Commission (CEA) and lecturer at the Paris Institute of Political Studies and on the Master's Programme in Ecology, Energy, and Society at the Université Paris Diderot. The author of numerous books on science and science-fiction, he is also president of the Utopiales international science-fiction festival in Nantes.

A loyal friend of Gandalf, **Loïc Mangin** is associate editor of the magazine *Pour La Science*, where he writes the "Art and Science" column. He is also the author of the blog Best of Bestioles (http://bit.ly/PL/BOB) and the book *Pollock, Turner, Van Goth, Vermeer . . . et la science* (Belin, 2018).

Originally from the Shire, **Jean-Sébastien Steyer** is a paleontologist at the French National Centre for Scientific Research (CNRS) and the Museum of Paris. A former freelance journalist for *Charlie Hebdo* (writing the column "La Science amusante"), he is a conference speaker, the author of several popular books, and a columnist for the journal *Espèces* (*Species*). He is passionate about using works of fiction to disseminate knowledge.

Researchers and educators at the Museum of Paris, **Christine Argot** and **Luc Vivès** are specialists in the history of the collections exhibited in the gallery of paleontology and comparative anatomy, sharing the results of this epistemological research with students and the public through conferences and publications.

Damien Bador is a paleontologist at the Centre for Paleontological Research—Paris (CR2P) and a data processing engineer at Sorbonne University. He studies evolutionary modules of morphology and their use to reconstruct relationships between organisms and identify the continuous mechanisms in the history of life.

Élodie Boucheron-Dubuisson is a lecturer at Sorbonne University, where she teaches plant biology. Subsequent to her research on the role of microgravity on the development of root systems, she is currently investigating the moss flora of Madagascar at the Museum of Paris.

After training in the plastic arts and then in protohistoric archaeology, **Cécile Breton** switched her focus to science and publishing, and founded two popular scientific journals: *Stantari, histoire naturelle & Culturelle de la Corse* in 2005, and *Espèces: Revue d'histoire naturelle* in 2011, for which she continues to serve as editor-in-chief.

An associate professor in Earth and Life Sciences (SVT) at the Université de Pau and a resident of the Shire (if only in his mind), **Jean-Philippe Colin** only emerges from his hobbit-hole to share his enthusiasm for the science of living things.

A lecturer at the Université Clermont-Auvergne, **Bruno Corbara** works on the ecology and ethology of social insects and the functioning of aquatic ecosystems. A former leader of international expeditions, including the famous Radeau des Cimes project investigating forest biodiversity, he is managing editor and editorial advisor for the journal Espèces.

The holder of a doctorate in chemistry, **Jean-Claude Crivello** is a researcher at the CNRS, assigned to the Institute of Materials Chemistry, Paris East. A specialist in chemical bonds in the solid phase, he studies the influence of chemical composition and crystallography in order to optimize materials for specific applications such as hydrogen storage.

A veterinarian and speleologist, **Sylvie-Anne Delaire** is passionate about exploration, both above and below ground. She organizes international expeditions with her team and has been an advocate for more than ten years for the protection and promotion of underground mining heritage.

A biologist and lecturer at Sorbonne University and in the Pierre-et-Marie-Curie Department of Sciences, **Sidney Delgado** teaches animal biology and evolution. A researcher on the Homology team at the Museum of Paris, he specializes in evolution and the development of mineralized tissues in vertebrates.

The holder of a PhD in French linguistics (pragmatics and syntax) and a graduate of the Université de Bretagne Occidentale and the Université de Paris IV, **Virginie Delgado-Bréüs** is a specialist in French army parlance.

An associate professor and PhD in philosophy, **Michaël Devaux** lectures on the philosophy of education at the Université de Caen Normandie. A member of the French National Research Agency (ANR) Condorcet, he teaches Tolkien at the Collège des Bernardins in Paris and has edited multiple books including *Tolkien, les racines du légendaire* (Éditions Ad Solem, 2003) and *J.R.R. Tolkien, l'effigie des Elfes* (Bragelonne, 2014).

Jean-Yves Dubuisson is a professor at the Sorbonne University, where he teaches biodiversity, the history of living things, (paleo)botany, and plant ecology. Focusing principally on the evolution of plants and working on models of ferns and orchids, his research is centered on the origin and history of plant communities in island environments.

Vincent Dupret is a Museum of Paris-certified paleontologist and specialist in the origin and diversification of jawed vertebrates including placoderms, or "ehaviou fish." He has worked all over the world and gives public lectures in France and abroad.

Camille Garrouste is a graduate of the Institut supérieur des arts appliqués. She is a digital artist and scientific illustrator.

A specialist in insects, **Romain Garrouste** is a researcher at the National Museum of Natural History (MNHN) in Paris and a member of the Society of French Explorers. He has taken part in numerous expeditions focused on (paleo)biodiversity in more than twenty countries and has co-discovered several fossil beds. He is passionate about the diffusion of knowledge and participates in many collaborative editorial projects.

A geoarchaeologist and topographer, **Stephen Giner** regularly designs museum exhibition spaces on geosciences in the southwest of France. Frequently engaged in fieldwork, he is also a member of the Scientific and Cultural Steering Committee for the UNESCO global geopark project in the Maures mountains.

An ethnomineralogist and specialist in paleolithophony, cave art, and megaliths, **Erik Gonthier** is the author of numerous books, director of film documentaries, and designer of exhibitions. An expert in precious stones, he has created several pieces for the Paris Mint and staged more than 130 exhibitions all over the world. His interest in semio-plastic art has led him to reclassify several collections, including in the Museum of Paris.

A doctor of cognitive science, biologist, and scientific sociologist, **Benoît Grison** is a lecturer and researcher at the UFR Collegium Sciences & Techniques at the Université d'Orléans. His current scientific interests include behavioral neuroscience, ethnozoology, and the epistemology of life sciences.

An associate professor in Earth and Life Sciences (SVT) at the Université de Picardie Jules-Verne, **Romaric Hainez** teaches biology liberally peppered with references to fictional worlds, to show that science is everywhere, even—and especially—in places where we least expect it.

A doctor of psychology and psychopathology, clinical psychologist, writer, essayist, and psychoanalyst, **Thierry Jandrok** is also a member of the Center for Literary Study and Research at the Université de Lorraine, focusing on the imaginary, and has contributed to numerous books on the subject.

An engineer and the holder of a PhD in materials science, **Jean-Marc Joubert** is Director of Research at the CNRS. He works in the Institute of Chemistry and Materials of Paris-East on metal alloys, focusing on the structure and properties of intermetallic components and their hydrides, and the determination and modelling of phase diagrams.

A doctor of paleontology specializing in crocodylomorphs (crocodiles and their close relatives), **Stéphane Jouve** is Head of Collections for geosciences at Sorbonne University.

Isabelle Kruta is a paleontologist with the Centre for Paleontological Research—Paris and a lecturer at Sorbonne University. Her research is focused on the exceptional preservation of fossil cephalopods in order to reconstruct the soft parts of the anatomy of these organisms and study their trophic interactions.

A paleontologist specializing in fossil birds at the CNRS and the geology laboratory of Lyon, passionate about birds and their observation, **Antoine Louchart** studies birds of different periods, with a particular focus on island birds that have gone extinct in recent centuries due to human activity.

Dan Lunt is a professor of climatology at the University of Bristol in the United Kingdom. The head of a climatic model comparison project (www.deepmip.org), he is also the main author of the sixth assessment of climate evolution by a group of intergovernmental experts (IPCC). Winner of the Philip Leverhulme Prize, he is also a fan of fantasy fiction.

François Marchal is a CNRS research fellow on joint research project (UMR) 7268, Biocultural Anthropology, Law, Ethics, and health. His research deals with the evolution of the locomotive system in hominids from a morphofunctional point of view, as well as the details of the sexual dimorphism of the human pelvis and their practical implications in terms of gender estimation.

A coordinator of cultural projects, **Cécile Michaux** received multidisciplinary training at the Sorbonne-Nouvelle, ranging from directing cultural projects to touristic development. Currently working in the field of performing arts, she continues to be interested and involved in all forms of culture and to work toward their transmission and the diffusion of knowledge about them.

Professor Emeritus in Literature at the École normale supérieure (Paris) and a member of the Institut d'histoire Moderne et Contemporaine, **Isabelle Pantin** works on the history of the book and the relationships between science, philosophy, and literature. She is the author of *Tolkien et ses legends: une experience en fiction* (CNRS Éditions, 2013).

Luc Perino is a physician, writer, and essayist certified in tropical medicine and epidemiology. After having worked in sub-Saharan Africa, tropical China, and rural France, he now teaches The Biology of Evolution and Medicine at the Université de Lyon. He is the author of publications on the vagaries of the health care system and the health care market, Darwin, and the science of evolution.

An associate professor in social sciences and high school teacher, **Thierry Rogel** is committed to spreading knowledge about his field. He has published several books on the subject, including *Introduction impertinente à la sociologie* (Éditions Liris, 1999 and 2004) and *La Sociologie des super-héros* (Éditions Hermann, 2012).

Stéphane Sarrade is director of research at the CEA in chemical engineering. Passionately dedicated to green chemistry, he is developing separation, decontamination, and extraction procedures using supercritical fluids and filtration membranes. Through his publications, he proposes new approaches to chemistry to make our world more sustainable.

A vulcanologist, geologist, and expert and consultant in geological risks, **Laurent Stieltjes** is a director of research for organizations including the French Geological Survey (BRGM). Dedicated to the sharing of knowledge on the culture of life, the Earth, and their evolution, he has produced talks, books, films, and conferences in which he offers his perspective as a geologist on various subjects.

Vivien Stocker is one of the directors of the Tolkiendil association, which promotes Tolkien's writings in the French-speaking world. He contributed to the *Dictionnaire Tolkien* (edited by V. Ferré, CNRS Éditions, 2012) and is co-author of *L'Encyclopédie du Hobbit et du Monde des Hobbits* (Éditions Le Pré aux Clercs, 2013 and 14).

Arnaud Varennes-Schmitt is the holder of a doctorate in paleontology certified by the Museum of Paris. His thesis focused on the evolution of elephants and their close relatives. He currently teaches biology in various higher education institutions.

ILLUSTRATION CREDITS